Single-Sided NMR

Federico Casanova · Juan Perlo · Bernhard Blümich
Editors

Single-Sided NMR

Springer

Editors
Dr. Federico Casanova
RWTH Aachen
Inst. Technische und
Makromolekulare Chemie
Worringerweg 1
52074 Aachen
Sammelbau Chemie
Germany
fcasanova@mc.rwth-aachen.de

Juan Perlo
RWTH Aachen
Inst. Technische und
Makromolekulare Chemie
Worringerweg 1
52074 Aachen
Sammelbau Chemie
Germany
jperlo@mc.rwth-aachen.de

Prof. Dr. Bernhard Blümich
RWTH Aachen
Inst. Technische und
Makromolekulare Chemie
Worringerweg 1
52074 Aachen
Sammelbau Chemie
Germany
bluemich@mc.RWTH-aachen.de

ISBN 978-3-642-16306-7 e-ISBN 978-3-642-16307-4
DOI 10.1007/978-3-642-16307-4
Springer Heidelberg Dordrecht London New York

Cover design: WMX Design GmbH, Heidelberg

Printed on acid-free paper

Springer is part of Springer Science+Business Media (www.springer.com)

Preface

The study of complex molecules requires high sensitivity and resolution for chemical analysis. This is why the NMR community pushes the development of ever more powerful superconducting magnets that generate stronger and more homogeneous magnetic fields every year. Today's NMR magnets are heavy giants installed in special NMR laboratories built to shield electromagnetic interference, control the temperature, and reduce magnetic field distortions in order to provide ideal experimental conditions in a small volume located inside the magnet. Besides the fact that samples must be taken to the magnet to be analyzed, they must fit in the limited space available in the magnet center. Large samples cannot be investigated in this way without cutting. This problem was already noted in the early days of NMR when the technique was identified as a potential tool to characterize rock formations in situ. Measuring properties of the pore structure as well as characterizing the saturating fluids downhole by NMR triggered the development of what today is called inside-out NMR. Here, instead of placing the sample inside the magnet, the apparatus is placed inside or at one side of the object.

The implementation of this concept requires the construction of sensors projecting considerable magnetic and radiofrequency (rf) fields outside the magnet and inside the sample under study. For more than 30 years well-logging tools used the earth's magnetic field as polarization field, but the low sensitivity and poor spatial selectivity limited their commercial success. Potential application of open NMR sensors to measure, for example, moisture in building materials or soil motivated engineers to use electromagnets to increase the magnetic field strength. However, these instruments were large, weighing up to 300 kg and operated at low frequencies like 3 MHz. Perhaps the most important step toward reducing the size of the magnets was the use of permanent magnets to generate the static magnetic field. This eliminated most of the power consumption of these NMR devices. During the late 1980s and early 1990s various magnet geometries have been proposed by different well-logging companies, and a number of research groups proposed small single-sided magnets generating magnetic fields of about 0.5 T external to the sensor. An example is the known U-shaped magnet geometry used by the NMR-MOUSE, which is obtained by opening the C-shaped geometry used in conventional closed magnets.

The big challenge faced when working outside the magnet is posed by the large inhomogeneities in both the magnetic and the radiofrequency fields. Under such experimental conditions even hard rf pulses act as selective pulses and off-resonance effects are considerable. Therefore, during the 1990s, several groups set their attention on understanding the response of the spin system to pulse sequences known from conventional NMR, like the Carr-Purcell-Meiboom-Gill (CPMG) sequence extensively used to measure the transverse relaxation time T_2, when they are applied in the presence of a strongly inhomogeneous magnetic field. Chapter 2 provides the tools needed to evaluate the evolution of the magnetization to such pulse sequences implemented in the magnetic field generated by open NMR sensors. Besides providing the timing of some well-known pulse sequences and the phase cycle for the rf pulses required to eliminate unwanted signals generated by resonance offsets, typical features that must be considered at the time of implementing them are discussed. Moreover, a general expression is derived for the sensitivity of NMR experiments in inhomogeneous fields. The conclusions drawn from this analysis set the rules for optimizing magnet and rf coils to maximize the signal-to-noise ratio of single-sided sensors.

In the presence of inhomogeneous magnetic fields, relaxation times and diffusion coefficients are the key NMR parameters to assess sample composition and dynamics. To extract maximum information from relaxation and diffusion measurements, a number of numerical tools have been proposed. One of the most powerful ones is based on calculating the inverse Laplace transform of the signal decay to obtain the relaxation time distribution from a relaxation measurement. Although initially this transformation has been applied to 1D data sets, recently the possibility to extend it to 2D experiments in reasonable computation times has been reported. It opened the door to measure powerful multi-dimensional diffusion–relaxation and relaxation–relaxation correlation spectra useful to disentangle overlapping signals in a way similar to conventional multi-dimensional Fourier NMR spectroscopy. This methodology is described in detail in Chap. 3.

Although initially magnets were optimized to maximize sensitivity by maximizing magnetic field strength (at the expense of a strong static gradient) or the excited volume (at the expense of field strength), later, harder constrains have been imposed to the magnet design in order to generate convenient magnetic field profiles. Several of these magnets, including suitable rf coils, are presented in Chap. 4. A particularly useful field profile is the one that defines a uniform gradient along the depth direction. In the presence of such a field profile a slice parallel to the sensor surface can be exited at variable positions inside the object. Achieving depth resolution is useful to elucidate the depth structure of large samples. Moreover, furnishing these magnets with a set of coils that generate pulsed gradient fields along the two lateral directions, 3D spatial localization has been achieved. An important step toward a functioning open tomograph required the adaptation of single-point imaging sequences to work in inhomogeneous fields where multi-echo acquisition is mandatory to achieve maximum sensitivity. The different methods available to encode position during CPMG-like sequences are described in Chap. 5, where also suitable pulsed field gradient sequences are discussed to measure molecular velocity in the presence of a strong background gradient.

Accepting the field inhomogeneities as unavoidable for open sensors, pulse sequences specially designed to minimize off-resonance effects have been proposed. A particularly challenging problem that attracted considerable attention over the years was the possibility to recover spectroscopic information from experiments performed in inhomogeneous fields. For a long time this was thought to be impossible because the static magnetic field inhomogeneities are usually orders of magnitude larger than those created by the microscopic structure of the molecules to be detected. In addition, terms in the spin Hamiltonian arising from chemical shifts or field inhomogeneities are formally identical. As a natural consequence, no radiofrequency pulse sequence was deemed capable of differentiating both types of interactions. A breakthrough was achieved in 2001 when an innovative alternative to recover a chemical-shift resolved spectrum in an inhomogeneous magnetic field was presented. The technique is based on matching the spatial dependence of the static and the rf magnetic fields, to generate nutation echoes whose phase is only sensitive to chemical-shift differences. This methodology and its implementation in a single-sided magnet are discussed in Chap. 6.

Although for decades magnetic field inhomogeneity was assumed to be a handicap inherent to these sensors, during the last years an important step ahead toward recovering spectroscopic resolution has been reported. In particular, it has recently been proven that highly homogeneous fields can be generated even outside a magnet by furnishing the main magnet with a shim unit built from small and adjustable permanent magnet blocks. Moreover, a way has been suggested to combine magnetic materials with different temperature coefficients to build a magnet with a temperature-compensated magnetic field. The strategy followed to homogenize the stray field of a single-sided magnet is discussed in Chap. 7.

Over the years, single-sided sensors have found applications in diverse fields such as the non-destructive testing of rubber and polymer products, food and life stock analysis, as well as the state assessment of objects of cultural heritage. Applications in the area of biological tissue are discussed in Chap. 8, while those in the area of material science and quality control are presented in Chap. 9. Finally, Chap. 10 intends to familiarize the reader with the particular hardware requirements of single-sided NMR.

Aachen, Germany
December 2010

Federico Casanova
Juan Perlo
Bernhard Blümich

Contents

1 Single-Sided NMR 1
Federico Casanova, Juan Perlo, and Bernhard Blümich
1.1 Development of Open NMR Sensors 2
1.1.1 Well-Logging Tools 2
1.1.2 Mobile Single-Sided Sensors 3
1.1.3 The NMR-MOUSE 4
1.2 Methods for Mobile NMR 6
References 7

2 NMR in Inhomogeneous Fields 11
Federico Casanova and Juan Perlo
2.1 Introduction 11
2.1.1 Evolution of the Magnetization During a Pulse Sequence . 12
2.1.2 Separation of the Magnetization into Coherence Pathways 14
2.1.3 Numerical Calculation of the NMR Signal 20
2.2 Pulse Sequence Analysis 21
2.2.1 Single rf Pulse 21
2.2.2 The Generation of Hahn Echoes 23
2.2.3 The CPMG Sequence 28
2.2.4 Inversion and Saturation Recovery 35
2.2.5 Diffusion Measurements 42
2.3 The SNR in Inhomogeneous Fields 47
2.3.1 The Reciprocity Principle 48
2.3.2 Numerical Calculations of the SNR 49
2.3.3 An Analytical Solution for the SNR 55
References 55

3 Ex Situ Measurement of One- and Two-Dimensional Distribution Functions 57
Martin D. Hürlimann
3.1 Introduction 57
3.1.1 Relaxation 57

3.1.2 Diffusion ... 58
3.1.3 Diffusion–Relaxation Distribution Functions ... 58
3.2 Pulse Sequences and Spin Dynamics in Inhomogeneous Fields ... 59
3.2.1 Relaxation Measurement: Carr-Purcell-Meiboom-Gill Sequence ... 59
3.2.2 Diffusion Measurements with Static Gradients ... 62
3.2.3 T_1 Measurements in Inhomogeneous Fields ... 64
3.3 One-Dimensional Distribution Functions ... 64
3.3.1 Data Inversion ... 65
3.3.2 Regularization ... 67
3.3.3 Systematic Errors ... 68
3.3.4 Uncertainties ... 68
3.4 Two-Dimensional Diffusion–Relaxation Distribution Functions ... 69
3.4.1 Two-Dimensional Diffusion-Relaxation Measurements ... 69
3.4.2 Data Analysis ... 71
3.4.3 Interpretation of Distribution Functions ... 73
3.5 Applications of Two-Dimensional Distribution Functions ... 74
3.5.1 Two-Component Systems ... 75
3.5.2 Wettability ... 76
3.5.3 Complex Miscible Fluid ... 77
3.5.4 Structured Fluid ... 78
3.5.5 Pore Geometry of Porous Media ... 79
3.6 Conclusion ... 81
References ... 82

4 Magnets and Coils for Single-Sided NMR ... 87
Juan Perlo
4.1 Magnets ... 88
4.1.1 B_0 Perpendicular to the Sensor Surface, the Bar Magnet Geometry ... 89
4.1.2 B_0 Parallel to the Sensor Surface, the U-Shaped Geometry 93
4.1.3 Magnets for Depth Profiling ... 95
4.1.4 Sweet-Spot Magnets ... 99
4.2 RF Coils ... 102
4.2.1 Coils for Depth Profiling ... 103
4.2.2 Coils for Sweet-Spot Magnets ... 105
4.3 Gradient Coils ... 107
References ... 108

5 Single-Sided Tomography ... 111
Federico Casanova
5.1 Depth Resolution Using the Static Field Gradient ... 111
5.2 Spatial Encoding by Fourier Imaging ... 113
5.3 Multi-Echo Acquisition Schemes ... 117

5.3.1 RARE-Like Imaging Sequence ... 117
5.3.2 CPMG-CP for Quadrature Detection ... 121
5.4 Performance of the Multi-Echo Detection Scheme ... 125
5.4.1 Sensitivity Improvement ... 125
5.4.2 Relaxation and Diffusion Contrast ... 126
5.4.3 3D Imaging ... 128
5.5 Displacement Encoding ... 129
5.5.1 PFG Methods in Inhomogeneous Fields ... 130
5.5.2 Measurement of Velocity Distributions ... 135
5.6 Spatially Resolved Velocity Distributions ... 137
5.6.1 2D Velocity Maps ... 137
References ... 140

6 High-Resolution NMR in Inhomogeneous Fields ... 143
Vasiliki Demas, John M. Franck, Jeffrey A. Reimer, and Alexander Pines
6.1 Introduction ... 143
6.2 Approaches Based on Spin Interactions ... 144
6.3 *Ex Situ* NMR: Spatially Dependent "z-Rotations" ... 145
6.3.1 *Ex Situ* Matching: Compensating Static Field Inhomogeneities via Spatially Matched rf ... 147
6.3.2 Shim Pulses: Corrections Based on Gradient Modulations During an Adiabatic Double Passage ... 157
6.3.3 Adjusted Chirp Shim Pulses ... 160
6.4 Summary ... 162
References ... 162

7 High-Resolution Spectroscopy in Highly Homogeneous Stray Fields . 165
Ernesto P. Danieli
7.1 Sensor Design ... 166
7.1.1 Main Unit ... 166
7.1.2 Shim Unit ... 167
7.2 Shimming Magnetic Fields with Movable Permanent Magnets ... 170
7.2.1 Generation of Linear Terms Along y ... 171
7.2.2 Generation of Linear Terms Along x and z ... 172
7.2.3 Generation of Quadratic Terms x^2 and z^2 ... 172
7.3 Experimental Results ... 175
7.4 Shimming the Magnet to Higher Order ... 177
7.4.1 Improving Resolution and Working Volume Size ... 178
7.5 Temperature Compensation ... 181
7.6 Conclusions ... 185
References ... 185

8 Applications in Biology and Medicine ... 187
Bernhard Blümich
8.1 Skin ... 188
8.2 Tendon ... 192

8.3 Mummies and Bones ... 194
8.4 Unilateral Imaging of Biological Matter ... 196
8.5 Conclusions ... 198
References ... 199

9 Applications in Material Science and Cultural Heritage ... 203
Jürgen Kolz
9.1 Elastomers ... 204
9.1.1 Crosslink Density ... 204
9.1.2 Aging ... 206
9.1.3 Imaging ... 209
9.2 Hard Polymers ... 212
9.2.1 Ingress of Solvents ... 215
9.3 Cultural Heritage ... 217
References ... 219

10 Spectrometer Hardware ... 221
Jörg Felder
10.1 Single-Sided vs. Conventional: Systematic Differences ... 222
10.2 Frontend Design ... 223
10.2.1 Matching and Balancing ... 223
10.2.2 Transmit-Receive Switching ... 226
10.3 Transmitter Design ... 228
10.3.1 Conventional Power Amplifiers ... 229
10.3.2 Alternative Amplifier Designs ... 230
10.4 Receiver Design ... 233
10.4.1 Low-Noise Amplifier ... 234
10.4.2 Frequency Generation and Mixing ... 235
10.5 Digital Hardware ... 237
10.5.1 Frontend Signal Processor Selection ... 237
10.5.2 Digital Phase Sensitive Detector ... 239
References ... 239

Index ... 241

Contributors

Bernhard Blümich Institut für Technische Chemie und Makromolekulare Chemie, RWTH Aachen University, D-52074, Aachen, Germany, bluemich@mc.rwth-aachen.de

Federico Casanova Institut für Technische Chemie und Makromolekulare Chemie, RWTH Aachen University, D-52074, Aachen, Germany, fcasanova@mc.rwth-aachen.de

Ernesto P. Danieli Institut für Technische Chemie und Makromolekulare Chemie, RWTH Aachen University, D-52074, Aachen, Germany, edanieli@mc.rwth-aachen.de

Vasiliki Demas University of California, Berkeley, CA 94720, USA; T2 Biosystems, Inc. Cambridge, MA 02141, USA, vdemas@t2biosystems.com

Jörg Felder Institut für Hochfrequenztechnik, RWTH-Aachen University, D-52074 Aachen, Germany; Institute of Neuroscience and Medicine 4, Medical Imaging Physics, Forschungszentrum Juelich GmbH, 52425 Juelich, Germany, j.felder@fz-juelich.de

John M. Franck University of California, Berkeley, CA 94720, USA, jfrank@chem.ucsb.edu

Martin D. Hürlimann Schlumberger-Doll Research, Cambridge, MA 02141, USA, hurlimann@slb.com

Jürgen Kolz Institut für Technische Chemie und Makromolekulare Chemie, RWTH Aachen University, D-52074, Aachen, Germany; ACT GmbH, 52074 Aachen, Germany, juergen.kolz@act-aachen.com

Juan Perlo Institut für Technische Chemie und Makromolekulare Chemie, RWTH Aachen University, D-52074, Aachen, Germany, jperlo@mc.rwth-aachen.de

Alexander Pines University of California, Berkeley, CA 94720, USA, pines@berkeley.edu

Jeffrey A. Reimer University of California, Berkeley, CA 94720, USA, reimer@berkeley.edu

Chapter 1
Single-Sided NMR

Federico Casanova, Juan Perlo, and Bernhard Blümich

Since its discovery in 1945 [1, 2], nuclear magnetic resonance (NMR) has developed into an inexhaustible research field. It is exploited in several areas in physics, chemistry, biology, and medicine to extract unique information at the molecular level [3–8]. In chemistry, for example, it is considered to be one of the most powerful analytical tools to elucidate molecular structure, and in medicine it is routinely used for diagnostic imaging. Driven by the fact that sensitivity and spectral resolution increase with the magnetic field strength and homogeneity, magnets are built larger and larger over the years. Today, magnets are heavy and static devices installed in special NMR laboratories designed to shield electromagnetic interference and reduce magnetic field distortions in order to provide ideal experimental conditions (Fig. 1.1a). Besides the fact that samples of interest must be taken to the magnet, they must fit into the limited space available in the bore of the magnet. These issues are certainly a limitation when arbitrarily large samples require non-destructive analysis.

The complications involved with bringing NMR devices to the sample were already noted in the early days of NMR when it was recognized that NMR could contribute to in situ studies of rock formations. Measuring the properties of fluids confined to rock pores downhole by NMR triggered the development of what is called the inside-out NMR concept, where instead of placing the sample inside the magnet, the apparatus is placed inside the object, the earth in this case (Fig 1.1b). The sensor is required to produce strong magnetic and rf stray fields that penetrate the sample under study. Applications in the field already appeared in the early 1950s driven by the incentive of oil companies [9–18]. The efforts undertaken by this community demonstrated that non-destructive NMR studies can successfully be carried out with mobile NMR instruments.

F. Casanova, J. Perlo, and B. Blümich (✉)
Institut für Technische Chemie und Makromolekulare Chemie, RWTH Aachen University, D-52074 Aachen, Germany

F. Casanova et al. (eds.), *Single-Sided NMR*,
DOI 10.1007/978-3-642-16307-4_1, © Springer-Verlag Berlin Heidelberg 2011

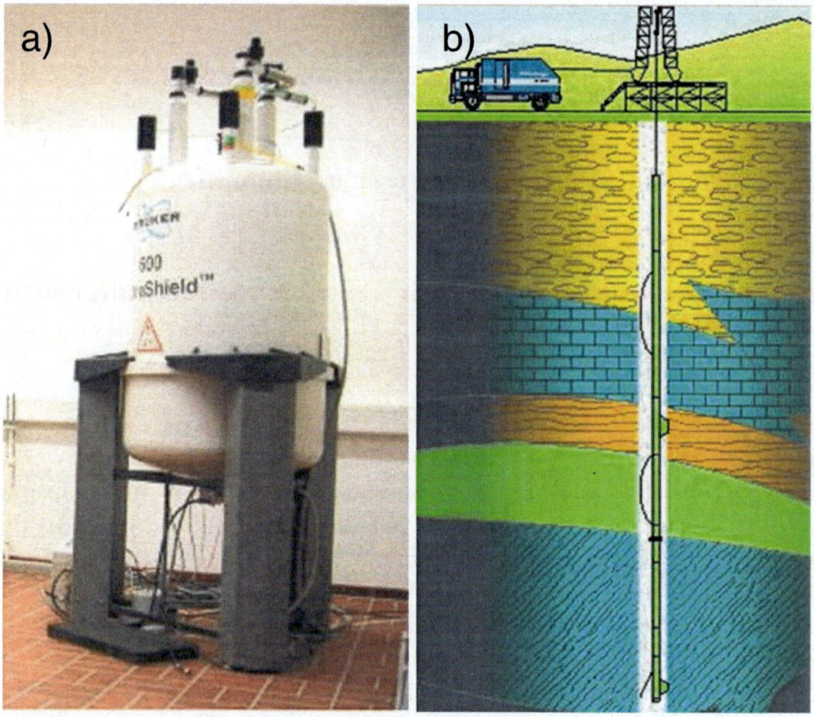

Fig. 1.1 NMR instrumentation. *Left*: 600 MHz NMR spectrometer for chemical analysis at the Institute of Technical Chemistry and Macromolecular Chemistry of RWTH-Aachen University. *Right*: Schematic illustration of well-logging NMR

1.1 Development of Open NMR Sensors

1.1.1 Well-Logging Tools

For more than 30 years well-logging tools used the earth's magnetic field as polarization field, but the low sensitivity and the poor selectivity to signals from the bore-hole fluid and the surrounding wall limited their commercial success. The break-through came with the incorporation of permanent magnets that produce magnetic fields 1,000 times stronger than the earth's magnetic field. The magnet geometry developed at Los Alamos and known as the Jackson geometry (Fig. 1.2a) was the first working prototype [11]. It was followed by two different ones developed independently by the companies Numar (now Huliburton) and Schlumberger, which nowadays are the two leading service companies for well-logging NMR (Fig. 1.2b, c).

Figure 1.2b, c shows schematics of both tools to illustrate the two different magnet concepts. While the Numar tool generates a magnetic field with a relatively large static gradient, the Schlumberger one generates a so-called sweet spot where several spatial derivatives of the magnetic field are zero in a point external to the magnet.

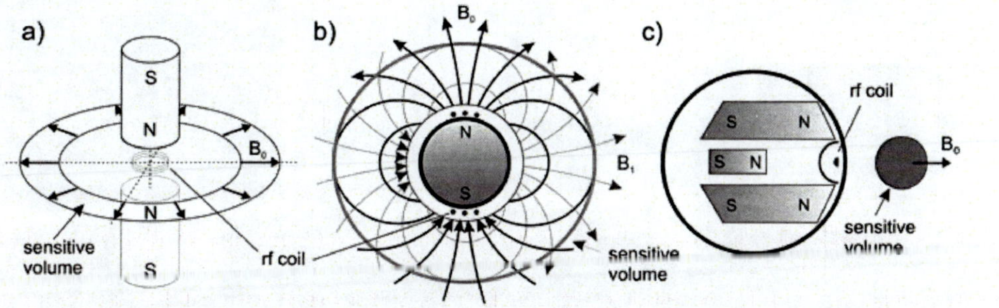

Fig. 1.2 Well-logging NMR sensors. (**a**) Jackson geometry. (**b**) Numar tool. (**c**) Schlumberger tool

Although the original idea was to generate a magnetic field relatively homogeneous over a large volume outside the magnet, to define a large signal-bearing volume, the Numar tool accepts a static magnetic field with a considerable gradient, but increases the signal-to-noise ratio by generating a stronger magnetic field over a volume extended along the tool. The scheme uses a long cylindrical magnet, magnetized transversal to its axis to generate a dipolar magnetic field in the x–y plane. The RF coil is wound around the magnet to generate a dipolar field perpendicular to $\mathbf{B}_0$. As the magnitude of B_0 is constant in rings and decays with the radial distance as $1/r^2$, by choosing different excitation frequencies, the signal originating from cylindrical shells of particular radius can be detected. In parallel Schlumberger developed a tool inspired in the Jackson geometry. They pushed the sensor to the wall of the borehole and exploits the magnetic stray field at this preferential side in order to increase the signal-to-noise by maximizing both coil filling factor and magnetic field strength. The magnet configuration uses two bar magnets magnetized transversal to the borehole axis that generates a point inside the earth formation where many spatial derivatives of the static magnetic field vanished. A third magnet is added to increase the field strength but still keeping the sweet spot (Fig. 1.2c). Although, at first sight, the tool sacrifices sensitive volume compared to the one generated by the Jackson geometry, because a single spot instead of a ring is detected, the volume can be regained by increasing the length of the magnet array.

1.1.2 Mobile Single-Sided Sensors

The community of scientists and engineers engaged in developing well-logging NMR also realized that similar single-sided sensors can be put to use outside the oil industry [11, 12, 14, 17]. The early interest was in the detection of moisture in building materials, soil, and food [14, 17, 18]. But other applications have been suggested as well, for example, in medicine, material science, and process control [11], and devices suitable for use of NMR in the production line have been produced by Southwest Research and others [12, 14, 15, 17, 19, 20]. Other applications were envisioned in nondestructive testing, process, and quality control [11, 12, 16]. Figure 1.3 depicts early instruments for moisture measurements in bridge decks

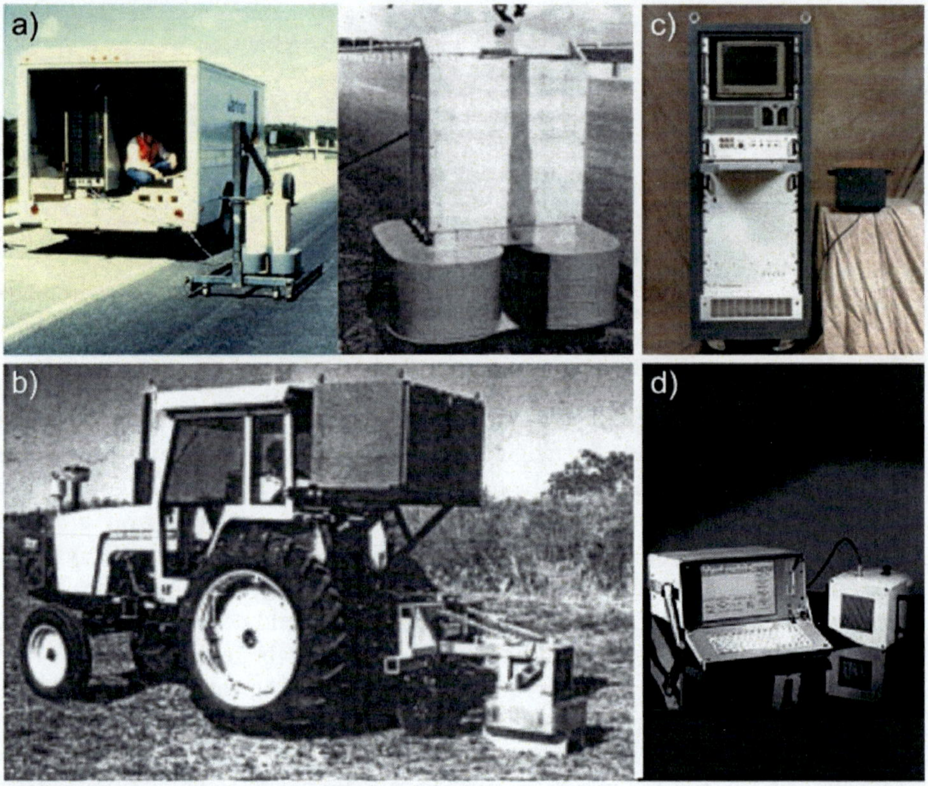

Fig. 1.3 NMR sensors for online detection. (**a**) Unilateral NMR device with electromagnets to measure moisture in bridge decks. (**b**) Tractor carrying an electromagnet to measure moisture in soil for calibration of satellite photography (Photos courtesy of G. A. Matzkanin). (**c**) One-sided access NMR (OSA). Spectrometer rack and sensor by Southwest Research. (**d**) Portable console and OSA sensor by Fraunhofer Institute for Nondestructive Testing

(Fig. 1.3a) [16] and in soil (Fig. 1.3b). These instruments were large, weighing up to 300 kg, employed mainly electromagnets, and operated at low frequencies like 3 MHz. An important step toward size reduction was the use of permanent magnets to generate the static magnetic fields. It not only reduced the size and weight but also eliminated most of the power consumption of these devices. An example on this realization is the one-sided access (OSA) sensor by Southwest Research (Fig. 1.3c) [12], which was subsequently fitted with a portable console by the Fraunhofer Institute of Nondestructive Testing (Fig. 1.3d).

1.1.3 The NMR-MOUSE

Although the idea of using this type of equipment has been formulated over two decades ago [9–13], systematic investigations of unilateral NMR did not appear in the scientific literature until the mid 1990's with the publication of the NMR-

MOUSE [21, 22]. In recognition of the fact, that most contrast filters used in imaging [8] do not need a homogeneous magnetic field to function, and that space encoding in imaging requires an inhomogeneous magnetic field, the need for expensive homogeneous magnetic fields in imaging and materials characterization was questioned, and the most simple NMR device was sought that would provide NMR information equivalent to that contained in a pixel of an NMR image. The NMR-MOUSE is a small unilateral NMR sensor that generates a polarizing magnetic field up to 0.5 T with an adjustable gradient. It has been used and explored in such diverse fields as the non-destructive testing of rubber and polymer products [23–30], food and livestock analysis [24, 31–34], as well as the state assessment of objects of cultural heritage [28, 35–46]. This in turn stimulates the development and understanding of NMR in highly inhomogeneous fields.

Today two principal geometries of the NMR-MOUSE are available (Fig. 1.4). One is the horseshoe geometry obtained by opening up the old-fashioned NMR magnet (left) [12, 21, 47] and the bar magnet (right), the most simple geometry with a suitable coil at one of its faces [48–50]. When limiting the size to a hand-held device and the weight to less than 2 kg, the maximum depth of the sensitive volume is about 10 mm. By tuning the rf frequency, the depth or distance to the sensor surface where the sensitive volume is generated can be shifted (Fig. 1.4). Optimizing the magnet and the rf coil geometries, a flat sensitive volume can be generated at variable distance from the sensor surface. The selection of a planar sensitive volume is convenient to gain depth resolution useful to resolve the depth structure of large samples. A sample profile is acquired by shifting the position of the slice through the object, which can be achieved by simply rerunning the resonance frequency or adjusting the distance between the sensor and the object.

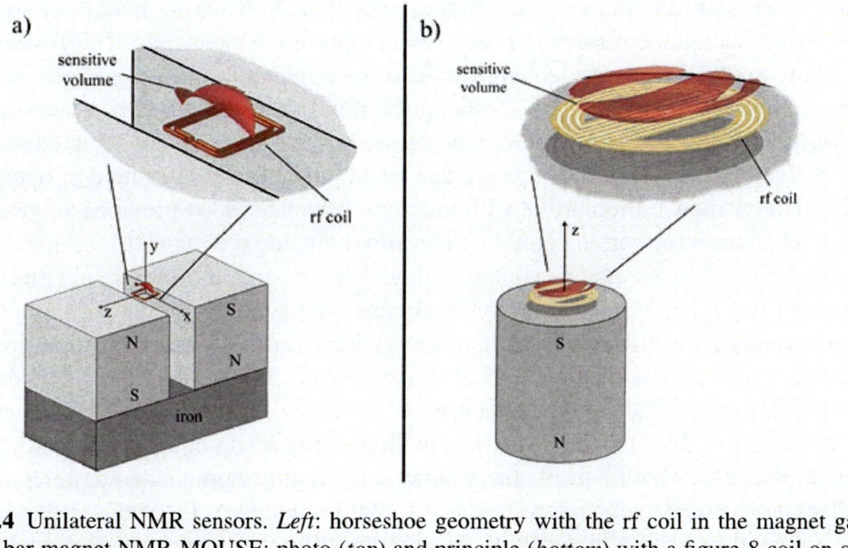

Fig. 1.4 Unilateral NMR sensors. *Left*: horseshoe geometry with the rf coil in the magnet gap. *Right*: bar magnet NMR-MOUSE; photo (*top*) and principle (*bottom*) with a figure-8 coil on one of the magnet faces

One major challenge has been the development of unilateral tomographs that provide lateral resolution across the sensitive volume. Although several ideas about different approaches to tackle field profiling, space encoding, and signal acquisition have appeared in the patent literature [49, 51–57], no practical realization of a unilateral or open tomograph has been reported so far. In recent years, sensors have been fitted with gradient coils to achieve spatial localization across the sensitive slice [58–60]. Later, it was shown that even velocity can be encoded in the presence of inhomogeneous fields, NMR [61, 62]. As encoding and detection are done in the presence of a static gradient, these methods use pure phase encoding combined with multi-acquisition schemes to maximize sensitivity.

By now different magnet geometries have been explored for unilateral NMR [49, 50, 58, 63]. The main goals in the design are often formulated as a region of homogeneous field remote from the sensor surface [9, 49, 64–66] or a field profile which varies linearly with the distance from the sensor surface [67, 68], and large penetration depth, although a plane of constant field strength at a given distance from the sensor surface is sufficient for slice selective imaging [69]. With these minimal constraints small devices can be constructed for nondestructive materials testing.

1.2 Methods for Mobile NMR

Initially, only the relaxation times T_1 and T_2 could be measured by single or multi-echo sequences like the Hahn-echo or Carr-Purcell-Meiboom-Gill (CPMG) sequences [70–72]. Once the acquisition of NMR signals in strongly inhomogeneous magnetic fields was better understood, methodologies to acquire more and more information with these tools were developed. Nowadays, in addition to relaxation times, a number of methods have been reported to measure self-diffusion coefficients, multiple-quantum coherences and their relaxation times, images, velocity distributions, velocity images, multi-dimensional maps correlating different relaxation times, and diffusion coefficients. Last but not least, it has been shown that even chemical shift-resolved spectra can be acquired in the stray field of open magnets. This demonstration broke with the myth accepted in the previous 50 years that spectroscopic information cannot be measured outside the magnet.

The arsenal of methods available today to interrogate the sample in situ is summarized in Fig. 1.5. It shows a multi-dimensional scheme similar to the one used in conventional NMR, where in its more general form the magnetization prepared during a first period then is labeled during an evolution interval, passes through a mixing period, and is finally measured, typically with a CPMG sequence, in a detection time. In fact, the direct acquisition domain can encode frequency in the sense of chemical shift in magnets generating highly homogeneous fields (single pulse) or transverse relaxation (Hahn or CPMG sequence). The indirect dimensions are sampled in a repetitive fashion varying the filter parameter in a systematic way to encode relaxation times, position, molecular diffusion, displacement, or chemical

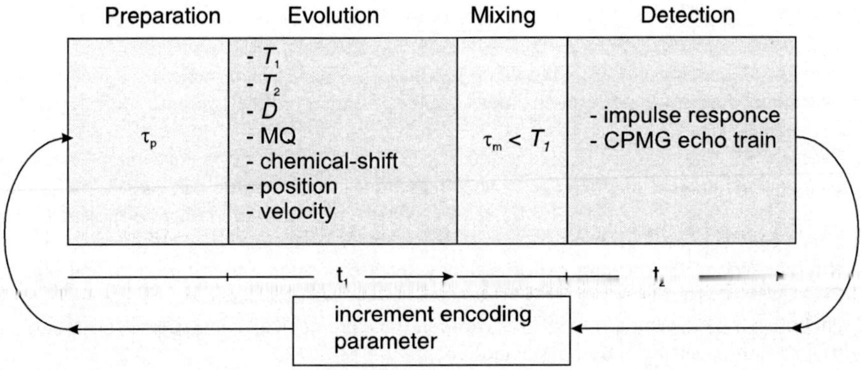

Fig. 1.5 Data acquisition in inhomogeneous fields. During the evolution period, the sequence is made sensitive to relaxation, diffusion, displacement, position, or even chemical shift and, then, the signal is acquired in the detection period with a single- or multi-echo train. A fixed mixing time can be used to connect evolution and detection periods

shift. Finally, the mixing time can be varied to study the magnetization transfer between preparation and detection periods. The different methods are described in detail in the different chapters of this book, which focus first on the understanding of NMR in inhomogeneous fields and describe the state of the art of the hardware, then describe the available methodologies, and finally discuss some of the applications explored over the last years.

References

1. Bloch F (1946) Nuclear induction. Phys Rev 70(7–8):460–474
2. Purcell EM, Torrey HC, Pound RV (1946) Resonance absorption by nuclear magnetic moments in a solid. Phys Rev 69(1–2):37–38
3. Abragam A (1983) Principles of nuclear magnetism. Oxford University Press, Oxford
4. Fukushima E, Roeder SBW (1986) Experimental pulse NMR: a nuts and bolts approach. Addison Wesley, New York, NY
5. Slichter CP (1990) Principles of magnetic resonance, 3rd edn. Springer, Berlin
6. Ernst RR, Bodenhausen G, Wokaun A (1990) Principles of nuclear magnetic resonance in one and two dimensions. Oxford University Press, Cambridge
7. Callaghan PT (1991) Principles of nuclear magnatic resonance microscopy. Clarendon Press, Oxford
8. Blümich B (2000) NMR imaging of materials. Clarendon Press, Oxford
9. Cooper RK, Jackson JA (1980) Remote (inside-out) NMR. 1. Remote production of a region of homogeneous magnetic-field. J Magn Resonance 41(3):400–405
10. Burnett LJ, Jackson JA (1980) Remote (inside-out) NMR. 2. Sensitivity of NMR detection for external samples. J Magn Reson 41(3):406–410
11. Jackson JA, Burnett LJ, Harmon JF (1980) Remote (inside-out) NMR. 3. Detection of nuclear magnetic-resonance in a remotely produced region of homogeneous magnetic-field. J Magn Reson 41(3):411–421
12. Matzkanin GA (1989) A review of nondestructive of composites using NMR, in: nondestructive characterization of materials. Springer, Berlin

13. Nordon A, McGill CA, Littlejohn D (2001) Process NMR spectrometry. Analyst 126(2):260–272
14. Paetzold RF, Delosantos A, Matzkanin GA (1987, Mar) Pulsed nuclear-magnetic-resonance instrument for soil-water content measurement – sensor configurations. Soil Sci Soc Am J 51(2):287–290
15. Paetzold RF, Matzkanin GA, Delosantos A (1985) Surface soil-water content measurement using pulsed nuclear magnetic-resonance techniques. Soil Sci Soc Am J 49(3):537–540
16. Hogan BJ (1985) One-sided NMR sensor system measures soil/concrete moisture. Design News, May 5
17. Rollwitz WL (1985) Using radiofrequency spectroscopy in agricultural applications. Agric Eng 66(5):12–14
18. Nicholls CI, Delosantos A (1991) Hydrogen transient nuclear-magnetic-resonance for industrial moisture sensing. Drying Technol 9(4):849–873
19. Rollwitz WL, Persyn GA (1971) On-stream NMR measurements and control. J Am Oil Chem Soc 48(2):59–66
20. Maciel GE (1994) NMR in industrial process control and quality control. In: Nuclear magnetic resonance in modern technology. Kluwer, Academic, Dordrecht
21. Eidmann G, Savelsberg R, Blümler P, Blümich B (1996, Sep) The NMR-MOUSE, a mobile universal surface explorer. J Magn Reson Ser A 122(1):104–109
22. Blümich B, Blümler P, Eidmann G, Guthausen A, Haken R, Schmitz U, Saito K, Zimmer G (1998, June) The NMR-mouse: construction, excitation, and applications. Magn Reson Imaging 16(5–6):479–484
23. Zimmer G, Guthausen A, Blümich B (1998) Characterization of technical elastomers by the NMR-mouse. Solid State Nucl Magn Reson 12:183–190
24. Guthausen G, Guthausen A, Balibanu R, Eymael F, Hailu K, Schmitz U, Blümich B (2000) Soft-matter analysis by the NMR-mouse. Macromol Mater Eng 276/277:25–37
25. Guthausen A, Zimmer G, Eymael R, Schmitz U, Blümler P, Blümich B (1998) Soft-matter relaxation by the NMR MOUSE. In: Spatially resolved magnetic resonance. Wiley-VCH, Weinheim
26. Guthausen A, Zimmer G, Blümler P, Blümich B (1998, Jan) Analysis of polymer materials by surface NMR via the mouse. J Magn Reson, 130(1):1–7
27. Blümich B (2001) NMR for product and quality control of elastomers. Kautschuk Gummi Kunststoffe 54:188–190
28. Blümich B, Bruder M (2003) Mobile NMR zur qualitätskontrolle. Kautschuk Gummi Kunststoffe 56:90–94
29. Blümich B, Anferova S, Casanova F, Kremer K, Perio J, Sharma S (2004, July) Unilateral NMR: principles and a applications to quality control of elastomer products. Kautschuk Gummi Kunststoffe 57(7–8):346–349
30. Blümich B, Anferova S, Kremer K, Sharma S, Herrmann V, Segre A. (2003) Unilateral nuclear magnetic resonance for quality control. Spectroscopy 18:18–32
31. Guthausen G, Todt H, Burk W (2002) Industrial quality control with time-domain NMR. Bruker Spin Rep 150/151:22–53
32. Pedersen HT, Ablett S, Martin DR, Mallett MJD, Engelsen SB (2003, Nov) Application of the NMR-mouse to food emulsions. J Magn Reson 165(1):49–58
33. Veliyulin E, van der Zwaag C, Burk W, Erikson U (2005, June) In vivo determination of fat content in atlantic salmon (salmo salar) with a mobile NMR spectrometer. J Sci Food Agric 85(8):1299–1304
34. Martin DR, Ablett S, Pedersen HT, Mallett MJD (2003) The NMR mouse: its applications to food science. Magn Resona Food Sci Latest Dev 286:54–61
35. Blümich B, Anferova S, Sharma S, Segre AL, Federici C (2003, Apr)Degradation of historical paper: nondestructive analysis by the NMR-mouse. J Magn Reson 161(2):204–209
36. Sharma S, Casanova F, Wache W, Segre A, Blümich B (2003, Apr) Analysis of historical porous building materials by the NMR-mouseo (r). Magn Reson Imaging 21(3–4):249–255

37. Proietti N, Capitani D, Pedemonte E, Blümich B, Segre AL (2004, Sep) Monitoring degradation in paper: non-invasive analysis by unilateral NMR. Part II. J Magn Reson 170(1):113–120
38. Casieri C, Senni L, Romagnoli M, Santamaria U, De Luca F (2004, Dec) Determination of moisture fraction in wood by mobile NMR device. J Magn Reson 171(2):364–372
39. Viola I, Bubici S, Casieri C, De Luca F (2004, July) The codex major of the collection altaempsiana: a non-invasive NMR study of paper. J Cult Heritage 5(3):257–261
40. Casieri C, Bubici S, Viola I, De Luca F (2004, Sep) A low-resolution non-invasive NMR characterization of ancient paper. Solid State Nucl Magn Reson 26(2):65–73
41. Casieri C, De Luca F, Fantazzini P (2005, Feb) Pore-size evaluation by single-sided nuclear magnetic resonance measurements: compensation of water self-diffusion effect on transverse relaxation. J Appl Phys 97(4):043901
42. Viel S, Capitani D, Proietti N, Ziarelli F, Segre AL (2004, July) NMR spectroscopy applied to the cultural heritage: a preliminary study on ancient wood characterisation. Appl Phys A Mater Sci Process 79(2):357–361
43. Poli T, Toniolo L, Valentini M, Bizzaro G, Melzi R, Tedoldi F, Cannazza G (2007, Apr) A portable NMR device for the evaluation of water presence in building materials. J Cult Heritage 8(2):134–140
44. Perlo J, Casanova F, Blümich B (2006) Advances in single-sided NMR. In: Modern magnetic resonance. Springer, Berlin
45. Presciutti F, Perlo J, Casanova F, Gloggler S, Miliani C, Blümich B, Brunetti BG, Sgamellotti A (2008, July) Noninvasive nuclear magnetic resonance profiling of painting layers. Appl Phys Lett 93(3):033505
46. Blümich B, Casanova F, Perlo J, Presciutti F, Anselmi C, Doherty B (2010, June) Noninvasive testing of art and cultural heritage by mobile NMR. Acc Chem Res 43(6):761–770
47. Anferova S, Anferov V, Adams M, Blümler P, Routley N, Hailu K, Kupferschlager K, Mallett MJD, Schroeder G, Sharma S, Blümich B (2002, Mar) Construction of a NMR-mouse with short dead time. Concepts Magn Reson 15(1):15–25
48. Blümich B, Anferov V, Anferova S, Klein M, Fechete R, Adams M, Casanova F (2002, Dec) Simple NMR-mouse with a bar magnet. Concepts Magn Reson 15(4):255–261.
49. Fukushima E, Jackson JA (2002) Unilateral magnet having a remote uniform field region for nuclear magnetic resonance. US Patent, 6489872
50. Rahmatallah S, Li Y, Seton HC, Mackenzie IS, Gregory JS, Aspden RM (2005, Mar) NMR detection and one-dimensional imaging using the inhomogeneous magnetic field of a portable single-sided magnet. J Magne Reson 173(1):23–28
51. Crowley CW, Rose FH (1994) Remotely positioned MRI system. US Patent, 5304930
52. Fukushima E, Rath AR, Roeder SBW (1988) Apparatus for unilaterally generating a region of uniform magnetic field. US Patent, 4721914
53. Kikis D (1995) Magnetic resonance imaging system. US Patent, 5390673
54. Krieg R (1999) Pulssequenz für ein kernspintomographiegerät mit vorgegebener, zeitlich konstanter inhomogenität in einer raumrichtung und vorrichtung zur ausführung der pulssequenz. DE Patent, 195 11 835 C 2
55. Pissanetzky S (1992) Structured coil electromagnets for magnetic resonance imaging and method for fabricating the same. US Patent, 5382904
56. Pulyer YM (1998) Planar open magnet MRI system. US Patent, 5744960
57. Westphal M, Knüttel B (1998) Magnet arrangement for an NMR tomography system in particular for skin and surface examinations. US Patent, 5959454
58. Prado PJ, Blümich B, Udo Schmitz U (2000) One-dimensional imaging with a palm-size probe. J Magn Reson 144:200–206
59. Casanova F, Blümich B (2003, July) Two-dimensional imaging with a single-sided NMR probe. J Magn Reson 163(1):38–45
60. Perlo J, Casanova F, Blümich B (2004, Feb) 3D imaging with a single-sided sensor: an open tomograph. J Magn Reson 166(2):228–235

61. Casanova F, Perlo J, Blümich B, Kremer K (2004, Jan) Multi-echo imaging in highly inhomogeneous magnetic fields. J Magn Reson 166(1):76–81
62. Casanova F, Perlo J, Blümich B (2004, Nov) Velocity distributions remotely measured with a single-sided NMR sensor. J Magn Reson 171(1):124–130
63. Prado PJ (2001, Apr) NMR hand-held moisture sensor. Magn Reson Imaging 19(3–4):505–508
64. Kleinberg RL (1996) Well logging. In: Encyclopedia of NMR. Wiley-Liss, New York
65. Marble AE, Mastikhin IV, Colpitts BG, Balcom BJ (2005) An analytical methodology for magnetic field control in unilateral NMR. J Magn Reson 174(1):78–87
66. Pulyer YM, Hrovat MI (2002) Generation of remote homogeneous magnetic fields. IEEE Trans Magn 38:1553–1563
67. Popella H, Henneberger G (2001) Design and optimization of the magnetic circuit of a mobile nuclear magnetic resonance device for magnetic resonance imaging. COMPEL 20:269–278
68. Glover PM, Aptaker PS, Bowler JR, Ciampi E, McDonald PJ (1999, July) A novel high-gradient permanent magnet for the profiling of planar films and coatings. J Magn Reson 139(1):90–97
69. Casanova F, Perlo J, Blümich B (2005) Depth profiling by single-sided NMR. In: NMR in chemical engineering. Wiley-VCH, Weinheim
70. Hahn EL (1950) Spin echoes. Phys Rev 80(4):580–594
71. Carr HY, Purcell EM (1954) Effects of diffusion on free precession in nuclear magnetic resonance experiments. Phys Rev 94(3):630–638
72. Meiboom S, Gill D (1958) Modified spin-echo method for measuring nuclear relaxation times. Rev Sci Instrum 29(8):688–691

Chapter 2
NMR in Inhomogeneous Fields

Federico Casanova and Juan Perlo

2.1 Introduction

Besides the redesigning of the hardware to excite and detect NMR signals from sample volumes external to the sensor, the field of single-sided NMR has required special effort for adapting or developing new measurement techniques suitable to work in the presence of inhomogeneous B_0 and B_1 fields [1–4]. When large samples are studied in the presence of an inhomogeneous B_0 field, the spectral bandwidth of the spin system easily exceeds the rf field strength B_1 and all rf pulses act as selective pulses. The spatial dependence of both the resonance frequency and the amplitude of the radiofrequency field makes it impossible to impart a uniform rotation to all excited spins, leading to a distribution of flip angles across the sample. Although the pulse can be set to define a 90° rotation in a certain part of the object chosen to be on-resonance, the magnetization in near voxels will experience an α rotation quickly departing from the desired angle. Consequently, instead of converting longitudinal magnetization into transverse magnetization (assuming the pulse is applied to the sample in thermal equilibrium) a pulse α takes part of the magnetization to the transverse plane, but leaves considerable magnetization along the longitudinal axis. These two types of magnetization evolve in completely different ways during a subsequent free evolution period and they are referred to as different coherence pathways. As we will see, pulse imperfections give rise to a number of coherence pathways that dramatically increase with the number of pulses. Each of these pathways exhibits a characteristic attenuation by relaxation and molecular self-diffusion and depends on the phase of the rf pulses according to the transitions caused by the pulses. While longitudinal magnetization relaxes with T_1 and is not affected by the presence of a field gradient, transverse magnetization relaxes with T_2 and is spread out by a B_0 gradient. Moreover, whereas the transverse magnetization generated by a single pulse has a phase proportional to the phase of the pulse, the remaining longitudinal magnetization has no memory of such phase. The unwanted interference

F. Casanova and J. Perlo (✉)
Institut für Technische Chemie und Makromolekulare Chemie, RWTH Aachen University, D-52074 Aachen, Germany

F. Casanova et al. (eds.), *Single-Sided NMR*,
DOI 10.1007/978-3-642-16307-4_2,

of the different pathways during signal detection has required re-examination of the spin system response to most pulse sequences known from NMR in homogeneous fields in order to find cycles for the phases of the rf pulses that lead to elimination of non-desired pathways.

This chapter provides the tools necessary to calculate the evolution of a non-interacting $\frac{1}{2}$ spin system in inhomogeneous B_0 and B_1 fields. To describe the dynamics of the magnetization in a vectorial picture, the effect of rf pulses and free evolution periods is first described in the $\{M_x, M_y, M_z\}$ basis set. Then, a basis set is described which is more convenient to separate the magnetization into the different coherence pathways created by an inhomogeneous and off-resonance rf field. This second formalism is more convenient to identify and filter non-desired signals. Both formalisms provide the dynamics of the spin system for a given B_0 and B_1 spatial distribution. In the next sections these numerical tools are used to calculate the response of a spin system to well-known sequences such as the Hahn echo [5]; CPMG (Carr-Purcel-Meiboom-Gill) [6, 7]; inversion and saturation recovery; and stimulated echo, which are the key sequences to measure longitudinal and transverse relaxation times; and molecular self-diffusion. The analysis is made first by including the off-resonance condition defined in an inhomogeneous B_0 field, but assuming a homogeneous B_1 field. Then, the complications introduced by the excitation with an inhomogeneous radiofrequency field are evaluated by including in the simulations the field generated by a surface rf coil. Finally, how to calculate the signal-to-noise ratio in a strongly inhomogeneous field is discussed. Based on the provided equations both magnet and rf coil geometry can be optimized in order to maximize the sensitivity of the sensor.

2.1.1 Evolution of the Magnetization During a Pulse Sequence

Every pulse sequence can be considered as a succession of two types of events, i.e., rf pulses and free evolution periods. The effect of each of these events on the magnetization can be described by a rotation represented by a 3 × 3 matrix. Thus, the evolution of the components of the magnetization vector $\mathbf{M}(t) = (M_x, M_y, M_z)$ under a pulse sequence composed of n events involves simply the product of n matrixes:

$$\mathbf{M}(t) = \prod_{i=1}^{n} R_{\hat{\eta}_i}(\epsilon_i)\mathbf{M}(0),$$

where $\mathbf{M}(0) = (0, 0, M_0)$ and $\mathbf{M}(t)$ are the equilibrium magnetization and the magnetization at a time t after the application of the sequence, respectively [8]. The $R_{\hat{\eta}_i}$ matrices represent rotations by an angle ϵ_i about an axis $\hat{\eta}_i$. In the case of applying an rf field with amplitude B_1, frequency ω_{rf}, and phase ϕ, the magnetization precesses about an effective field of amplitude

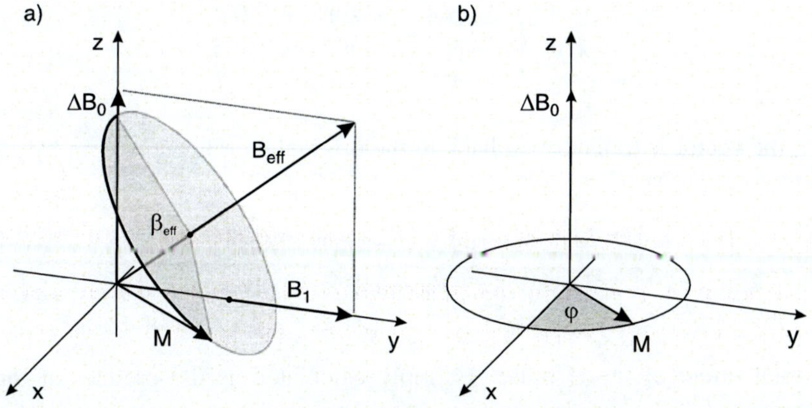

Fig. 2.1 (**a**) Precession of the magnetization vector about the tilted effective magnetic field B_{eff} during the application of an rf field with amplitude B_1, phase $\pi/2$, and offset resonance $\Delta\omega_0 = \gamma\Delta B_0$. (**b**) Precession about the off-set field ΔB_0 during a free evolution period, which is simply a rotation about the z-axis

$$B_{\text{eff}} = \sqrt{B_1^2 + \Delta B_0^2}, \tag{2.1}$$

which is tilted with respect to the z-axis by an angle

$$\theta = \tan^{-1}(B_1/\Delta B_0), \tag{2.2}$$

where $\Delta B_0 = B_0 - \omega_{\text{rf}}/\gamma$. Figure 2.1a shows the cone described by the magnetization precessing about the effective field for an rf pulse applied with phase $\phi = \pi/2$. The effective angle nutated about B_{eff} during a pulse of duration t_{p} is given by:

$$\beta_{\text{eff}} = \gamma B_{\text{eff}} t_{\text{p}}. \tag{2.3}$$

The components of the magnetization after the rf pulse can be calculated by rotating the vector about the effective field by an angle β_{eff}. To perform this rotation it is convenient to transform the magnetization before the pulse to a local frame with the z-axis along B_{eff} and the x-axis lying in the plane defined by the vectors B_{eff} and B_1. Depending on the phase ϕ of the rf pulse the general rotation is given by

$$\mathbf{A} = \begin{pmatrix} \cos\theta\cos\phi & \cos\theta\sin\phi & -\sin\theta \\ -\sin\phi & \cos\phi & 0 \\ \sin\theta\cos\phi & \sin\theta\sin\phi & \cos\theta \end{pmatrix}.$$

Once the magnetization vector is in the new basis, it is rotated by an angle β_{eff} about the effective field by applying the matrix

$$\mathbf{B} = \begin{pmatrix} \cos\beta_{\text{eff}} & -\sin\beta_{\text{eff}} & 0 \\ \sin\beta_{\text{eff}} & \cos\beta_{\text{eff}} & 0 \\ 0 & 0 & 1 \end{pmatrix}.$$

Then, the vector is transformed back to the original coordinate system by applying the inverse rotation

$$\mathbf{A}^{-1} = \begin{pmatrix} \cos\theta\cos\phi & -\sin\phi & \sin\theta\cos\phi \\ \cos\theta\sin\phi & \cos\phi & \sin\theta\sin\phi \\ -\sin\theta & 0 & \cos\theta \end{pmatrix}. \tag{2.4}$$

The total effect of the rf pulse is simply calculated as the product of the three matrices

$$\mathbf{P} = \mathbf{A}^{-1}\mathbf{B}\mathbf{A}. \tag{2.5}$$

During free evolution periods, the magnetization simply precesses about the offset field ΔB_0 with a frequency $\Delta\omega_0$ (see Fig. 2.1b), and the components of the magnetization after a time τ can be calculated as $\mathbf{M}(t+\tau) = \mathbf{E}\mathbf{M}(t)$, where

$$\mathbf{E} = \begin{pmatrix} e_2(\tau)\cos\Delta\omega_0\tau & e_2(\tau)\sin\Delta\omega_0\tau & 0 \\ e_2(\tau)\sin\Delta\omega_0\tau & e_2(\tau)\cos\Delta\omega_0\tau & 0 \\ 0 & 0 & e_1(\tau) \end{pmatrix}. \tag{2.6}$$

The effects of longitudinal and transverse relaxation are included during free evolution periods via the attenuation factors $e_1(\tau) = \exp(-\tau/T_1)$ and $e_2(\tau) = \exp(-\tau/T_2)$, but they are usually neglected during rf pulses. The factor e_1 includes T_1 attenuation but does not take into account the magnetization created during each free evolution period. To include the new magnetization after a free evolution τ, a vector $\mathbf{M}_{\text{new}}(\tau) = (0, 0, 1 - \exp(-\tau/T_1))$ must be added to the resultant magnetization at the end of the evolution period. In this way, after a free evolution τ, the magnetization is given by $\mathbf{M}(t+\tau) = \mathbf{E}\mathbf{M}(t) + \mathbf{M}_{\text{new}}(\tau)$. Using this formalism, the evolution of the magnetization under defined B_0 and B_1 fields can be calculated at any time of the sequence by applying the right number of rotations to the initial magnetization vector. The strategy to integrate the total signal response of a spin system to a particular sequence requires partitioning of the sample into small voxels where a uniform resonance frequency and rf amplitude can be defined. In Sect. 2.2 this approach to calculate the dynamics of the magnetization will be used to analyze the response to different pulse sequences.

2.1.2 Separation of the Magnetization into Coherence Pathways

The formalism presented in Sect. 2.1.1 describes the evolution of the magnetization in the vector basis set $\{M_x, M_y, M_z\}$. This basis set is convenient from a graphical point of view, because it provides the trajectory of the magnetization vector during the pulse sequence. However, when a number of pulses is applied with arbitrary phases, separation of the magnetization into different pathways with defined evolution becomes extremely complicated, if not impossible. In this basis set, longitudinal and transverse magnetization can be separated, but dephasing and rephasing transverse magnetization cannot be identified in a simple way. An example where these two types of coherence are generated is in a spin echo sequence. In this case, magnetization dephasing during the time between the first and second rf pulses is converted into refocusing magnetization that leads to the echo formation at a defined time after the second pulse. However, if the second pulse does not define a perfect 180° rotation, part of the dephasing magnetization keeps dephasing after the second pulse. This coherence is known to be responsible for one of the four echoes generated after the application of a third rf pulse. A basis set where the magnetization is split into the three mentioned coherence states is the one defined as

$$\begin{aligned} M_{+1} &= M_x + iM_y, \\ M_{-1} &= M_x - iM_y, \\ M_0 &= M_z, \end{aligned} \tag{2.7}$$

where M_{+1} is the dephasing, M_{-1} the rephasing, and M_0 the longitudinal magnetization, which are designated as $q = +1, -1$, and 0, respectively [9].

Following the notation used by Hürlimann [4], the evolution under a free precession period of duration τ is described by the matrix

$$\mathbf{\Gamma} = \begin{pmatrix} e_2 e^{(i\Delta\omega_0\tau)} & 0 & 0 \\ 0 & e_2 e^{(-i\Delta\omega_0\tau)} & 0 \\ 0 & 0 & e_1 \end{pmatrix}, \tag{2.8}$$

where, as defined in Sect. 2.1.1, $e_1(\tau) = \exp(-\tau/T_1)$ and $e_2(\tau) = \exp(-\tau/T_2)$ are the factors that describe attenuation by relaxation during a free evolution period. As this matrix has only diagonal elements, coherence pathways are not mixed during free precession, M_{+1} and M_{-1} are just multiplied by a phase term that depends on the offset frequency. Radiofrequency pulses, on the contrary, mix all the states according to the rotation

$$\mathbf{\Lambda} = \begin{pmatrix} \Lambda_{+1,+1} & \Lambda_{+1,-1} & \Lambda_{+1,0} \\ \Lambda_{-1,+1} & \Lambda_{-1,-1} & \Lambda_{-1,0} \\ \Lambda_{0,+1} & \Lambda_{0,-1} & \Lambda_{0,0} \end{pmatrix}. \tag{2.9}$$

The complex matrix elements of $\mathbf{\Lambda}$ for a pulse of duration t_p, offset frequency $\Delta\omega_0$, amplitude B_1, effective nutation frequency $\Omega = \gamma B_{\text{eff}}$, and phase ϕ are [4, 10]

$$\begin{aligned}
\Lambda_{+1,+1} =& \frac{1}{2}\left\{\left(\frac{\omega_1}{\Omega}\right)^2 + \left[1+\left(\frac{\Delta\omega_0}{\Omega}\right)^2\right]\cos(\Omega t_\mathrm{p})\right\} + i\left(\frac{\Delta\omega_0}{\Omega}\right)\sin(\Omega t_\mathrm{p}),\\
\Lambda_{0,0} =& \left(\frac{\Delta\omega_0}{\Omega}\right)^2 + \left(\frac{\omega_1}{\Omega}\right)^2\cos(\Omega t_\mathrm{p}),\\
\Lambda_{+1,0} =& \frac{\omega_1}{\Omega}\left\{\frac{\Delta\omega_0}{\Omega}[1-\cos(\Omega t_\mathrm{p})] - i\sin(\Omega t_\mathrm{p})\right\}\exp(+i\phi),\\
\Lambda_{0,+1} =& \frac{1}{2}\frac{\omega_1}{\Omega}\left\{\frac{\Delta\omega_0}{\Omega}[1-\cos(\Omega t_\mathrm{p})] - i\sin(\Omega t_\mathrm{p})\right\}\exp(-i\phi),\\
\Lambda_{+1,-1} =& \frac{1}{2}\left(\frac{\omega_1}{\Omega}\right)^2[1-\cos(\Omega t_\mathrm{p})]\exp(+i2\phi),\\
\Lambda_{-1,-1} =& \Lambda^*_{+1,+1},\\
\Lambda_{-1,0} =& \Lambda^*_{+1,0},\\
\Lambda_{0,-1} =& \Lambda^*_{0,+1},\\
\Lambda_{-1,+1} =& \Lambda^*_{+1,-1}.
\end{aligned} \tag{2.10}$$

As the three states are mixed only during the application of rf pulses, Kaiser et al. [9] proposed a simple indexing where a particular coherence pathway is described by an array of numbers $q_0, q_1, q_2, \ldots, q_N$, where q_k denotes the coherence after the kth pulse. For example, a coherence pathway after applying N rf pulses separated by free evolution periods of duration t_k is calculated as

$$\begin{aligned}
M_{q_0,q_1,q_2,\ldots,q_N} =& \prod_{k=1}^{N} \Lambda^k_{q_k,q_{k-1}} \times \exp\left(i\Delta\omega_0 \sum_{k=1}^{N} q_k t_k\right)\\
&\times \exp\left(-\sum_{k=1}^{N}\left(\frac{q_k^2}{T_2} + \frac{1-q_k^2}{T_1}\right)t_k\right),
\end{aligned} \tag{2.11}$$

where $q_0 = 0$ when the system is assumed to be initially in thermal equilibrium.

In this expression, we can identify the inherent beauty of this formalism. First, each particular pathway depends on the phase of the rf pulses via the coefficients of the matrix Λ^k that communicate the state before the kth pulse with the state after the pulse. For example, if $q_{k-1} = -1$ and $q_k = +1$, the kth matrix element in the product of Eq. (2.11) $\Lambda^k_{q_k,q_{k-1}} = \Lambda^k_{+1,-1}$, which depends on the phase of the kth pulse as $\exp(+i2\phi_k)$ (see Eq. 2.10). After multiplying the Λ^k coefficients defining the pathway, a final phase factor is obtained that depends on the phase of the rf pulses in a characteristic way. As we will see, this phase can be exploited to define a proper phase cycle that cancels unwanted terms. The second advantage of this formalism comes from the fact that the phase accumulated during free precession periods is simply added or subtracted depending on the sign of q_k, if during the kth interval the magnetization is stored as longitudinal magnetization $q_k = 0$ and no phase is accumulated. If after the N pulses there is a time t_N when the condition

$$\sum_{k=1}^{N} q_k t_k = 0 \tag{2.12}$$

is fulfilled, the coherence pathway generates an echo. If the times t_k are much larger than T_2^*, the free induction decay (FID) generated by the pulse k will be completely dephased during t_k and will not contribute to the signal detected in posterior free evolution times. In general, for single-sided sensors, this condition is well fulfilled and pathways which do not satisfy Eq. (2.12) are discarded. It must be noted that magnetization brought to the plane for the first time by the pulse N does not fulfill Eq. (2.12) during the period t_N. However, it should be retained for the signal calculation during t_N, because it does generate a signal that starts dephasing after the pulse and, depending on the timing of the sequence, may interfere with the desired signal. Finally, the last factor in Eq. (2.11) is the attenuation that the magnetization suffers due to relaxation.

A last important fact to be noted when using this formalism is that the following relation between pathways is fulfilled:

$$M_{-q_0,-q_1,-q_2,\ldots,-q_N} = M^*_{q_0,q_1,q_2,\ldots,q_N}. \tag{2.13}$$

This means that, except for the pathway where the magnetization evolves always as longitudinal magnetization ($M_{0,0,0,\ldots,0}$), all the others have a twin pathway containing basically the same information. Twin pathways do not need to be taken into account in the signal calculation and it is done by selecting from the paths generating echoes only those ending with $q_N = +1$. As a convention, it is held that only coherence $+1$ contributes to the signal. Then, the total signal can be calculated as

$$S_{\mathrm{Re}} + i S_{\mathrm{Im}} = \sum_{q_1,\ldots,q_{N-1}} \left(\mathrm{Re}(M_{q_0,q_1,q_2,\ldots,+1}) + i\,\mathrm{Im}(M_{q_0,q_1,q_2,\ldots,+1})\right). \tag{2.14}$$

To illustrate how to proceed at the time of selecting a particular pathway, a pulse sequence composed of three arbitrary rf pulses with phases ϕ_1, ϕ_2, and ϕ_3 is analyzed in the following. To do this, the pathways generating signal during each evolution time are analyzed. The expanding tree of coherence pathways under the application of three rf pulses is shown in Fig. 2.2 [9]. The first pulse generates 3^1 pathways, but only M_{+1} gives a signal and corresponds to the FID generated by the first pulse (FID_1). When the second pulse is applied after a time t_1, it splits the three initial pathways into $3^2 = 9$ paths. To find out which of these paths generate a signal during t_2 is not relevant if they generated a signal during t_1. The strategy is to identify out of the total number of pathways those with $q_2 = +1$. So, the potential candidates to generate a signal during t_2 are $M_{0,+1,+1}$, $M_{0,-1,+1}$, and $M_{0,0,+1}$. The first one is magnetization brought to the transverse plane by the first pulse and is not affected by the second one. If $t_1 \gg T_2^*$, this path does not generate any signal during t_2 because at the time the second pulse is applied, the coherence is fully dephased. The second path describes magnetization that dephases during t_1 and refocuses during t_2 to generate the Hahn echo. The third pathway describes the magnetization that remains along the z-axis after the first pulse but is brought to the transverse plane by the second pulse to generate the FID observed after the second pulse (FID_2). These two last paths interfere during t_2; to separate them, the phase of the pulses must be cycled. As it can be easily observed, by explicitly writing Eq. (2.11) for these two particular pathways, the phase of the Hahn echo depends on the phase of the rf pulses as $-\phi_1 + 2\phi_2$. Moreover, the phase of the FID_2 only depends on the phase of the second pulse as ϕ_2. A typical phase cycle used to cancel the FID_2 combines two experiments where the phase of the first pulse is incremented from 0 to π. In this way, the echo changes its phase also in π, but the FID_2 remains unchanged. Changing the receiver phase also from 0 to π, the echoes are added constructively and the FID_2s are canceled. This is typically known as the add–subtract phase cycle.

Finally, the third pulse splits the nine pathways generated by the first two pulses into $3^3 = 27$ paths. From the nine pathways ending in $q_3 = +1$, only five can generate a signal [5]. Figure 2.2 shows the paths giving a signal if $t_2 > t_1$, and indicates their dependence on the phases of the rf pulses. The first echo in the list is the so-called *direct echo* (DE) formed by magnetization that is always in the transverse plane (defined as direct echo by [1]). The second echo is that formed by the magnetization dephased during the first two evolution periods and rephased during the third one (E_{13}). The third one is the stimulated echo (STE) formed by magnetization that dephases during the first free precession period, is stored as longitudinal magnetization during the second one, and rephases during t_3 after being

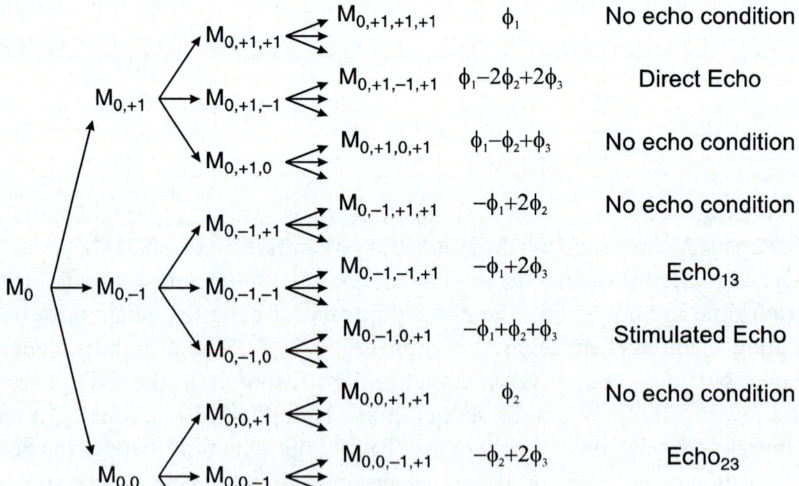

Fig. 2.2 Coherence pathways generated by the application of three pulses separated by times t_1 and t_2. Only coherences with $q_3 = +1$ are shown after the third pulse. From all these potential candidates to generate signal after the third pulse only those fulfilling the echo condition of Eq. (2.12) are listed together with their respective dependence on ϕ_1, ϕ_2, and ϕ_3

Table 2.1 Phase cycle to select the stimulated echo generated by a three-pulse sequence

ϕ_1	ϕ_2	ϕ_3	ϕ_{DE}	$\phi_{E_{13}}$	ϕ_{STE}	$\phi_{E_{23}}$	ϕ_{F3}	ϕ_{rec}
0	0	0	0	0	0	0	0	0
π	0	0	π	π	π	0	0	π
0	π	0	0	0	π	π	0	π
π	π	0	π	π	0	π	0	0

converted in transverse magnetization by the third pulse. The fourth echo is generated by the magnetization brought to the transverse plane by the second pulse which is refocused after the third pulse (E_{23}). The last path describes the signal generated by the third pulse (FID_3). Table 2.1 exemplifies the phase cycle needed to select the stimulated echo out of the five interfering signals. It also provides the phase of each signal for each combination of phases chosen for the rf pulses. As the stimulated-echo has to be added constructively, the receiver phase is cycled following the phase shift of this particular pathway. In this way, it can be seen that the other signals are averaged out after the four scans. The selection of other paths can be achieved by changing the phase cycle.

Besides allowing a straightforward identification of interfering signals and simplifying the design of proper phase cycles to disentangle them from the desired pathway, a really powerful advantage of this formalism over the vectorial one described in Sect. 2.1.1 is the possibility to include the effect of diffusion attenuation for each particular pathway. Even though relaxation can be easily included in both formalisms as simple multiplicative factors attenuating the magnetization during the free evolution periods, in order to account for the effect of diffusion we need to know the type of coherence during previous evolution periods to identify if the generated echo has evolved as a direct echo or as a stimulated echo. It is well known that the diffusion attenuation experienced by the direct echo is different from the one experienced by a stimulated echo [5, 11–13]. Thus, each pathway must be weighted by a different diffusion factor. This strategy requires subdivision of a given pathway into segments that form sequences of the form $(-1, +1)$, $(-1, -1, +1, +1)$, or higher known as single, double, or higher Hahn echoes and sequences of the form $(-1, 0, \ldots, 0+1)$, $(-1, -1, 0, \ldots, 0+1, +1)$, or higher, known as single, double, or higher stimulated echoes, respectively. The diffusion attenuation factors for each of these segments can be easily calculated once the type of echo and the timing are defined, hence the total attenuation after N pulses is just the product of these attenuation factors [4, 10].

Please note that the fresh magnetization created during a free evolution period is not included in this calculation. The new magnetization created during the evolution time t_k is $\mathbf{M}^k = (0, 0, 1 - \exp(-t_k/T_1))$ and has to be taken into account only from the $(k+1)$th rf pulse on (see [4]). In most cases, the magnetization created in free evolution intervals set after the first rf pulse is not desired and is eliminated by the add/substract phase cycle of the first rf pulse (exceptions are sequences used to measure T_1, where the magnetization created during the evolution period subsequent to the preparation interval is the desired one).

2.1.3 Numerical Calculation of the NMR Signal

The NMR signal detected by a given sensor depends exclusively on the geometry of the magnet and surface coil generating the static and rf fields. The strategy to calculate the total signal induced in the coil is based on the subdivision of the object into small voxels where defined $\mathbf{B}_0$ and $\mathbf{B}_1$ fields can be assumed. The sequence response is then calculated for every voxel using the rotations defined in the previous sections, and the total signal is obtained as the integration over the volume, weighting the contribution of each voxel by the detection efficiency of the rf coil. This calculation requires the knowledge of the spatial distribution of both fields.

The magnetic field generated by an array of permanent blocks and iron yokes or pole shoes can be calculated by means of finite element methods (FEM). On the other hand, the rf field produced by the rf coil can be obtained by integrating the Biot–Savart equation for the defined coil geometry. This solution does not consider possible distortion of the rf field due to the presence of metallic components in the vicinity of the coil, but it can be applied in most cases. A more exact solution would require the use of finite element calculations also for the rf coil. Once both magnetic field vectors are known at every voxel, the evolution of the magnetization in each voxel can be calculated. As both fields $\mathbf{B}_0$ and $\mathbf{B}_1$ may change in magnitude and direction from voxel to voxel, the component of the rf field perpendicular to the static field ($\mathbf{B}_0$ is assumed to define the local z-axis in the voxel)

$$(B_1)_{xy} = \mid \mathbf{B}_0 \times \mathbf{B}_1 \mid / B_0$$

must be used to calculate the effect of the rf pulse at each voxel.

Besides playing a primary role during excitation, the rf coil defines the sensitivity of the sensor to detect the NMR signal from each volume element. This fact leads to a reduction of the contribution from voxels with weak $(B_1)_{xy}$ in a twofold way. First, just part of the magnetization is brought to the transverse plane and, second, it is detected with less efficiency. The contribution from each voxel to the total signal can be calculated considering that the EMF induced in the rf coil C by the oscillating magnetization component M_{xy} placed at r is, according to the reciprocity principle [14]

$$\xi = -\partial/\partial t\{(B_1/i)_{xy} M_{xy} \exp(i\omega_0 t)\}, \tag{2.15}$$

where it becomes obvious that the efficiency during the detection of the signal is proportional to $(B_1)_{xy}$ per unit of current. Then, the total complex signal in the rotating frame can be obtained in arbitrary units by integrating the contribution of the transverse magnetization $M_{xy} = M_x + iM_y$ at each voxel as

$$S(t) \propto (B_1/i)_{xy} \int_{V_s} M_{xy}(r)\mathrm{d}r^3, \tag{2.16}$$

where the components of the magnetization are calculated at the time of the acquisition by applying to the initial magnetization $\mathbf{M}_0 = (0, 0, 1)$ the rotations describing the pulse sequence.

2.2 Pulse Sequence Analysis

The formalism given in the last section allows one to easily calculate the evolution of the magnetization in response to different pulse sequences as a function of both the off-resonance and B_1 inhomogeneity. For simplicity, in this section the analysis is performed by assuming a magnetic field decreasing linearly along the depth and a homogeneous radiofrequency field. In this situation, the frequency distribution is constant, and all frequencies are excited by the same rf amplitude, condition that simplifies the understanding of the off-resonance effect. Then, in those cases where the B_1 inhomogeneity is expected to introduce an appreciable effect, the response is calculated by including in the simulation the field of a surface coil. To emulate one of the most common spatial dependencies found in single-sided sensors, the sample is assumed to be in the presence of a magnetic field B_0 parallel to the surface of the sensor, and it is excited by the rf field generated by a circular single-loop rf coil. For the calculations presented here a coil diameter of 10 mm is used and it is assumed that the on-resonance condition is met 2.5 mm above the coil. Thus the excited planar slice is placed in a region where the rf field is expected to be approximately constant in an area of the size of the coil and then decays quickly as the lateral position increases beyond the radius of the coil. This can be considered as a moderated condition where the rf field is not extremely inhomogeneous.

2.2.1 Single rf Pulse

The simplest pulse sequence to be studied is the response to a single rf pulse. Understanding the dynamics of the magnetization under the application of a selective pulse is the key to visualize why off-resonance excitation is an important source of error during most sequences. Figure 2.3a shows the magnetization components after applying a single rectangular rf pulse as a function of the offset $\Delta\omega_0$. The simulations were conducted by assuming a homogeneous rf field with an amplitude B_1 such that $\omega_1 t_p = \gamma B_1 t_p = \pi/2$, set to define a flip angle of 90° for on-resonance magnetization. It can be observed that, as a consequence of the nutation about a tilted effective field, the magnetization does not necessarily remains on the $x - z$ plane under the application of an rf pulse along the y-axis. Figure 2.3b illustrates the position of the magnetization as a function of the offset frequency for an rf pulse applied along the y-axis.

The magnetization brought to the x–y plane by the rf pulse, which is a measure of the signal detectable for each frequency, can be calculated as $M_{xy} = \sqrt{M_x^2 + M_y^2}$, where M_x and M_y are the components plotted in Fig. 2.3a. The amplitude of M_{xy}

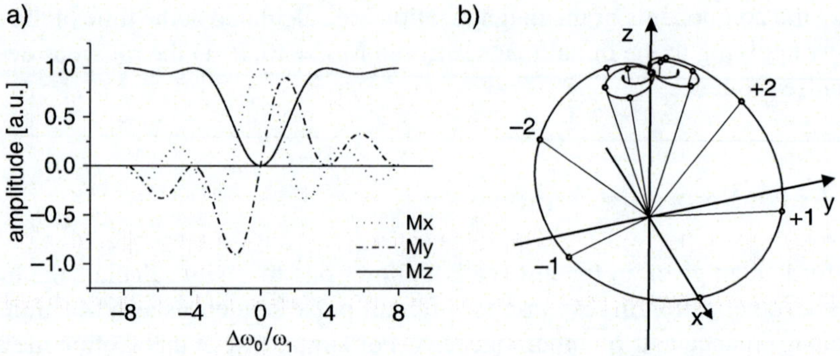

Fig. 2.3 (**a**) Magnetization components as a function of the off-resonance after the application of an rf pulse set to define a 90°rotation for on-resonance spins. (**b**) Orientation of the magnetization vector after the rf pulse as a function of the offset frequency for the rf pulse set in (**a**). The *dots* mark the final position of the magnetization vectors for offsets $\Delta\omega_0/\omega_1 = 0, \pm 1, \ldots, \pm 8$

as a function of the offset frequency is plotted in Fig. 2.4a. As it can be observed, full amplitude is obtained up to offset frequencies of the order of the nutation frequency ω_1. However, the amplitude quickly decreases for larger offsets showing that efficient excitation can only be achieved in a defined bandwidth. As a rule of thumb, it is usually said that an rf pulse of length t_p excites a bandwidth $\Delta\nu_0 = 1/t_p$. Considering that for a 90° pulse $\nu_1 = 1/(4t_p)$, this rule defines a region in Fig. 2.4a between $\Delta\omega_0/\omega_1 = \pm 2$ (indicated with vertical lines). At the border of this band the magnetization in the transverse plane is $M_{xy} = 0.8$. Finally, the dashed line in Fig. 2.4a depicts the sinc function corresponding to the Fourier transform of the rectangular shape of the rf pulse, which predicts the frequency response when working in the linear response limit [8]. Except predicting the position of the nodes, it should be noted that the difference between both curves is not negligible (in particular, in the range marked as the excitation bandwidth of the pulse), a fact that must serve to encourage the reader to use the real spin response for predicting the evolution of the magnetization under defined sequences.

As a consequence of precessing in conical trajectories about effective fields tilted at different angles, not only the magnitude but also the phase of the transverse magnetization varies with the offset. Figure 2.4b shows the phase of the magnetization calculated as $\phi = \arctan(M_y/M_x)$ as a function of the excitation frequency. It can be observed that ϕ varies almost linearly in the whole range, a fact that will be of extreme importance in the next section at the time of refocusing the magnetization to generate an echo. The spreading of the magnetization on the x–y plane during the application of the rf pulse leads to important signal attenuation at the end of the pulse. Moreover, as the gradient is continuously present, the magnetization further spreads after the pulse and the signal is averaged out in a free evolution time of the order of the pulse length, which in most cases is of the order of the dead time of the rf probe. Consequently, under typical conditions no FID can be readily detected after applying a selective rf pulse when large samples are in the presence of a static gradient.

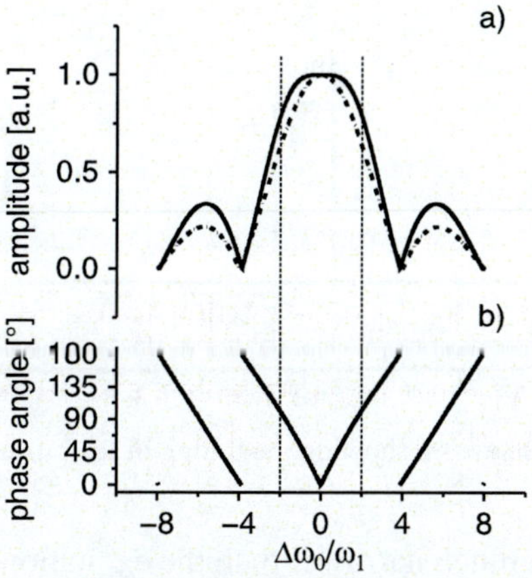

Fig. 2.4 (**a**) Magnetization component brought to the plane by the rf pulse plotted as a function of the off-resonance. While a perfect 90° rotation is achieved for on-resonance spins, the efficiency of the excitation quickly decreases with the offset. The *vertical lines* indicate the bandwidth corresponding to $1/t_p$, usually pointed as the excitation bandwidth of a rectangular pulse. (**b**) Phase of the transverse magnetization. The phase acquired by the magnetization during the pulse is highly linear with the offset frequency, a feature with an important implication at the time of refocusing the spreading during the pulse

2.2.2 The Generation of Hahn Echoes

The presence of a distribution of resonance frequencies results in a dephasing of transverse magnetization during and after the application of the rf pulse. The possibility to refocus the phase spreading of the magnetization was discovered by Hahn in 1950 and has become one of the most important realizations for the field of single-sided NMR [5]. The sequence requires the application of a second rf pulse generating a 180° rotation about an axis lying on the x–y plane. Although in the original work a 90° pulse was used, it was quickly realized that a 180° is needed to efficiently refocus the phase spreading [15]. This rotation inverts the sign of the phase acquired by the magnetization during the time previous to the 180° pulse, leading to its cancelation in a posterior free evolution after the application of the refocusing pulse. Although a free evolution period τ is expected to cancel the phase spreading acquired during the time τ between the 90° and the 180° pulses, the phase spreading developed during the 90° also needs to be cancelled to acquire full signal. As the phase acquired by the magnetization during the first rf pulse depends almost linearly on the offset, it can be refocused by letting the magnetization evolve an extra time after the application of the inversion pulse. The dephasing during the pulse over a broad frequency range is about the half of the dephasing acquired during free precession, so that the extra time required after the inversion pulse should be about $t_p/2$. However, close to on-resonance, the slope of the dephasing curve shown in Fig. 2.4 is slightly larger than the slope defined by half of the offset frequency, requiring a longer delay for magnetization rephasing. If signal coming from

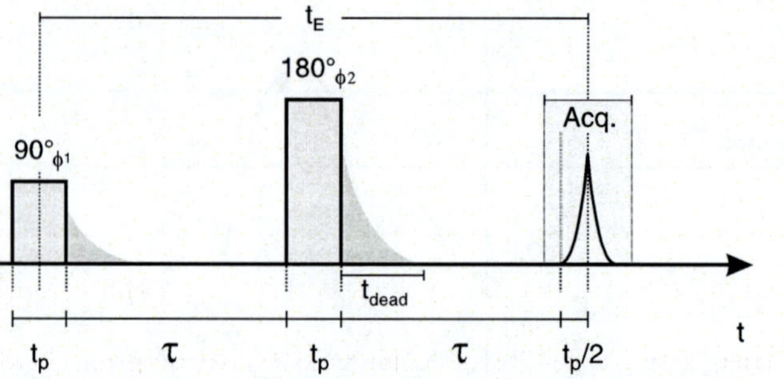

Fig. 2.5 Hahn echo pulse sequence showing the timing of the pulses and the echo formation

a frequency bandwidth much narrower than the excitation bandwidth of the pulse is going to be retained, the time for optimum refocusing is $2t_\mathrm{p}/\pi$ instead of $t_\mathrm{p}/2$. This time correction is reported by Hürlimann in [16]. However, as the optimum delay approaches $t_\mathrm{p}/2$ for bandwidths of the order of the excitation bandwidth of the pulse and because minimal sensitivity improvement is achieved with the optimum delay instead of $t_\mathrm{p}/2$, the later will be used in the following.

Figure 2.5 shows the timing of a Hahn echo sequence, where the echo is generated at a time $\tau + t_\mathrm{p}/2$ after the application of the refocusing pulse. The fact that the dephasing during the first rf pulse is almost half of the one acquired during the same time but under free evolution leads to the definition of zero time for the sequence at the center of the 90° pulse. Then, defining the echo formation as echo time t_E, the sequence appears symmetric with respect to the center of the 180° pulse. The need for an extra free evolution time after the refocusing pulse is comparable to the inversion of gradient polarity after the application of selective pulses used for slice excitation in conventional NMR imaging, which is used to refocus the phase spreading developed during the application of the soft rf pulse [17].

In conventional experiments, where the on-resonance condition is fulfilled, 180° rotations can be well defined. However, as we learned in Sect. 2.2.1, the effect of an rf pulse applied in the presence of a static gradient depends on the resonance offset; a fact that makes impossible to define a 180° rotation for every resonance frequency. The effect of the off-resonance on the refocusing efficiency can be easily calculated assuming a homogeneous B_1 and setting the second rf pulse to define a 180° pulse for the on-resonance spins. At this point, two possibilities are available: either the amplitude of the pulse is doubled keeping its length constant or the amplitude is kept constant and the length is doubled. For the simulations presented in this book we prefer the first option. The total evolution that has to be applied to M_0 to obtain the final state of the magnetization at the time t_E when the echo is formed can be written as

$$M(t_\mathrm{E}) = E(\tau + t_\mathrm{p}/2) \rightarrow P(180°, \phi_2) \rightarrow E(\tau) \rightarrow P(90°, \phi_1) \rightarrow M_0. \quad (2.17)$$

Figure 2.6a shows the magnetization at the echo maximum plotted as a function of the offset. It can be observed that the amplitude of the components is modulated by a well-defined frequency. The oscillations appear due to the fact that at the moment when the second pulse is applied, the magnetization is strongly spread in the transverse plane. As the pulse is applied along a particular direction, in this case the x-axis, those vectors lying along the direction of the rf field at the time when the pulse is applied, remain in the plane untouched, at least for small offsets. However, vectors lying along y-axis are rotated along the effective field corresponding to the particular offset. For a small offset in the resonance frequency, the rotation is close to be an ideal 180° and the magnetization is placed along the y-axis after the pulse. However, as the offset increases, the trajectory followed by the magnetization increasingly departs from the ideal one, a fact that leads to the reduction of the component lying in the xy-plane. Since longer τ delays lead to larger phase spread, smaller offsets can complete a rotation about the z-axis, and the oscillation frequency increases with τ. These oscillations cannot be observed in real experiments due to the destructive interference between anti-phase magnetization component at the time of the echo formation. Hence, it becomes necessary to consider in the simulations the fact that in a real experiment signals with different frequencies are detected simultaneously and are thereby subject to strong interference in the time domain. A way to avoid interference would be to use a very thin sample on a magnet generating a uniform gradient along the depth direction, then the signal amplitude for each frequency could be sampled independently to reproduce the result of this simulation.

The numerical tools used up to here to calculate the evolution of the magnetization can be easily applied to calculate the NMR signal generated by the pulse sequence. In contrast to the previous analysis, the calculation of the signal requires knowledge of the real static and rf fields. For the present analysis, a static magnetic field varying linearly along the depth direction and a homogeneous rf field are assumed. In this simplified situation, the real and imaginary signals at a defined time are calculated as the addition of the M_x and M_y magnetization components for a large number of offset values lying in a bandwidth much larger than the excitation bandwidth of the rf pulse.

Figure 2.6b compares the magnetization generated along the x-axis by setting an ideal 180° rotation for every frequency with the one obtained by applying a real rf pulse set to define a 180° rotation for on-resonance spins. Whereas in the first case a spectrum that reproduces the one defined by the first pulse (solid line) is recovered, the resonance offset present during the refocusing pulse yields an oscillatory behavior as a function of frequency (dotted line) that reduces the effective bandwidth measurable in the spectrum obtained as the Fourier transform of the echo signal (dashed line) by a factor of about 2. As the length of the rf pulses defines the excitation bandwidth of the sequence, the echo width is expected to depend on the pulse duration; the longer the pulse is the broader the echo. As a rule of thumb, the echo width is about the pulse length.

Considering that points acquired during the acquisition time can be added to improve the signal-to-noise ratio of the detected echo, the question that arises is how to set the length of the acquisition window to maximize the sensitivity? To answer this question we need to know the shape of the echo for a given t_{p}. This analysis

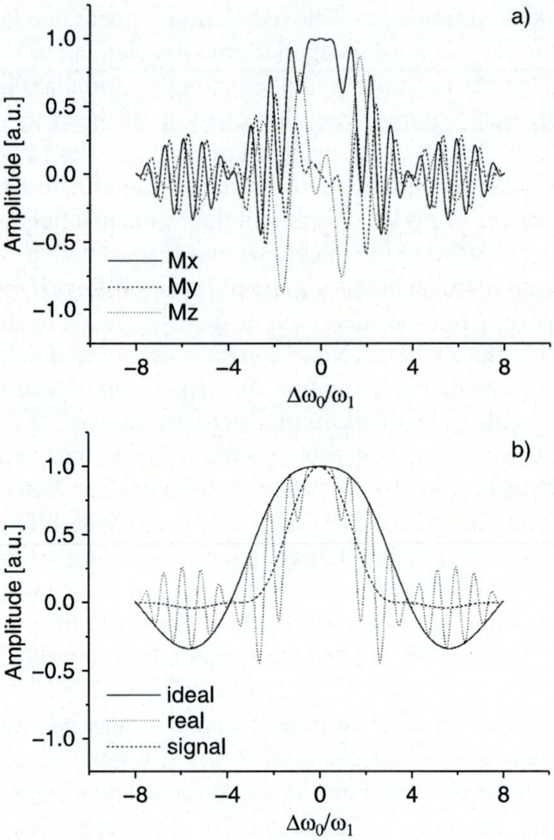

Fig. 2.6 (**a**) Magnetization components at the time of the echo formation plotted as a function of the resonance offset. (**b**) Comparison between the excitation bandwidth of a single rf pulse and the one defined by a Hahn echo sequence

uses the shape of the Hahn echo generated by assuming a uniform distribution of frequencies and a homogeneous rf field. Figure 2.7 shows the signal-to-noise ratio calculated as the addition of the acquired points (total signal) divided by the square root of the number of points added for different acquisition windows (noise). Since the noise power is the same for all points and the signal amplitude decreases when departing from the echo center, maximum sensitivity is obtained when the echo is acquired during a time equal to the length of the rf pulse used for excitation (a more detailed discussion can be found in [8], p. 152).

The timing needed to acquire the echo centered in the acquisition window requires starting acquisition at a time $\tau + t_p/2 - \mathrm{Acq}/2$ after the application of the 180° pulse, with Acq being the total time to be acquired. However, because the signal is excited and detected by a resonance circuit with a finite bandwidth, a short delay shift (sh) proportional to the quality factor Q of the circuit is introduced. The value of sh, which must be added to the delay between the second pulse and the acquisition window, needs to be determined for every rf probe, and it can be a few

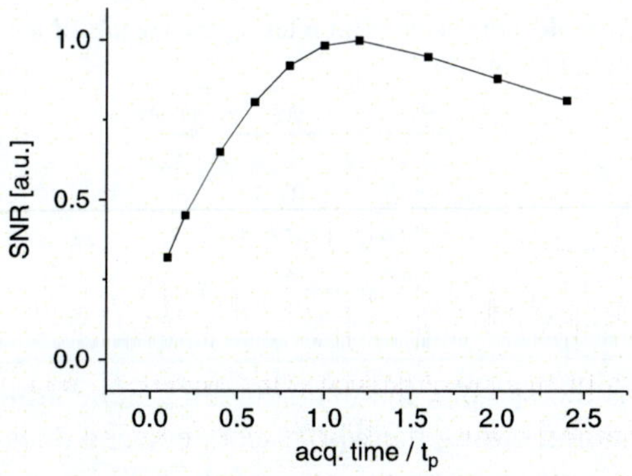

Fig. 2.7 Signal-to-noise ratio calculated as a function of the acquisition time normalized to the pulse length t_p for the echo generated by the Hahn echo sequence

microseconds. Under this condition the shortest τ that can be reached is determined by the dead time (t_{dead}) of the circuit, the acquisition time, and the pulse length ($\tau > t_{dead} + \mathrm{Acq}/2 - t_p/2 - \mathrm{sh}$). To shorten the minimum τ half of the echo can be acquired by starting the acquisition at the maximum of the echo. Nevertheless, it is achieved at the expense of sensitivity.

The phase of the second pulse influences the phase of the echo, but does not affect the efficiency to refocus the phase spread of the magnetization. In this example we have chosen to apply the 90° pulse along the y-axis and the 180° pulse along the x-axis, thereby generating the echo along the x-axis. A phase cycle of at least two steps (add/substract) is generally used. It requires applying the 90° pulse first along $+y$-axis and then along $-y$-axis keeping the phase of the 180° pulse along the $+x$-axis in both experiments. Cycling the phase of the first pulse changes the phase of the echo and requires cycling the phase of the receiver from $+x$ to $-x$ to constructively add the two signals.

The effect of this phase cycle is twofold. First, it serves to cancel magnetization brought to the transverse plane by an imperfect 180° pulse (path $M_{0,0,+1}$). Second, it cancel spurious signals coming from the dead time of the resonance circuit and possible acoustic ringing generated by the application of the 180° pulse. These unwanted signals are expected to be the same for both experiments because the 180° pulse is applied with the same phase. Thus, as soon as they reach a voltage level that can be digitized by the receiver they are add-subtracted by cycling the phase. It must be noted here that for very short echo times, ringing generated by the first rf pulse might by present at the time the echo is formed. In this case the phase cycle shown in Table 2.2 must be used. This phase cycle eliminates both the signal generated by the first pulse (path $M_{0,+1,+1}$), which for short echo times may still be present in systems with a small static gradient, and the ringing generated also by this rf pulse.

The Hahn echo sequence is of particular interest to measure extremely short relaxation times of the order of a few dead times of the resonance circuit, a situation where multi-echo techniques can only generate a few echoes and are inaccurate to determine the T_2 of the sample. Although resonance offset leads to a distribution

Table 2.2 Phase cycle to select the Hahn echo filtering the signals of the first and second rf pulses as well as acoustic ringing present after the pulses

ϕ_1	ϕ_2	ϕ_{rec}
$+\pi/2$	0	0
$-\pi/2$	0	π
$+\pi/2$	$+\pi/2$	π
$-\pi/2$	$+\pi/2$	0

of flip angles across the sample, the magnetization generating the echo evolves as transverse magnetization during the whole sequence and decays purely with T_2 (no T_1 contamination is present as in the case of the CPMG decay). In this case, the application of imperfect rotations to refocus the phase spread introduced by the field inhomogeneities do not affect the characteristic time decay measured by the sequence, which is expected to be the same as the one measured in homogeneous fields. A exception is for the case of liquids measured in a strong gradient. In this case, the echo amplitude is also attenuated by the self-diffusion of the molecules in the presence of the static gradient. In the case of unrestricted Brownian motion, this attenuation is

$$S(t_E) = S_0 \exp\left(-\frac{t_E}{T_2} - \frac{1}{12}\gamma^2 G_0^2 D(t_E)^3\right). \tag{2.18}$$

For sensors generating a uniform static gradient, the Hahn echo sequence is a very sensitive method to measure diffusion coefficients. Since gradients generated by single-sided sensors are very strong, they can be used to measure very small diffusion coefficients. Moreover, the Hahn echo sequence has been extensively used in low-field NMR to study polymers, where the echo decay is usually fitted by a number of exponentials or even the addition of exponential and gaussian functions.

2.2.3 The CPMG Sequence

The success of single-sided sensors mainly relies on the extraordinarily good performance of multi-echo sequences that successively refocus the phase spread due to the static field inhomogeneities and lead to the generation of echo trains. For most samples, the minimum echo time that can be set in a Hahn echo sequence is much shorter than T_2, and a large number of echoes can be generated by applying a train of refocusing pulses with the right phase. It was first Carr and Purcell [6] who proposed the use of repeated refocusing to measure T_2 in a single shot. However, as the original sequence uses refocusing pulses with the same phase as the first pulse, it suffers from accumulative distortions due to slight errors in the definition of the flip angle of the refocusing pulses. A modification to the CP sequence, proposed later by Meiboom and Gill [7], is known as the CPMG sequence. The timing of the pulse sequence is depicted in Fig. 2.8.

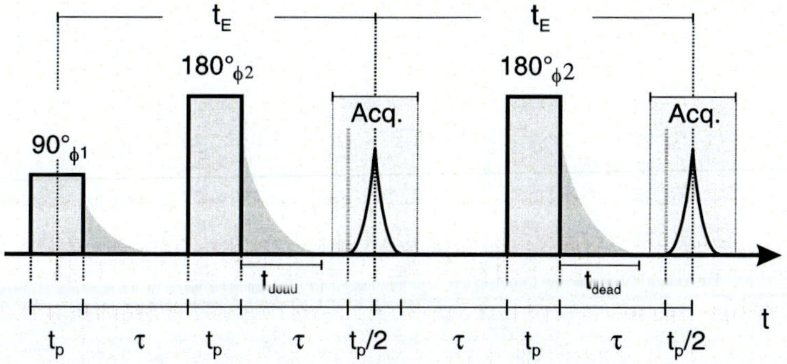

Fig. 2.8 Carr-Purcell-Meiboom-Gill pulse sequence. A 90° phase shift between the first pulse and the train of 180° pulses eliminates accumulative distortions introduced by imperfection of the rf pulse and resonance offset. The time between the first two pulses is not exactly half the time between 180° pulses, but $t_p/2$ shorter. As discussed in Sect. 2.2.2, the sensitivity can be improved by varying this extra delay time [16], but $t_p/2$ can be used with similar performance

The modification introduced by Meiboom and Gill consists of shifting the phase of the refocusing train by 90° with respect to the first pulse. By doing so the distortions coming from the mentioned imperfections are strongly reduced. The sequence is so robust that it can be implemented in strongly inhomogeneous fields with outstanding performance. Figure 2.9 shows the behavior of the first echoes generated by this sequence. For this calculation, an echo time $t_E = 0.2$ ms, rf pulses of 10 μs, and infinite T_1 and T_2 were assumed. After a transient of three echoes, the signal amplitude remains constant showing that the field inhomogeneities do not lead to a signal loss as in the case of the CP sequence. The CPMG sequence is, in general, implemented using a simple add-subtract phase cycle. It requires cycling ϕ_1 and the receiver phase from 0 to π keeping ϕ_2 unchanged.

The dynamics of the magnetization during the CPMG sequence have been extensively studied in the past [1–3]. Different authors have demonstrated that the echo signals detected during this echo train are complicated superposition of coherence pathways where the magnetization stays alternatively along the longitudinal direction or in the transverse plane. Whereas, for example, the first echo is formed by magnetization that evolved only in the transverse plane (pathway $M_{0,-1,+1}$), the second echo is the superposition of signal that was in the plane (direct echo) plus the stimulated echo generated by magnetization stored along the z-axis by the first 180° pulse and brought back to the transverse plane by the second pulse (pathways $M_{0,+1,-1,+1}$ and $M_{0,-1,0,+1}$, respectively). As the echo number increases, the number of pathways contributing to the echoes also does it. In strongly inhomogeneous fields, this superposition of pathways quickly reaches a steady state and after three or four echoes the signal amplitude remains constant (in absence of relaxation). The typical behavior shows that the amplitude of the first echo is the smallest and that of the second is the largest. Then, subsequent echoes reach a steady state.

This behavior can also be observed by calculating the bandwidth of each echo. Figure 2.10 compares the excitation bandwidth of a single pulse, shown as a

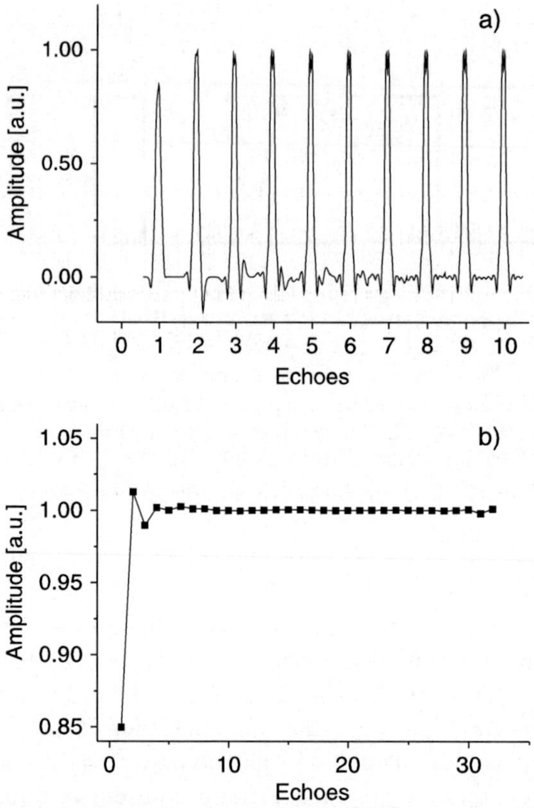

Fig. 2.9 (**a**) First echoes generated by the CPMG sequence. Like in the previous section the rf field was assumed to be homogeneous and the signal was calculated as the addition of 2500 signals for frequencies distributed in a bandwidth of $\pm 2/t_p$. The number of frequencies used to discretize the frequency bandwidth depends on the total time T of the sequence. As a rule of thumb the total bandwidth must be divided by a frequency step $1/T$. (**b**) Echo amplitudes obtained by integration of the echo during a time t_p

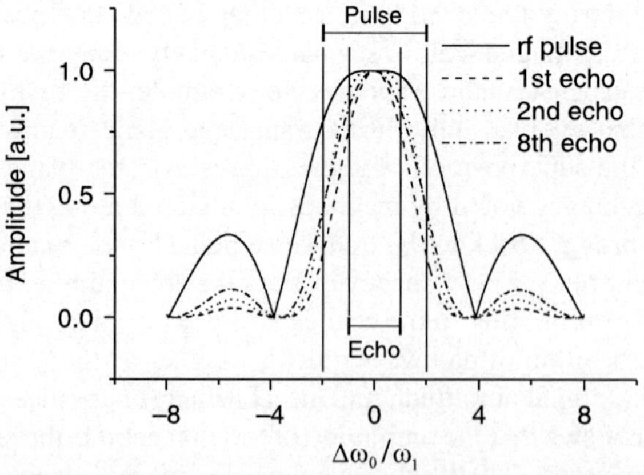

Fig. 2.10 Excitation bandwidth for different echoes of a CPMG sequence. The bandwidth of a single pulse is shown for reference. Whereas the first echo has a bandwidth of about half of that of a single pulse, thanks to the mixture of coherence pathways, the steady-state echo bandwidth is slightly broader than the one of the first echo (Hahn echo)

reference, with that of the first, the second, and the eighth echo. The bandwidth of the first echo is clearly the narrowest. Then, it increases to a maximum for the second echo due to the superposition of the direct echo with the stimulated echo. Thereafter, it slightly decreases to reach the steady state. In this case, the spectra were obtained as the Fourier transforms of the echo signals shown in Fig. 2.9. The transient depends on the B_0 and B_1 distributions and the acquisition bandwidth. For an acquisition window set longer than the pulse length, off-resonance signals can be filtered, and almost no oscillation in the first echoes is observed. The amplitudes of the transient can be determined theoretically or experimentally to be used later as a calibration for correcting the data before numerical analysis.

An important consequence of the mixture of different coherence pathways comes from the fact that magnetization stored along the z-axis relaxes with T_1, while magnetization that remains on the plane relaxes with T_2. This leads to an effective echo train decay time $T_{2\mathrm{eff}}$, which is a mixture of T_1 and T_2, instead of only T_2 [1]. The effective time decay is directly obtained from the simulations by including the attenuation factors $e_1(\tau)$ and $e_2(\tau)$ in the rotation matrix of the free evolution (Eq. 2.6). The increasing superposition of coherence pathways leads to a signal decay that departs from a single exponential decay. Figure 2.11 shows the results of numerical simulations assuming $T_1 = T_2$ and $T_1 = 6T_2$. The calculations were done by assuming first a homogeneous B_1 field and then including the real rf field generated by a surface coil (inhomogeneous B_1). As expected, for $T_1 = T_2$ an exponential signal decay is obtained, but for $T_1 = 6T_2$ the superposition of pathways leads to a non-exponential decay that is apparent for sufficiently long times. When a inhomogeneous rf field is included in the simulations the mixture of coherence

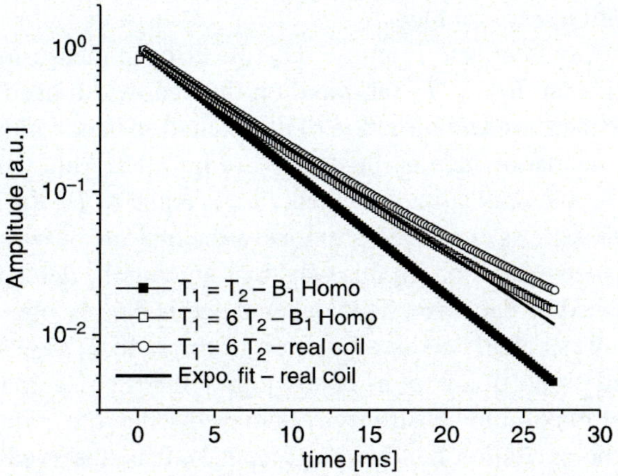

Fig. 2.11 Signal decay during a CPMG sequence calculated for $T_1 = T_2 = 5\,\mathrm{ms}$ and $T_1 = 6$, $T_2 = 30\,\mathrm{ms}$. The simulations used the following parameters, $t_E = 100\,\mu\mathrm{s}$, $t_p = 10\,\mu\mathrm{s}$, acquisition bandwidth equal to the excitation bandwidth of the rf pulse ($1/t_p = 100\,\mathrm{kHz}$), and $N_{\mathrm{echoes}} = 300$. It can be observed how the T_1 contamination influences the value of the slope ($T_{2\mathrm{eff}}$) and defines a non-exponential decay. The presence of an inhomogeneous B_1 field, included in the simulation by integrating over a plane $15 \times 15\,\mathrm{mm}^2$ at 2.5 mm above a circular loop with 10 mm in diameter, does not lead to appreciable enhancement of T_1 contamination compared to the result obtained for a homogeneous B_1 field. The fitting shows that the a single exponential can fit well the data if the decay is considered up to one-third

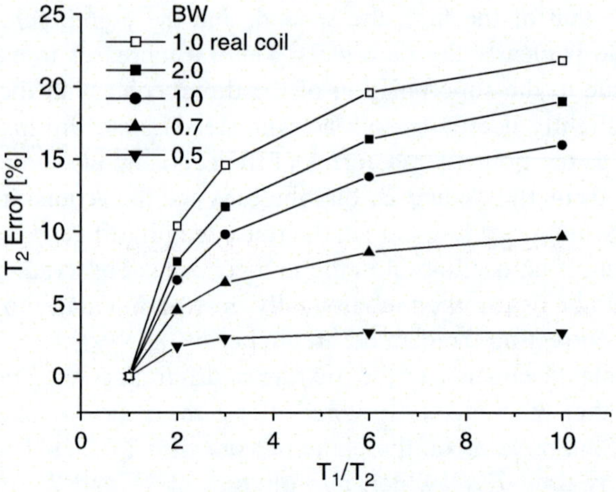

Fig. 2.12 Deviation of the decay time of the echo train generated by a CPMG sequence simulated as a function of the T_1/T_2 ratio and for different detection bandwidths. The decay times where obtained by fitting the signal decays up to one-third of its initial amplitude (one T_2). The deviation obtained in the presence of the inhomogeneous B_1-field generated by the surface coil was calculated only for an acquisition bandwidth matching the excitation bandwidth, condition where maximum sensitivity is expected. The simulations where done assuming the parameters listed in Fig. 2.11

pathways is larger. However, only a slightly longer $T_{2\text{eff}}$ is obtained. In practice, the signal decay can be well fitted to a single exponential if a maximum time of the order of T_2 is considered, which is the case even in the presence of a inhomogeneous B_1 field (Expo. fit in Fig. 2.11).

As shown by Goelman and Prammer [1], the error in measuring T_2 via a CPMG sequence depends on the T_1/T_2 ratio and on the bandwidth used for the detection, which should be measured as a fraction of the excitation bandwidth. The dependence of the error on this parameters is illustrated in Fig. 2.12. Although the error is zero only when T_1=T_2, the only situation when $T_{2\text{eff}}$ is equal to T_2, the error grows as the ratio T_1/T_2 increases. As explained previously, the mixture between T_2 and T_1 is due to the off-resonance excitation. In this way, by reducing the detection bandwidth the mixture is expected to decrease. This dependence is clearly observed in Fig. 2.12, where the error was calculated as a function of the ratio T_1/T_2, for different bandwidths measured proportional to the bandwidth of the rf pulse. In the last section we have shown that maximum sensitivity is achieved when the acquisition bandwidth is set equal to the excitation bandwidth (Fig. 2.7). It is observed that when setting the detection bandwidth to achieve maximum sensitivity, the error in measuring T_2 remains lower than 15% even for a T_1 10 times longer than T_2.

These simulations were conducted assuming a homogeneous B_1 field, but this is not the situation found in real sensors. In the presence of an inhomogeneous radiofrequency field, even on-resonance magnetization is spread to coherence pathways that include evolution periods when it is stored along the longitudinal axis and is contaminated by T_1. Figure 2.12 shows the error in the T_2 measurement obtained when a surface rf coil is used for excitation and detection. In this case, the acquisition window was set to match the acquisition bandwidth with the bandwidth of the rf pulse. Although the error is larger than the one obtained for a homogeneous B_1 field, it is still about 20% for a T_1 10 times longer than T_2. This confirms that the CPMG sequence can be used to measure the transverse relaxation time even in the presence of a strong static gradient excited with an inhomogeneous B_1 field.

When the CPMG sequence is applied to measure the relaxation time of liquids in the presence of a static gradient, the signal decay is attenuated by molecular Brownian displacement. In slightly inhomogeneous fields, where only direct echoes contribute to the signal, the characteristic time decay of the echoes is given by

$$S(mt_{\mathrm{E}}) = A\exp\{-(1/T_2 + 1/12(\gamma G_0 t_{\mathrm{E}})^2 D)mt_{\mathrm{E}}\}. \tag{2.19}$$

This equation shows that in the presence of molecular self-diffusion, the signal decay measured by a CPMG sequence is shortened depending on the strength of the static gradient. A way to reduce this distortion is to reduce the echo time t_{E}, but this is a realistic solution only up to certain limit. Whereas for a gradient of 1 T/m, an echo time of 0.1 ms suffices to measure a T_2 of about 1 s with an error of 10%, in a gradient of 20 T/m, a time decay of about 20 ms would be measured for this echo time. Actually, for off-resonance excitation the equation given above is not valid anymore and the particular attenuation of each of the many coherent pathways contributing to the signal must be calculated considering their particular time dependencies.

The effect of diffusion during a CPMG was considered by Goelman and Prammer [1] who separated the signal into direct and indirect echoes and calculated expressions for the first three echoes. Later Hürlimann presented a formalism that decomposed the signal into different pathways and calculated the diffusion decay for the first 15 echoes, where he estimated a superposition of 10^6 pathways. Finally, a formalism that allows the incorporation of diffusion attenuation for the complete echo train was presented by Song [10], who classified the coherence pathways and showed that only a fraction of the total number of pathways significantly contributes to the signal. He observed that long pathways can be decomposed into a combination of short segments like stimulated (−1,0,+1) and spin echoes (−1,+1). Using this approach, he demonstrated that direct and stimulated echoes contribute to 95% of the CPMG echo signal, thereby showing that despite the large variety of contributing pathways, the final behavior remains simple.

For material characterization, numbers need to be derived from the echo envelope decay (Fig. 2.11). These numbers can be obtained either from a fit of the experimental data with a model function or by calculating a relaxation weight similar to spin echo imaging, for example, $S(t_1)/S_0$. In the case of exponential relaxation, the fit parameters are the amplitude $S_0 = S(t = 0)$ corresponding to the number of spins in the sensitive volume and the relaxation time $T_{2\mathrm{eff}}$. But usually the decay is not mono-exponential. A good empirical fit function is the scaled stretched exponential function $S(t) = S_0\exp(-(t/T_{2\mathrm{eff}})^b/b)$, where the factor $1/b$ has been introduced to arrive at an exponential function for $b = 1$ and a Gaussian function for $b = 2$. If a bi-exponential function with a long relaxation time $T_{2\mathrm{eff,long}}$ and a short relaxation time $T_{2\mathrm{eff,short}}$ is fitted with this function, $T_{2\mathrm{eff}}$ is close to $T_{2\mathrm{eff,long}}$, which from experience is the relaxation time which varies more strongly with the material properties. Compared to the bi-exponential function this stretched exponential function has one less fit parameter and therefore provides a more reproducible fit in the presence of noise-contaminated signals. Before the CPMG decay is fitted, the first echoes must be corrected for the transient effect. This characteristic oscillation can be extracted by measuring a sample with long T_2 or they can be simply eliminated if a large number of echoes are available.

Instead of fitting the experimental data with a model function, they can be transformed for analysis. A useful transformation is the regularized inverse Laplace transformation which transforms the echo envelope into a distribution of relaxation times. This approach is routinely used in well logging, where relaxation time spectra are interpreted in terms of hydrocarbon content, viscosity of fluids in porous rocks, and pore-size distribution [18]. Alternatively, the initial amplitude of the echo envelope can be diffusion encoded, and 2D data sets be measured and transformed into correlation and exchange maps of distributions of relaxation times and diffusion coefficients [19, 20]. Similar to 2D spectroscopy, this approach is useful in separating the contributions to the relaxation time distribution in multi-component systems such as oil and water in well logging. This approach is described in detail in Chap. 3.

When a CPMG sequence is implemented on a real sensor two parameters must be set carefully, the optimum rf pulse duration and the minimum echo time. In a numerical simulation the pulse amplitude or length can be easily set to define a desired flip angle in a defined voxel. However, in a real experiment, where the measured signal amplitude is a value averaged over the sensitive volume, this is not an option. The criterium adopted in this case to set the pulse length is simply the maximization of the signal amplitude. Figure 2.13 shows the dependence of the signal amplitude with the rf strength for both a homogeneous rf field and the field distribution generated by a surface coil. For a homogeneous B_1 field the maximum signal is obtained when the rf amplitude is set to define a rotation slightly larger than 90° for on-resonance spins; however, in the presence of a B_1 distribution this value is generally larger. In the present simulation, where a single loop is used to excite the spins at a depth defined at half of its radius, the maximum is obtained when the strength in the center is about 20% larger than the one needed to define the 90° rotation for on-resonance spins. Thus, the maximum signal is obtained when the optimal rotation is defined for spins lying away from the coil axis. The position of the maximum depends on the rf field distribution generated by the rf coil, which even for the same coil is expected to vary when a different working depth is set.

Regarding the echo time to be set in a CPMG sequence, it has to be noted that to maximize the sensitivity, the maximum number of echoes has to be generated during the echo train. It is achieved by setting the shortest echo time without signal contamination resulting from dead time. It can be found simply by setting an acquisition time shorter than the echo width to ensure that signal is present at all acquired points. Then, the echo time is reduced until the first acquired point becomes distorted. This

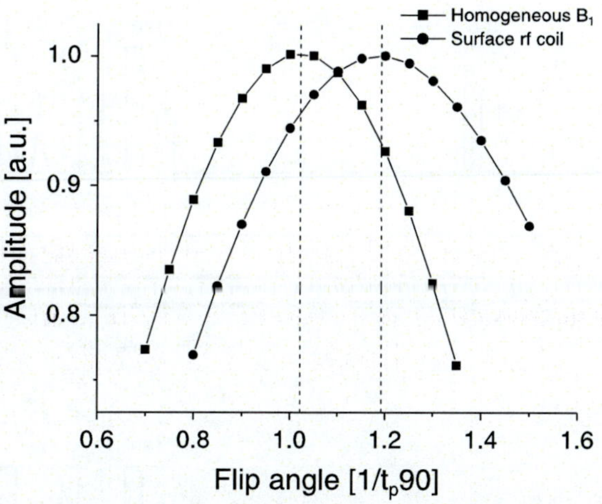

Fig. 2.13 Signal amplitude detected in a CPMG sequence calculated as a function of the strength of the rf pulses. The rf strength is normalized in the plot to the one required to define a 90° rotation (for the first pulse) for on-resonance spins. In the case of a homogeneous B_1 the maximum is obtained for an rf strength only 5% larger than the optimal one. However, for a real coil the maximum signal is achieved when the strength is 20% larger than that required to define a 90° rotation for on-resonance spins in the axis of the coil. The maximum of the distribution is achieved when the optimal flip angle is defined for spins lying at a certain radius from the coil axis where the rf strength is smaller. The position of the maximum strongly depends on the coil geometry, distance to the coil, and detection bandwidth. For this simulation, the field of a single circular loop 10 mm in diameter was calculated at a depth of 2.5 mm, and the detection bandwidth was matched to the excitation bandwidth of the pulses

is the evidence that the dead time is longer than the delay between the 180° pulses and the acquisition windows. The minimum echo time is the shortest one without signal distortion.

2.2.4 Inversion and Saturation Recovery

The spin-lattice relaxation time T_1 is the characteristic time needed by the spin system to exchange energy with its surrounding (the "lattice"). Pulse sequences designed to measure T_1 consist mainly of three stages: (i) a first preparation period during which one or more rf pulses are applied to modify the longitudinal magnetization of the system, taking it away from its equilibrium state, (ii) a free evolution period when the spin system is allowed to relax to the equilibrium state, and (iii) a detection period when the state of the system is measured by converting the longitudinal magnetization into transverse magnetization. The measurement requires a 2D experiment in which the duration of the evolution interval is varied to cover a time range of the order of several T_1s.

Figure 2.14 shows the two main alternative pulse sequences known as inversion recovery [21] and saturation recovery [22], which are adapted for use in

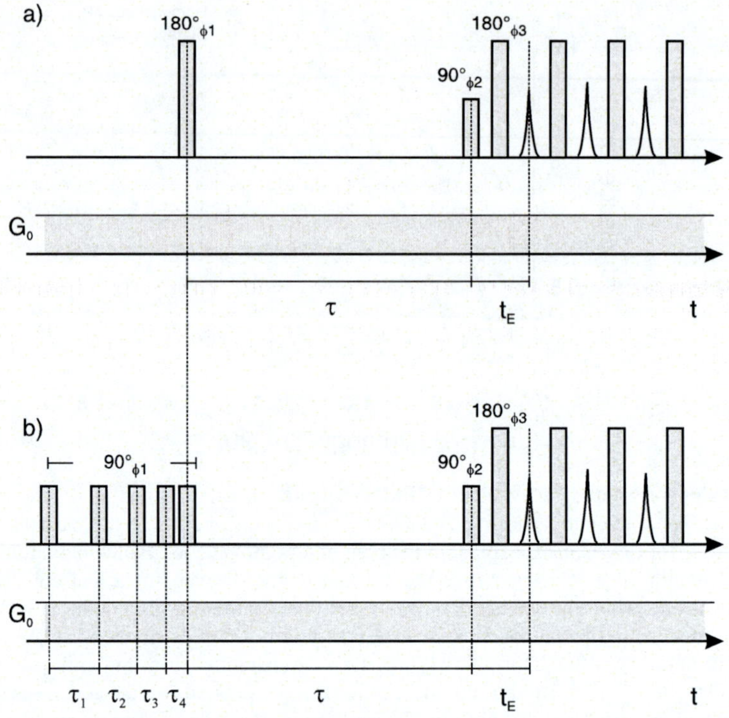

Fig. 2.14 Inversion- (**a**) and saturation-recovery (**b**) pulse sequences adapted to measure the spin-lattice relaxation time T_1 in inhomogeneous fields

inhomogeneous magnetic fields. Implementing these pulse sequences in inhomogeneous fields is difficult mainly because it is impossible to achieve full magnetization inversion or saturation across the full sample. A detailed description of this problem is presented below. It should also be noted here that instead of the single 90° pulse typically used in the detection period to excite an FID, a CPMG sequence is implemented to generate a long echo train that can be co-added to improve sensitivity.

In homogeneous B_0 and B_1 fields, the inversion-recovery sequence uses a hard 180° pulse in the preparation period to invert the equilibrium magnetization M_0 from z to $-z$. Under this ideal condition, the magnetization recovers during the free evolution period τ as $M(\tau) = M_0(1 - 2e^{-\tau/T_1})$. This result is actually the addition of two contributions: one being the magnetization stored along $-z$-axis by the inversion pulse that decays as $M(\tau) = M_0 e^{-\tau/T_1}$ (described by the pathway $M_{0,0,-1,+1}$), and the second, the fresh magnetization created during the evolution time τ that grows as $M(\tau) = M_0(1-e^{-\tau/T_1})$ (described by the pathway $M^1_{0,0,-1,+1}$). However, as described in Sect. 2.2.1, rf pulses act only in a finite bandwidth defined by the pulse length t_p, so their performance strongly depends on the off-resonance frequency (see Fig. 2.3). In this way, when the 180° pulse is applied in the presence of a strong static gradient, it cannot invert the magnetization over the whole frequency range across the object.

As a consequence of the partial inversion defined by the 180° pulse, the magnetization at $\tau = 0$ is not $-M_0$, but just a fraction of it. If M_{max} is the maximum magnetization detected during the CPMG sequence (detection period) in a certain bandwidth and for long evolution times, it is observed that the measurable magnetization at $M(\tau = 0)$ depends on the frequency bandwidth. For example, if just a very narrow portion of the spectrum close to the on-resonance condition is integrated, the initial magnetization can be made $-M_{max}$ (if B_1 is assumed to be homogeneous). However, this is not a practical solution in terms of sensitivity because just a very small fraction of the total excited spins is detected. If the signal is integrated over a bandwidth $1/t_p$ for optimum sensitivity (see Fig. 2.7), the initial value is about $-M_{max}/2$ (Fig. 2.15a). Thus, the dynamic range reduction of the T_1 curve,

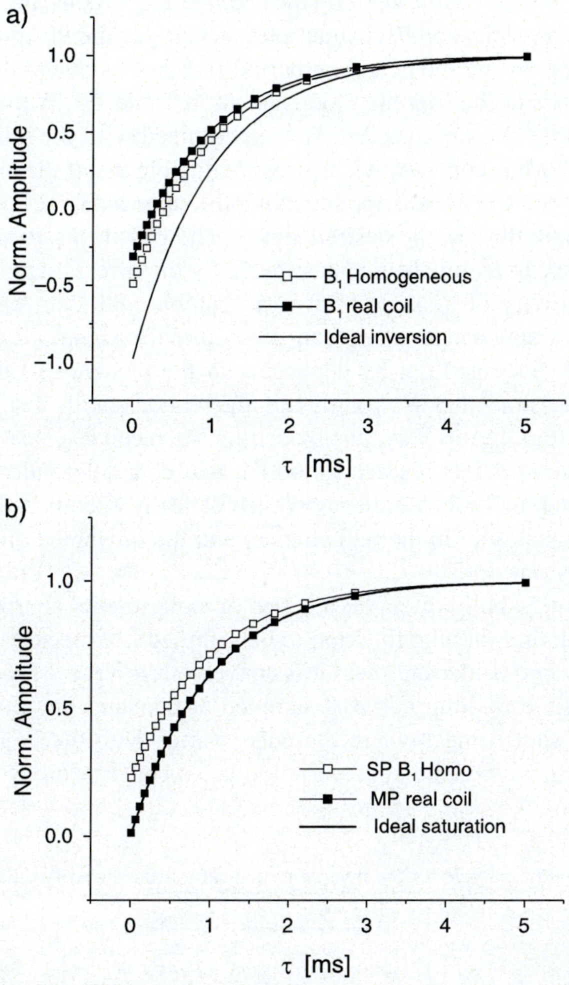

Fig. 2.15 T_1 buildup curves measured by the inversion- (**a**) and saturation-recovery pulse sequence (**b**) in inhomogeneous magnetic fields. The curves are normalized to M_{max}

which is more pronounced in the presence of a inhomogeneous B_1 field (Fig. 2.15a), reduces the accuracy of the T_1 determination, but does not lead to the measurement of a wrong relaxation time.

A further complication observed when implementing the sequence in inhomogeneous fields arises from the fact that the flip angle distribution generated by the inversion pulse leaves considerable transverse magnetization that may interfere during the detection period with the coherence pathways encoding the T_1 information. To filter these spurious signals, a proper phase cycle is needed. As the pathways evolving as transverse magnetization during the evolution period depend on the phase of the inversion pulse, but those of interest for an inversion-recovery measurement ($M_{0,0,-1,+1}$ and $M^1_{0,0,-1,+1}$ describing the magnetization inverted by the first pulse and the fresh magnetization created during the evolution period, respectively) do not, the unwanted ones can be eliminated by cycling the phase of the inversion pulse by 180° while keeping the receiver phase constant. Combining these two steps with the typical add/substract phase cycle for the 90° pulse of the CPMG sequence applied during the detection period (needed to cancel the signal after the 180° pulses) leads to the four-step loop shown in Table 2.3. Although the phase of the inversion pulse does not need to be synchronized with the phases of the pulses applied in the CPMG sequence, it is interesting to note that if the inversion pulse and the 90° pulse of the CPMG are applied along the same axis, the spurious signals are generated in quadrature to the desired signal. The possibility to separate these two signals allows one to eliminate the phase cycle for the inversion pulse. Then, in cases where the sensitivity of a single scan is acceptable and short experimental times are mandatory, a single phase step can be used if the signal after the 180° pulses of the CPMG is averaged out by dephasing in the presence of the static gradient (a condition that is fulfilled in strongly inhomogeneous fields if rf pulses are shorter than the dead time). In this way, just by setting the receiver phase in a conventional CPMG sequence to detect the echoes on the real channel would ensure that when using the first step in Table 2.3 the coherence pathway encoding the T_1 information will generate signal only in the real channel and the unwanted ones will contribute to the imaginary one.

To measure a T_1 buildup curve, a series of experiments is repeated varying the delay τ. Considering that the function to be sampled is expected to be an exponential (at least for liquid-like samples) it is convenient to vary τ logarithmically. This guarantees that the buildup curve is sampled at constant amplitude steps, setting more points at short times, where the curve varies with the largest rate, and less points at the end, where almost no variation is expected. Moreover, as experiments

Table 2.3 Phase cycle for the inversion- and saturation-recovery pulse sequences

ϕ_1	ϕ_2	ϕ_3	ϕ_{rec}
0	0	$\pi/2$	0
π	0	$\pi/2$	0
0	π	$\pi/2$	π
π	π	$\pi/2$	π

with longer τ are more time consuming to measure, this sampling strategy leads to a reduction in the experimental time. If τ is set to vary up to $5T_1$ the time steps can be calculated as

$$\tau_i = -T_1 \ln\left(1 - \frac{(1-\exp(-5))}{(N_{\rm pts}-1)} \times i\right), \tag{2.20}$$

where it is assumed that the function $e^{-\tau/T_1}$ varies from 1 to exp(−5), and $N_{\rm pts}$ is the number of points to be measured in the T_1 curve. As the magnetization is affected during the detection period, a recycle delay of at least 5 T_1 must be included between experiments in order to ensure that the system is in equilibrium before the next experiment starts. A wrong setting of the recycling delay would lead to a smaller initial magnetization (distorting the T_1 curve), requiring a good guess of T_1 before it can actually be measured. This fact leads to time-consuming overestimations. It can be calculated that the measurement time to complete a inversion-recovery experiment is $7.5 \times T_1 \times N_{\rm pts}$ for a linear τ increment, and $6 \times T_1 \times N_{\rm pts}$ for a logarithmic one.

Another variant used to measure T_1 is the saturation-recovery pulse sequence [22]. In homogeneous fields it uses a first 90° pulse to saturate the magnetization instead of inverting it, consequently, during the evolution period τ the longitudinal magnetization grows from zero to its equilibrium value M_0 with the the time-dependence $M(\tau) = M_0(1-e^{-\tau/T_1})$. As this sequence starts nullifying the longitudinal magnetization, it does not require any recycle delay in between experiments. This not only leads to a considerable shortening of the measurement time, but it also eliminates the need to know in advance an estimation of the T_1 of the sample. In other words, a wrong setting of the time range for τ just leads to sample the buildup curve in a non-optimal way. Nonetheless, the characteristic time extracted from the fit is the correct one (just a poorer accuracy is expected from the experiment because the signal variation does not cover the full dynamic range).

When the sequence is implemented in an inhomogeneous magnetic field, saturation is only achieved for the on-resonance spins (assuming a homogeneous B_1 field), but just partial saturation is achieved for off-resonance ones. Consequently, some remnant magnetization stays along the z-axis after the saturation period. Hence, the T_1 curve at $\tau = 0$ starts from an offset value, which, for a detection bandwidth matched to the bandwidth of the pulse, is about 1/4 of the maximum signal measured for $\tau \gg T_1$. To visualize the offset, a T_1 measurement was simulated by assuming the system is in equilibrium before the next saturation pulse is applied (Fig. 2.15b). Experimentally this condition is fulfilled by including a long recycle delay between scans. As for the inversion-recovery sequence, the T_1 extracted from this buildup curve is the exact one, without being distorted by the off-resonance effects. However, in the real experiment, where no recycle delay is used, the situation is more complicated because the offset signal measured in a given scan depends on the longitudinal magnetization at the end of the previous one. The remnant magnetization at the end of one measurement is determined by both the evolution time and the saturation effect of the detection period during this scan. Whereas single

Hahn echo detection may leave considerable longitudinal magnetization, a long CPMG echo train can act as an effective saturation period. Then, the offset would be present in the first scan, but already the second scan would show a negligible offset thanks to the saturation provided by the CPMG detection of the previous scan. Important improvement can be achieved when the T_1 curve is sampled in an inverse way, starting from the long τ values. Then, if the first point is measured for a recovery time $\tau \gg T_1$, the value measured is close to M_0 and is independent of the offset left after the saturation. When τ is shortened to measure the second point, the pre-saturation achieved by the first scans reduces the longitudinal magnetization, attenuating the offset by a considerable factor.

Even though a CPMG detection combined with inverse sampling of the evolution times can considerably reduce the offset in the curve, it must be noted here that the measurement of a given point depends on the efficiency of previous scans to saturate the magnetization, which also depends on the B_1 homogeneity. Moreover, it must be pointed out that the recycle delay between scans is not a precise delay and it may depend on interruptions of the spectrometer between experiments. If this happens, unexpected recovery of the magnetization can induce a larger offset in the measured point. A way to improve decoupling between scans uses multiple 90° pulses applied at irregular times in order to achieve more efficient saturation during the preparation period (see Fig. 2.14b). Optimum performance is achieved by applying an odd number of 90° pulses separated by increasing time delays longer than the T_2^* of the FID. In this way, the natural inhomogeneity of the magnet is used as crusher gradient to spoil the magnetization in between pulses. In addition to the mixing achieved by resonance offset, the fact that the rf field is also inhomogeneous leads to a more efficient cancelation of the magnetization. This is true, because during the train of saturation pulses the magnetization across the sample nutates with different frequency. This spreading defined by the inhomogeneous B_1 field is efficient to mix even on-resonance magnetization.

This approach is useful for magnets with poor homogeneity and becomes very efficient in single-sided systems, where really short delays between pulses of the order of the rf pulse length are enough to strongly mix magnetization components. It should be noted that at the end of the saturation period, the longitudinal magnetization is not zero for each frequency, but it quickly oscillates from positive to negative values as a function of the resonance offset. The anti-phase longitudinal magnetization created during the preparation period gives no signal when it is brought to the transverse plane during the detection period. Thus, only the magnetization recovered during the evolution period is expected to contribute to the signal. Using multi-pulse saturation and inverse sampling, a full dynamic range of the magnetization (from 0 to M_0) is obtained even in inhomogeneous fields. The simulations shown in Fig. 2.15b were performed using the B_1 field generated by a surface coil, indicating that the good performance of the sequence is not affected by the presence of radiofrequency inhomogeneity. It must be mentioned that to reduce the offset, the delays between pulses may require small adjustment depending on the sensor where it is implemented in order to achieve the cancelation of the longitudinal magnetization stored in anti-phase (same positive and negative magnetization stored along the

z-axis). In spectrometers with rf modulation capabilities more complicated schemes based on the use of soft pulses or frequency sweeps can be implemented to improve the degree of saturation.

Finally, the important fraction of magnetization brought to the transverse plane by the saturation pulses applied during the preparation period of this sequence needs to be canceled to avoid interference with the magnetization created during the evolution time τ, which is brought to the plane during the detection period. In contrast to the inversion-recovery sequence, even working under ideal conditions, full magnetization is left in the transverse plane by the preparation period. A simple solution implemented in systems equipped with gradient coils uses a crasher gradient to spoil the residual magnetization right after the saturation pulses. Although in inhomogeneous fields the background gradient acts also as a spoil gradient during the evolution period, the same gradient is present during the detection period when refocusing pulses are applied. Then, under certain conditions (evolution time shorter than the duration of the detection sequence), the magnetization spread out during the evolution period is refocused and may interfere with the desired magnetization encoding the T_1 information. The same phase cycle described for the inversion recovery sequence removes the signal coming from this non-desired coherence pathways, but its use requires similar magnetization before the saturation period for all the scans involved in the phase cycle loop. This condition is satisfied up to certain point when the T_1 curve is sampled in a inverse way (from long to short τ values) and the longest τ is larger than T_1. It is interesting to note that applying the multiple saturation pulses with the same phase as the 90° pulse of the CPMG sequence, with the signal of the desired coherence pathways can be generated 90° out of phase with respect to the spurious signal. Hence, the signals can also be separated by phasing the receiver as described for the inversion-recovery sequence.

In the absence of a recycle delay, and neglecting the detection time (duration of the CPMG sequence), the saturation-recovery sequence requires an experimental time of about $2.5 T_1 N_{\mathrm{pts}}$ for a linear stepping of τ, and just $T_1 N_{\mathrm{pts}}$ for a logarithmic one. In this way, this sequence is six times faster than the inversion-recovery sequence. Considering that the T_1 curve measured with the inversion-recovery sequence has a range of about $1.5 M_0$ (this range may be smaller if B_1 is inhomogeneous) and that the range measured by saturation recovery is M_0, two scans of the saturation recovery are needed to approximately compensate for the sensitivity difference. At the same sensitivity, the saturation recovery is three times faster than the inversion-recovery variant.

While a single-echo detection requires, in general, a large number of scans that lead to extremely long experimental times, multi-echo detection achieved by applying a train of refocusing pulses after the T_1 editing sequence considerably reduces the measurement time. A critical issue that must be taken into account when a long echo train is added is the possible presence of multiple T_1 times in the sample. If the different T_1s show similar T_2s then the echo train addition leads to a T_1 build up curve where the fractions of the T_1s are preserved. However, if spins with a short T_1 have also a short T_2 and the length of the echo train is set to sample the long T_2, then the signal amplitudes are weighted to the limit that the short T_1 may even disappear from the buildup curve. To preserve the amplitude ratio between the two or more T_1 components, the number of echoes to be added must be small enough so that one cannot observe an appreciable signal decay of the shortest T_2 component during the echo train. If the number of echoes is not properly set, it is expected to affect the signal amplitude ratio, but the T_1 values extracted from the curve remain unaffected.

2.2.5 Diffusion Measurements

Magnetic field inhomogeneity has been pointed out to be disadvantageous in single-sided NMR because resonance offsets across the sample complicate the performance of conventional pulse sequences [1–4] and reduce the sensitivity during the detection period [23]. However, some applications like high-resolution sample profiling or the measurement of diffusion coefficients benefit from strong field gradients.

A difficulty faced in the early days of single-sided NMR when trying to exploit the static gradient of open magnets for these types of measurements was the generation of uniform gradients. However, this requirement has been successfully achieved with a number of magnet geometries in the last years (see Chap. 4). In this section, we describe how the echo formation in the presence of a strong and uniform static gradient can be exploited to encode self-diffusion, a parameter that can be used to reveal molecular dynamics and sample microstructure. The effect of molecular self-diffusion on the amplitudes of Hahn and stimulated echoes generated in the presence of a static magnetic field gradient has been analyzed theoretically and experimentally since the very early days of NMR, and the working principle is much older than MRI itself [5, 12, 13]. The use of strong static gradients, such as the one found in the stray field of superconducting magnets, allows measurements of root-mean square molecular displacements as small as 20 nm and self-diffusion coefficients as small as 10^{-16} m^2/s. Large gradients simplify measurements of the diffusion coefficient in heterogeneous materials such as porous materials and biological systems, since it reduces the relative contribution from background gradients due to susceptibility variation across the sample. This is true because the background gradients are proportional to the magnitude of the applied static field B_0, but essentially independent of the field gradient. From this point of view, measurements with unilateral low-field NMR sensors, which produce relatively strong field gradients, offer an interesting advantage over conventional methods.

The pulse sequences typically used are based on the generation of Hahn (SE) [5, 12] and stimulated echoes (STE) [13], both operating in the presence of a steady gradient (see Fig. 2.16). To improve the sensitivity of these experiments, a CPMG sequence is applied after the main diffusion-editing period to generate an echo train that takes advantage of long transverse relaxation times of liquid samples.

Thanks to the sensitivity improvement achieved by adding the echo train, complete diffusion curves can be measured in times shorter than a minute for protonated

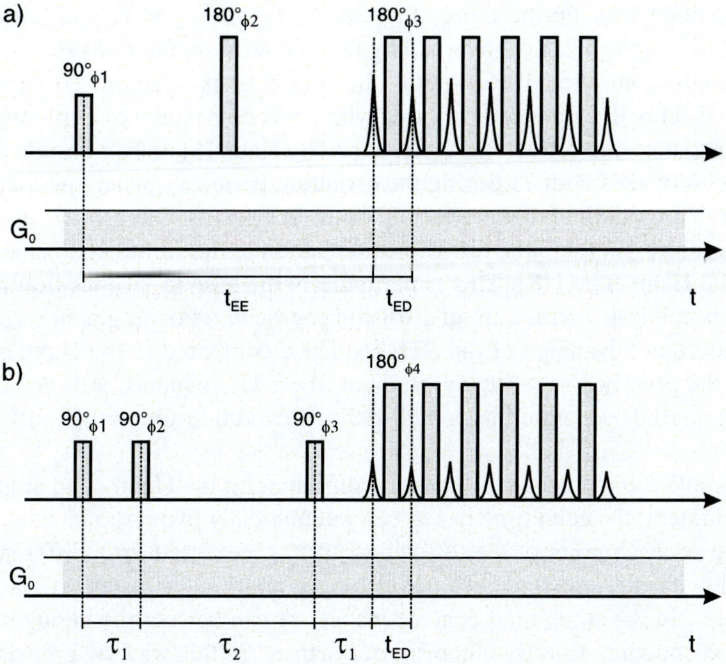

Fig. 2.16 Diffusion editing pulse sequences based on the generation of a Hahn echo (**a**) and a stimulated echo (**b**) in the presence of a static gradient. After the generation of the echo with an amplitude attenuated by diffusion, a train of refocusing pulses is applied to generate a train of echoes that can be added to improve sensitivity. To set the right timings the center of the SE and STE echoes defines the zero time or center of the 90° pulse in a conventional CPMG sequence

solvents. The normalized signal attenuations for the SE and the STE are given by [17, 24]

$$\ln\left(\frac{S}{S_0}\right) = -\frac{1}{12}\gamma^2 G_0^2 t_{\mathrm{EE}}^3 D - \frac{t_{\mathrm{EE}}}{T_2} \tag{2.21}$$

and

$$\ln\left(\frac{S}{S_0}\right) = -\gamma^2 G_0^2 \tau_1^2 \left(\tau_2 + \frac{2}{3}\tau_1\right) D - \frac{2\tau_1}{T_2} - \frac{\tau_2}{T_1}. \tag{2.22}$$

The time delays t_{EE}, τ_1, and τ_2 are defined in Fig. 2.16. The normalization signal S_0 corresponds to the amplitude of the echoes at very small times t_{EE} and τ_1 for SE and STE, respectively. The amplitudes of the Hahn and stimulated echoes are attenuated by the transverse and longitudinal relaxation times (see Eqs. (2.21) and (2.22)). For large values of D, like in the case of less viscous liquids, and for strong magnetic field gradients, the diffusion terms in Eqs. (2.21) and (2.22) dominate over the relaxation terms. Furthermore, if t_{EE}, $\tau_1 \ll T_2$, and $\tau_2 \ll T_1$, the echo amplitude in Eqs. (2.21) and (2.22) is governed by the first term on the right side.

Although in absence of relaxation, the spin echo is a factor two larger than the stimulated echo, there are certain cases where it is convenient to implement the STE sequence to measure diffusion. As described by Hahn [5], the amplitude of the stimulated echo relaxes with T_1 during the evolution period τ_2, when the magnetization is stored as longitudinal magnetization. Thus, in systems where $T_1 > T_2$, it might be advantageous to measure the diffusion attenuation of the stimulated-echo rather than that of the Hahn echo [13]. This is particularly the case in viscous liquids, where a short T_2 is combined with a small diffusion coefficient. For single-sided sensors, the most important advantage of the STE sequence compared to the Hahn echo is that even in the presence of a steady gradient, the STE sequence still consists of two encoding periods separated by a well-defined evolution time when diffusion takes place.

To sample the attenuation caused by diffusion on the Hahn echo amplitude in a steady gradient, the echo time has to be systematically increased in a defined range. By doing so, each point of the diffusion curve is measured for a different diffusion time, allowing the molecules to travel longer and longer distances for increasing echo times. In the stimulated echo sequence, τ_2 can be set long enough in order to require an encoding time τ_1 much shorter than τ_2. In this way, by increasing τ_1, the diffusion curve can be sampled without appreciably varying the total diffusion time $(\tau_1 + \tau_2)$, which in this case is defined by τ_2.

The possibility to define the diffusion time is extremely useful when studying molecular diffusion in the presence of restrictions or physical barriers like in the case of water or oil diffusing in the porous matrix of rocks. Extensive work done by Mitra and Sen [25, 26] showed that the dependence of the effective diffusion coefficient $D(L)$ on the molecular mean displacement L in the time τ_2 is linear for small L, and the slope is proportional to the surface-to-volume ratio of the porous material. A measurement of D vs. L simply requires measuring diffusion curves as a function of the evolution time (τ_2). In order to sample the diffusion curve properly, the variation range of the encoding time τ_1 has to be adapted as the evolution time is increased. Since the minimum τ_1 is positive (actually it is at least the pulse length t_p plus few microseconds needed by the spectrometer in between rf pulses), the maximum evolution time accessible to the technique is either limited by T_1 or by the dynamic range of τ_1 (for excessively long τ_2 the minimum τ_1 already leads to important signal attenuation). For example, for water measured with a G_0= 20 T/m and a t_p = 5 μs, the maximum evolution time τ_2 is about 100 ms. To reach longer evolution times it is convenient to work under a lower G_0. On the other hand, to reach extremely short evolution times (free diffusion limit), a strong gradient is required. Thus, in most cases a combination of two different sensors may be convenient. To access shorter diffusion times, the direct echo of the CPMG sequence can be exploited [27], in this case the diffusion time is set by the echo time of the sequence.

As we have already discussed in this chapter, the application of a series of rf pulses generates a number of coherence pathways weighted by different relaxation and diffusion factors. To make sure that the signal measured during the CPMG detection sequence has acquired a defined diffusion attenuation during the diffusion editing period (Hahn or STE sequence) a proper phase-cycle has to be used to select only the pathway generating the Hahn- or the stimulated echo. We have already described how to exploit the particular dependence of each pathway on the phase of the rf pulses in order to select a given signal. However, we have to point out here an important difference to the procedure explained in Sect. 2.2. As we know, after the second pulse of the Hahn echo sequence, there are nine coherence pathways and only two generate a signal. Thus, a simple add/subtract phase cycle suffices to separate them. However, in the present case, as a number of rf pulses is applied after the generation of the Hahn echo, it is not enough to implement a phase cycle that eliminates the pathways generating non-desired signals after the second pulse. We have to make sure that also those pathways that either do not fulfill the echo condition after the second pulse ($M_{0,+1,+1}$ or $M_{0,-1,-1}$) or are stored as longitudinal magnetization by the second pulse ($M_{0,\pm1,0}$) are also eliminated. It is so because they can simply generate signals when the CPMG is applied. An example is the pathway $M_{0,-1,-1}$, although it does not generate signal after the second rf pulse, a fraction of it is converted into $M_{0,-1,-1,+1}$ after a third pulse is applied. Hence, the strategy to design a proper editing sequence requires cancelation of every pathway except the one encoding the desired information. Table 2.4 provides the phase cycle required to select the coherence pathway generating the Hahn echo and which eliminates the rest.

The add/subtract phase cycle of the first pulse eliminates the pathways where the magnetization stays as longitudinal magnetization after the first pulse. From the pathways that evolve as transverse magnetization after the first pulse ($M_{0,\pm1}$), we need to eliminate the ones stored as longitudinal magnetization and the ones that are not affected by the second pulse. This is achieved by exploiting their different dependence on ϕ_2. Cycling the phase of the second pulse from 0 to π, the first group is canceled after the first four phase steps. Then, repeating these four steps but with a ϕ_2 incremented by $\pi/2$, the magnetization that is not affected by the second pulse is separated from the desired pathway. Although it is true that the unwanted echoes are generated at different times, and may not overlap for a particular t_{EE} and t_{ED}, in a diffusion experiment t_{EE} is varied over a defined range and for certain values

Table 2.4 Phase cycle to select only the direct echo in a Hahn echo sequence

ϕ_1	ϕ_2	ϕ_3	ϕ_{rec}
0	$+\pi/2$	$\pi/2$	π
π	$+\pi/2$	$\pi/2$	0
0	$-\pi/2$	$\pi/2$	π
π	$-\pi/2$	$\pi/2$	0
0	0	$\pi/2$	0
π	0	$\pi/2$	π
0	π	$\pi/2$	0
π	π	$\pi/2$	π

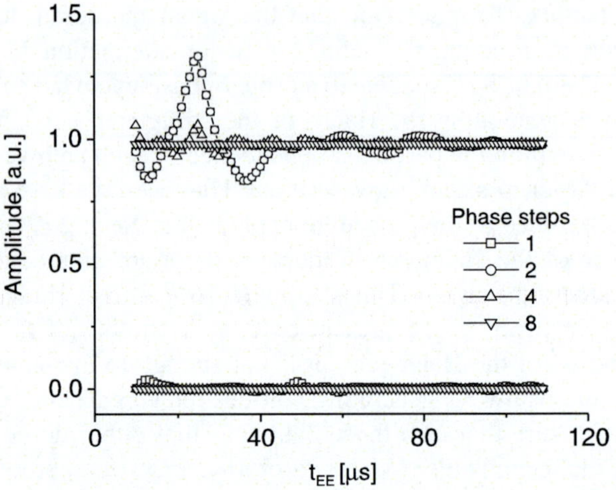

Fig. 2.17 Signal amplitude obtained by addition of 32 echoes generated by the multi-echo train applied after the Hahn echo sequence of Fig. 2.16 with a $t_{ED} = 50\,\mu s$ plotted as a function of the encoding time t_{EE} (in absence of diffusion and relaxation). When a single scan is used, the full coherence pathways interfere during the echo train and give rise to variations in the amplitude that depend on t_{EE}. The first step (two scans) eliminates the signal coming from pathways that stay as longitudinal magnetization after the first rf pulse. In this case, this signal is in the real channel (the desired pathway is along the imaginary channel). Cycling the phase of the second rf pulse in steps of π, the stimulated echo is eliminated. Finally, the coherence pathways converted into transverse magnetization by the first pulse, but untouched by the second one are eliminated by shifting the phase of the second pulse by $\pi/2$ and repeating the first four steps

the signals can overlap. Moreover, when a CPMG train is used for detection, the unwanted coherence pathways superimpose sooner or later during the train. Then, when the train is added for sensitivity improvement, the distortions are observed as oscillations in the diffusion curve.

The effect of cycling the phases of the different pulses can be observed in Fig. 2.17 where the real and imaginary echo train integrals are plotted as a function of the encoding time t_{EE} for the different phase steps (with this receiver phase, the desired signal is generated along the imaginary channel, and no signal is expected in the real channel). The largest distortion is introduced by the signal generated by the stimulated echo, but the other two are not negligible and, in general, require elimination.

The same goal needs to be achieved for the stimulated echo sequence. In this case, the pathway generating the stimulated echo must be separated from the remaining 26 pathways. Although in Sect. 2.2 we describe a phase cycle to filter the signals generated after the third pulse, the four steps listed in Table 2.1 do not suffice to ensure that other pathways will not contribute to the signal during the CPMG detection. This is the reason why the 16 steps phase cycle shown in Table 2.5 must be used [19].

Table 2.5 Phase-cycle to select only the stimulated-echo and eliminate the rest of the pathways after three rf pulses

ϕ_1	ϕ_2	ϕ_3	ϕ_4	ϕ_{rec}
0	0	0	$\pi/2$	π
π	0	0	$\pi/2$	0
0	π	0	$\pi/2$	0
π	π	0	$\pi/2$	π
0	0	π	$\pi/2$	0
π	0	π	$\pi/2$	π
0	π	π	$\pi/2$	π
π	π	π	$\pi/2$	0
$+\pi/2$	0	0	0	$+\pi/2$
$-\pi/2$	0	0	0	$-\pi/2$
$+\pi/2$	π	0	0	$-\pi/2$
$-\pi/2$	π	0	0	$+\pi/2$
$+\pi/2$	0	π	0	$-\pi/2$
$-\pi/2$	0	π	0	$+\pi/2$
$+\pi/2$	π	π	0	$+\pi/2$
$-\pi/2$	π	π	0	$-\pi/2$

2.3 The SNR in Inhomogeneous Fields

A common fact to all techniques developed for single-sided NMR is the intrinsic poor sensitivity associated with low and strongly inhomogeneous magnetic fields. From the methodological side it has been shown that the sensitivity can significantly be enhanced by multi-echo detection methods based on the application of CMPG-like pulse trains. By adding the echoes of a train, the measuring times can be reduced by up to two orders of magnitude compared to single echo methods (Hahn echo like) [3, 28]. However, how to optimize the hardware is less obvious mainly because of the lack of a theoretical formulation to quantify the signal-to-noise ratio (SNR) in inhomogeneous fields. Hoult's formulation for the SNR [14] cannot be applied in a straightforward way to single-sided sensors. These sensors are used to study samples much larger than the sensor itself and only a small portion of the sample can be reached in a single experiment. This fact leads to the definition of the sensitive volume, which depends on the magnet, rf-coil geometry, and the pulse sequence applied. The optimization of a magnet requires a delicate balance between field and gradient strength. Clearly a stronger field leads to higher sensitivity, but if it is achieved at expenses of the gradient a smaller sensitive volume is excited, then what happen with the SNR? On the other hand, another typical question that cannot be answered is how much does the SNR improve when replacing a 300 W rf amplifier by 1 kW one?

In this section we present a theoretical formulation for the SNR applicable to single-sided NMR. The size of the sensitive volume is defined considering the rf-pulse excitation bandwidth and the resonance-circuit bandwidth. Furthermore, single- and multi-echo detection sequences are addressed. Finally, an example of how to optimize a single-sided sensor in terms of sensitivity is discussed.

2.3.1 The Reciprocity Principle

The SNR of an NMR experiment was initially formulated by Abragam [29], and the analysis was then extended by Hoult and Richards [14]. Although Hoult and Richards give an analytical expression for the SNR, it has been obtained assuming certain aspects that are not valid in the limit of highly inhomogeneous fields. Therefore, a more general expression for the SNR that takes into account off-resonance excitation, inhomogeneous rf fields, and a more general detection scheme has to be derived. In the following, the steps followed in [14] are revised and the assumptions that are not valid for our case are highlighted.

The principle of reciprocity [14] states that a magnetic dipole **d** placed at a point **r** will induce an EMF in a coil given by

$$\xi = -\partial/\partial t\{\mathbf{B}/i \cdot \mathbf{d}\}, \tag{2.23}$$

where $\mathbf{B}/i$ is the magnetic field per unit of current produced by the coil at the point **r**. In the case of NMR, $(B_1/i)_{xy}$ is the component of rf field perpendicular to the magnetic field $\mathbf{B}_0$, and **d** is the detectable component (M_{xy}) of the magnetization per unit of volume M_0 generated after an excitation sequence. If the sample has a volume V_s, a spatial integration is necessary:

$$\xi = -\int_{V_s} \partial/\partial t\{(B_1/i)_{xy} M_{xy}\}\mathrm{d}r^3 \tag{2.24}$$

In a conventional NMR experiment, $(B_1/i)_{xy}$ can be assumed to be reasonably homogeneous over the whole sample volume and the excitation is considered to be on resonance; therefore both $(B_1/i)_{xy}$ and M_{xy} can be moved outside the integral [14]. Under these assumptions a 90° pulse tips the full equilibrium magnetization, therefore M_{xy} can be replaced M_0 and the integral is simply the sample volume. However, for the case of a single-sided sensor these conditions are not fulfilled and the integral has to be kept. To solve the integral, the signal response $\Gamma(B_0, B_1, \Delta f_L)$ for a particular pulse sequence (single pulse, Hahn echo, CPMG, etc.) has to be considered. Δf_L is the rf-circuit bandwidth. The signal response is a complex function that describes phase and magnitude of the transverse magnetization generated by the pulse sequence. The output is measured in units of M_0, thus Γ is nondimensional and counts the fraction of the equilibrium magnetization that can be detected. The spatial dependence of Γ (sensitive volume) is given by the variations of B_0 and B_1 with position. The effect of the rf-circuit bandwidth is to act as a filter for both excitation and detection and to determine the dead time which imposes a limit on the minimum echo time that can be used in echo sequences.

Equation (2.24) is then written as follows:

$$\xi = -\int_{V_s} \partial/\partial t\{(B_1/i)_{xy} M_0 e^{-i\omega_0 t}\Gamma(B_0, B_1, \Delta f_L)\}\mathrm{d}r^3. \tag{2.25}$$

In the high-temperature limit the equilibrium magnetization can be approximated by

$$M_0 = N\gamma^2\hbar^2 I(I+1)B_0/3kT, \tag{2.26}$$

where N is the number of spins per unit of volume, γ is the gyromagnetic ratio, and T is the temperature of the sample. The RMS noise power generated at the terminals of the rf coil is, in principle, determined only by the thermal noise (Johnson noise). In practice, there are other sources that could induce noise as well. Typical external noise sources are FM radio stations, gradient and shim coils, mechanical devices, NMR spectrometers, and even the sample itself. In the optimum case, the noise induced by external sources can be neglected so that the time-domain RMS noise per unit of frequency, Δf, is purely given by the equation

$$\sigma = \sqrt{4kT\Delta f R}, \tag{2.27}$$

where R and T are the resistance and temperature of the coil, respectively. By combining Eqs. (2.25), (2.26) and (2.27) the signal-to-noise ratio can be expressed as

$$\Psi = \left\{ \frac{N\gamma^3\hbar^2 I(I+1)}{6\sqrt{2}(kT)^{\frac{3}{2}}} \right\} \frac{B_0{}^2}{\sqrt{\Delta f R}} \int_{V_s} (B_1/i)_{xy} \Gamma(B_0, B_1, \Delta f_L) \mathrm{d}r^3. \tag{2.28}$$

The expression can be divided into two terms, the first one within the brackets is only determined by the sample properties while the term outside the brackets depends on the hardware and the applied pulse sequence. In the case of homogeneous B_0 and B_1 and a single 90° pulse for excitation, the signal response becomes quite simple. Γ is 1 over the complete volume of the sample and the integral is replaced by $(B_1/i)_{xy} V_s$, leading to Hoult's equation [14].

2.3.2 Numerical Calculations of the SNR

Based on Eq. (2.28) the SNR can be predicted for different coil and magnet geometries, and in this way a sensor can be optimized in terms of sensitivity. However, from Eq. (2.28) it is quite difficult to visualize the dependence on, for example, the gradient strength, the bandwidth of the resonance circuit, or the rf power. In the following, some common features to single-sided sensors are deduced by means of numerical simulations. First, a relationship between the excited volume and the excitation bandwidth of rf pulses is introduced. The main consequence of this relationship is the decoupling of magnet and rf coil parameters. Second, it is shown that the SNR is maximized when the rf circuit bandwidth matches the excitation bandwidth of the rf pulses. These two main results are used to solve Eq. (2.28) obtaining a more straightforward equation for the SNR.

2.3.2.1 Magnet Homogeneity: Excited Volume vs. Excitation Bandwidth

A common observation when working with single-sided sensors is the fact that the higher the rf power output W_{rf}, the shorter the rf pulses, and consequently the larger the excited volume, resulting in an increment of the SNR. However, there is a limit ($W_{rf} \to \infty$) where the excitation bandwidth is not longer determined by the rf pulse length but by the rf circuit bandwidth (See next section). In spite of that one can still talk in general about the excitation bandwidth, Δf_{exc}, and its relationship with the excited volume V_{exc}. For example, in the case of sensors generating a main gradient G_0 along the depth direction, the relationship is given approximately by

$$V_{exc} = l^2 \Delta z = \frac{l^2}{\gamma G_0} \Delta f_{exc}, \tag{2.29}$$

where l is the lateral size of the rf coil. It is assumed that the rf field does not vary appreciable along the depth direction within the excited slice; typical slice thickness is on the order of few 100 μm. Moreover, the rf field is considered to be constant along the lateral directions over a region of the size of the coil and drops quickly down at the borders of the coil. In general, the lateral selection is an important requirement to this kind of sensors, which are used to obtain depth information [28, 30].

This relationship is quite straightforward to obtain when assuming a 1D problem; however, at first sight, the relation does not have to hold for sensors generating a magnetic field with a sweet spot. In this case, the first-order components of the gradient field (linear terms) are zero and strong quadratic and higher order terms dominate the spatial variations resulting in a complicated "3D" sensitive volume. A quantification of the excited volume considering its shape and dimensions is impractical. Instead, one can compute the magnetic field over a region around the center of the sweet spot and then measure the excited volume simply as the amplitude of a field histogram. In general, the field distribution is centered at the field value corresponding to the middle of the sweet spot. Thus, a simple plot of the maximum value (cm^3) vs. bin size (kHz) shows the relation between excitation bandwidth and excited volume. Figure 2.18a shows the results obtained for a typical magnet geometry generating a sweet spot, the so-called barrel magnet [31]. The same idea, histogram maximum vs. bin size can be used for sensors generating a strong gradient (Fig. 2.18b). While in Fig. 2.18b the proportionally between V_{exc} and Δf_{exc} holds perfectly, in Fig. 2.18a is still a reasonably good approximation.

It is worth mentioning that in both cases B_1 was considered to be homogeneous within the sensitive volume. The assumption of lateral inhomogeneities only modify the proportionality constant (l^2) in Eq. (2.29) but not the kind of relationship between the excited volume and the excitation bandwidth, i.e., they remain proportional. Variations along the depth direction are negligible within few 100 μm. In the case of sweet spot magnets, the lateral selection is not an important requirement allowing in general the use of rf coils larger than the size of the sweet spot. Therefore B_1 variations can be considered to be smooth.

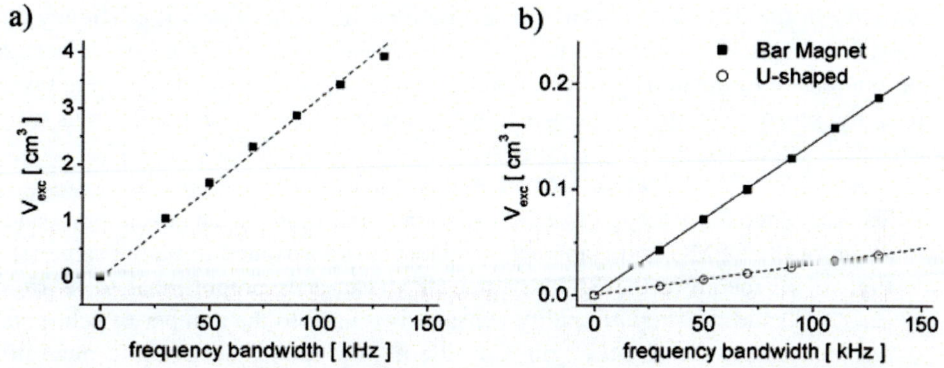

Fig. 2.18 (**a**) Excited volume as a function of the frequency bandwidth for the Barrel magnet [31]. The result is obtained by plotting the maximum value of a field histogram vs. the bin size. (**b**) Same as (**a**) but for magnets having a strong gradient along the depth direction

For both types of geometries the relation

$$V_{\text{exc}} = V^{*}\Delta f_{\text{exc}} \tag{2.30}$$

holds, where V^{*}, the magnet homogeneity, is the slope of the linear fits in Fig. 2.18. The magnet homogeneity indicates how many cubic centimeters are contained in a frequency range of 100 kHz, or the other way around how large is the frequency spread over a volume of 1 cm^3. Equation (2.30) is a fundamental step that allows ones to decouple the magnet parameters from the rf coil parameters. Then, assuming a uniform B_1 over the excited volume, the integral of Eq. (2.28) can be replaced by $(B_1/i)_{xy}V_{\text{exc}}$ obtaining

$$\Psi = \left\{ \frac{N\gamma^3\hbar^2 I(I+1)}{6\sqrt{2}(kT)^{\frac{3}{2}}} \right\} B_0^2 V^{*} \frac{(B_1/i)_{xy}\Delta f_{\text{exc}}(B_1, \Delta f_{\text{L}})}{\sqrt{\Delta f R}}. \tag{2.31}$$

A direct consequence of it is that one can already compare the sensitivity of different magnets by knowing V^{*} and B_0. In the case of sensors with a strong gradient, the inhomogeneity is only along the depth direction (1D problem) and the SNR can be quantified as B_0^2/G_0.

2.3.2.2 Optimum rf Circuit Bandwidth

The dependence of the SNR on the rf coil parameters is quite complicated in inhomogeneous fields, while in homogeneous fields it is simply $(B_1/i)_{xy}/\sqrt{R}$. As already mentioned, the problem is the quantification of the excitation bandwidth, that can be determined either by the length of rf pulses, the circuit bandwidth, or a combination of both. The circuit bandwidth also determines the dead time, which limits the maximum number of echoes that can be acquired in a CPMG train for sensitivity enhancement. Moreover, the coil efficiency $(B_1/i)_{xy}$ must be considered not only in the detection but also for excitation; the higher the coil efficiency the shorter the rf pulses. In consideration of the complexity we decided to run numerical simulations to evaluate the SNR. The magnet and rf coil characteristics were taken from the sensor described in [28]. The following assumptions and definitions were used:

- The sequence for detection is a CPMG train of duration T_2. The excitation bandwidth of the CPMG and Hahn echo sequences are assumed to be the same (see Fig. 2.10), therefore the SNR for the CPMG case is proportional to Ψ from Eq. (2.31) and the proportionality constant is related to the number of echoes n_e sampled during the CPMG train. Considering the echoes decay up to one-third of the initial amplitude the SNR can be calculated as

$$\Psi_{\mathrm{CPMG}} = 2/3\sqrt{n_\mathrm{e}}\ \Psi, \tag{2.32}$$

$$\Psi_{\mathrm{CPMG}} = 2/3\sqrt{T_2/(2t_\mathrm{d} + t_\mathrm{acq})}\ \Psi, \tag{2.33}$$

  where the number of echoes n_e is the ratio between T_2 and the echo time t_E. The echo time is chosen to be the shortest one; i.e., $t_\mathrm{E} = 2t_\mathrm{d} + t_\mathrm{acq}$, with t_d the dead time and t_acq the acquisition time.
- The dead time is defined as the time needed for the voltage induced in the coil during the application of an rf pulse V_pulse to reach the RMS voltage generated by the thermal noise V_noise in the rf coil

$$t_\mathrm{d} = (L/2R)\ln(V_\mathrm{pulse}/V_\mathrm{noise}), \tag{2.34}$$

  with L and R the coil inductance and resistance, respectively. Although the ratio $V_\mathrm{pulse}/V_\mathrm{noise}$ may change from coil to coil and also depends on the applied rf power, its ln can be considered to be constant (~25). The quantity $\ln(x)$ changes only 10% per decade near $x \sim 10^{11}$.
- The 180° pulse condition

$$\gamma B_1 t_{180} = \pi, \tag{2.35}$$

  with

$$B_1 = 1/2(B_1/i)_{xy}\sqrt{W_\mathrm{rf}/R}. \tag{2.36}$$

- The rf circuit bandwidth is

$$\Delta f_\mathrm{L} = R/\pi L. \tag{2.37}$$

The SNR is calculated taking into account the following steps. The 90° and 180° pulses are calculated using Eqs. (2.35) and (2.36) and choosing the same length for both pulses and half of the amplitude for the 90° pulse. The effects of the finite circuit bandwidth are considered as a frequency filter acting on $B_1(\omega)$ for excitation and for detection. Then the echo signal is calculated as the addition of signals coming from different voxels over a volume much larger than the rf coil size. The acquisition time t_{acq} is calculated from the echo signal as the time needed to sample the echo from half- to half-amplitude. The acquisition time and the dead time (Eq. 2.34) define the number of echoes n_e. The frequency bandwidth used to calculate the time domain RMS noise (Eq. 2.27) is calculated as $1/t_{acq}$.

The rf coil resistance is usually modified by including additional resistors in the circuit in order to increase its frequency bandwidth and reduce the Q-factor. In the following the coil resistance R is used as a variable to control the rf circuit bandwidth. Figure 2.19a shows the dependence of the echo amplitude, i.e., excited volume, as a function of the coil resistance. In the limit $R \to 0$ the excited bandwidth is fully determined by the circuit bandwidth, therefore, an incremental change of the resistance is translated one to one (echo amplitude $\propto R^1$) to an increment of the excited volume. In the other limit, $R \to \infty$, the excited bandwidth is determined by length of the rf pulses, which decreases in proportion with $R^{-1/2}$ (Eq. 2.36). The echo amplitude shows a maximum in the intermedius regime where the rf circuit bandwidth is comparable to the excitation bandwidth of rf pulses. The echo width (Fig. 2.19b) behaves similar to the echo amplitude, in the log–log plot they are horizontally mirrored which reflects the fact that their product is constant. The product is proportional to the spectral spin density, that in this case (strong G_0) is the spatial spin density. Figure 2.19c shows the dependence of the echo time of the CPMG train as a function of R. In the limit $R \to 0$ the echo time is determined by both the dead time and the acquisition time. The dead time (Eq. 2.34) as well as the echo width (Fig. 2.19b) is proportional to R^{-1}, therefore the asymptotic behavior of the echo time is also R^{-1}. In the opposite limit ($R \to \infty$) the dead time is negligible and the echo time is dominated by the acquisition time which grows as the $R^{1/2}$.

The SNR for the Hahn echo sequence (Fig. 2.19d) does not depend on the coil resistance in the limit ($R \to 0$) where the circuit bandwidth determines the excitation bandwidth. This fact can be understood by combining the results in the same limit, of Fig. 2.19a and b and considering the detection sensitivity to be proportional to $(B_1/i)_{xy}/\sqrt{R}$. The SNR is proportional to the echo amplitude (echo amplitude $\propto R^1$), proportional to the square-root of the acquisition time (acquisition time $\propto R^{-1}$), and proportional to the detection sensitivity ($\propto R^{-1/2}$), hence all different contributions cancel each other. In the opposite limit $R \to \infty$, the contribution due to the echo amplitude is proportional to $R^{-1/2}$, the one of the acquisition to $R^{+1/4}$, and the detection sensitivity remains unchanged leading to an asymptotic dependence of the SNR proportional to $R^{-3/4}$.

Figure 2.19d also shows the behavior for different rf power values. From these results one can conclude that the only influence of increasing the applied rf power is to expand the plateau of maximum SNR toward higher R values. It is worth mentioning that if the intrinsic coil resistance (no Q-damping) is already in the plateau there is no gain in SNR by increasing the applied power when using a single echo sequence. The situation is different for the case of the CPMG sequence. In this case the echo time determines the maximum number of echoes that can be acquired in the

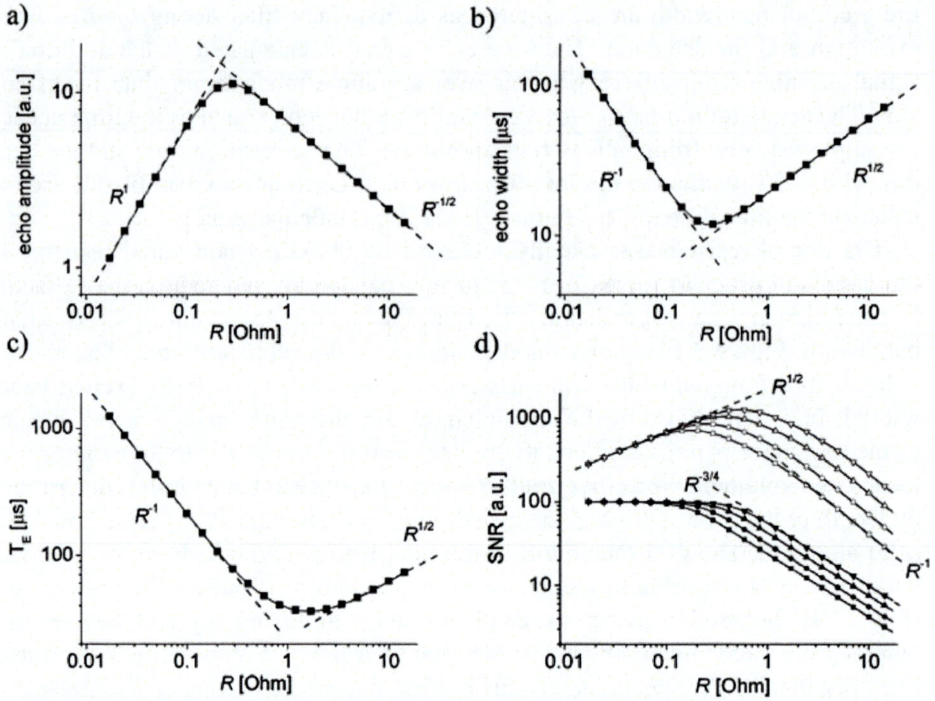

Fig. 2.19 Dependence of the echo amplitude ($\propto$ excited volume) (**a**), echo width (or acquisition time) (**b**), and minimum echo time of the CMPG sequence (**c**) on the coil resistance R. (**d**) SNR for the Hahn echo sequence and for the CPMG sequence in *solid* and *open symbols*, respectively. From *square* to *triangles* down W_{rf} is 100, 300, 1,000, and 3,000 W

CPMG (Eq. 2.34). The asymptotic behavior can be simply calculated as the one of the Hahn echo sequence times ${T_{\mathrm{E}}}^{-1/2}$ leading to SNR $\propto R^{1/2}$ for $R \to 0$ and SNR $\propto R^{-1}$ for $R \to \infty$. The SNR shows a maximum with a value and a position that depends on the applied rf power. At the maximum there is a compromise between, detection sensitivity, excitation bandwidth, and number of echoes acquired in the CMPG. Interestingly, for the resistance value R_{max} that maximize the signal SNR, the circuit bandwidth and excitation bandwidth of the rf pulses are matched, thus

$$\Delta f_{\mathrm{L}} = 1/t_{180} \tag{2.38}$$

which, combined with Eqs. (2.35) and (2.37), results in

$$R_{\mathrm{max}} = \left(\frac{1}{2}\gamma L (B_1/i)_{xy}\right)^{2/3} {W_{\mathrm{rf}}}^{1/3}. \tag{2.39}$$

The condition of Eq. (2.38) for the maximum sensitivity can be introduced in the general expression for the SNR (Eq. 2.28). The ratio between the SNR measured with the Hahn and CMPG sequences is defined in Eq. (2.33), which, considering

$t_{acq} = t_{180}$, can be rewritten as

$$\Psi_{CPMG} = \sqrt{\pi T_2/k_d}\sqrt{\Delta f_L}\ \Psi \tag{2.40}$$

with

$$k_d = 9/4\left(\ln(V_{pulse}/V_{noise}) + \pi\right). \tag{2.41}$$

2.3.3 An Analytical Solution for the SNR

Based on the issues discussed above, Eq. (2.28) can be combined with Eqs. (2.30), (2.38), and (2.39) to write the SNR as

$$\Psi_{CPMG} = \left\{\frac{N\gamma^3\hbar^2 I(I+1)}{\pi 6\sqrt{2}(kT)^{\frac{3}{2}}}\sqrt{\frac{\pi T_2}{k_d}}\right\} V^* {B_0}^2 (B_1/i)_{xy}\left(\frac{\gamma(B_1/i)_{xy}}{2L^2}\right)^{1/3} W_{rf}^{1/6} \tag{2.42}$$

The term between the brackets {} depends on the sample properties with the exception of k_d while the terms outside the bracket depend on the hardware characteristics. The magnet homogeneity V^* tells one, for example, how many cubic centimeters are contained in a frequency range of 100 kHz. Notice that for sensors with a strong gradient, the inhomogeneity is only along the depth direction (1D problem); therefore, it can be quantified simply as $1/G_0$. The term ${B_0}^2$ accounts for polarization and induction. The dependence of R on the frequency (due to the skin depth effect) is not present because the resistance was already optimized to match the circuit and excitation bandwidth (Eq. 2.39). The coil efficiency $(B_1/i)_{xy}$ appears twice, once to the power of one to account for the detection sensitivity, as in conventional NMR, and once to the power of 1/3 to account for the selective excitation. The SNR depends also on the inductance of the coil and the applied rf power, parameters that play no role when working in homogeneous fields.

Equation (2.42) can be used to optimize in terms of sensitivity the magnet and rf coil. The magnet sensitivity $B_0^2 V^*$ can be obtained in a straightforward way by making a field histogram from a 3D field map around the region of interest as described in Fig. 2.18. The rf coil sensitivity, $(B_1/i)_{xy}^{4/3}/L^{2/3}$, is maximized as a function of the coil size and number of turns with the constrain that the coil resistance R fulfills the condition $R < R_{max}$ (Eq. 2.39). The final resistance of the circuit is set to R_{max} by adding a resistor to the resonance circuit.

References

1. Goelman G, Prammer MG (1995, Mar) The CMPG pulse sequence in strong magnetic-field gradients with applications to oil-well logging. J Magn Reson Ser A 113(1):11–18
2. Balibanu F, Hailu K, Eymael R, Demco DE, Blümich B (2000, Aug) Nuclear magnetic resonance in inhomogeneous magnetic fields. J Magn Reson 145(2):246–258
3. Hürlimann MD, Griffin DD (2000, Mar) Spin dynamics of carr-purcell-meiboom-gill-like sequences in grossly inhomogeneous b-0 and b-1 fields and application to NMR well logging. J Magn Reson 143(1):120–135

4. Hürlimann MD (2001, Feb) Diffusion and relaxation effects in general stray field NMR experiments. J Magn Reson 148(2):367–378
5. Hahn EL (1950) Spin echoes. Phys Rev 80(4):580–594
6. Carr HY, Purcell EM (1954) Effects of diffusion on free precession in nuclear magnetic resonance experiments. Phys Rev 94(3):630–638
7. Meiboom S, Gill D (1958) Modified spin-echo method for measuring nuclear relaxation times. Rev Sci Instrum 29(8):688–691
8. Ernst RR, Bodenhausen G, Wokaun A (1990) Principles of nuclear magnetic resonance in one and two dimensions. Oxford University Press, New York, NY
9. Kaiser R, Bartholdi E, Ernst RR (1974) Diffusion and field-gradient effects in NMR fourier spectroscopy. J Chem Phys 60(8):2966–2979
10. Song YQ (2002, July) Categories of coherence pathways for the CPMG sequence. J Magn Reson 157(1):82–91
11. Stejskal EO (1965) Use of spin echoes in a pulsed magnetic-field gradient to study anisotropic restricted diffusion and flow. J Chem Phys 43(10P1):3597–3603
12. Stejskal EO, Tanner JE (1965) Spin diffusion measurements – spin echoes in presence of a time-dependent field gradient. J Chem Phys, 42(1):288–292
13. Tanner JE (1970) Use of stimulated echo in NMR-diffusion studies. J Chem Phys 52(5):2523–2526
14. Hoult DI, Richards RE (1976) Signal-to-noise ratio of nuclear magnetic-resonance experiment. J Magn Reson 24(1):71–85
15. Hahn EL (1953) Free nuclear induction. Phys Today 6:65–70
16. Hürlimann MD (2001, Apr) Optimization of timing in the carr-purcell-meiboom-gill sequence. Magn Reson Imaging 19(3–4):375–378
17. Callaghan PT (1991) Principles of nuclear magnatic resonance microscopy. Clarendon Press, Oxford
18. Coates GR, Xiao L, Prammer MG (1999) NMR logging principles and applications. Halliburton Energy Services. Houston, TX
19. Hürlimann MD, Venkataramanan L (2002, July) Quantitative measurement of two-dimensional distribution functions of diffusion and relaxation in grossly inhomogeneous fields. J Magn Reson 157(1):31–42
20. Song YQ, Venkataramanan L, Hürlimann MD, Flaum M, Frulla P, Straley C (2002, Feb) T-1-t-2 correlation spectra obtained using a fast two-dimensional laplace inversion. J Magn Reson 154(2):261–268
21. Vold RL, Waugh JS, Klein MP, Phelps DE (1968) Measurement of spin relaxation in complex systems. J Chem Phys 48(8):3831–3832
22. Markley JL, Horsley WJ, Klein MP (1971) Spin-lattice relaxation measurements in slowly relaxing complex spectra. J Chem Phys 55(7):3604–3605
23. McDonald PJ (1997, Mar) Stray field magnetic resonance imaging. Prog Nucl Magn Reson Spectrosc 30:69–99
24. Kimmich R (1997) NMR: tomography, diffusometry, relaxometry. Springer, Berlin
25. Mitra PP, Sen PN, Schwartz LM (1993, Apr) Short-time behavior of the diffusion-coefficient as a geometrical probe of porous-media. Phys Rev B 47(14):8565–8574.
26. Mitra PP, Sen PN, Schwartz LM, Ledoussal P (1992, June) Diffusion propagator as a probe of the structure of porous-media. Phys Rev Lett 68(24):3555–3558
27. Zielinski LJ, Hürlimann, MD (2005, Jan) Probing short length scales with restricted diffusion in a static gradient using the CPMG sequence. J Magn Reson 172(1):161–167
28. Perlo J, Casanova F, Blümich B (2004, Feb) 3D imaging with a single-sided sensor: an open tomograph. J Magn Reson 166(2):228–235
29. Abragam A (1983) Principles of nuclear magnetism. Oxford University Press, New York, NY
30. Perlo J, Casanova F, Blümich B (2005, Sep) Profiles with microscopic resolution by single-sided NMR. J Magn Reson 176(1):64–70
31. Fukushima E, Jackson JA (2002) Unilateral magnet having a remote uniform field region for nuclear magnetic resonance. US Patent, 6489872

Chapter 3
Ex Situ Measurement of One- and Two-Dimensional Distribution Functions

Martin D. Hürlimann

3.1 Introduction

Measurements of diffusion [1–4], relaxation [5–7], and distribution functions between relaxation and diffusion [8–10] by nuclear magnetic resonance have become important techniques to study the structure of materials and porous media ranging from biological systems to hydrocarbon bearing sedimentary rocks. These measurements probe the dynamics of molecules on the molecular level and are sensitive to the local environment. The techniques are also particularly well suited for the characterization of heterogeneous systems.

The experimental implementation of relaxation and diffusion measurements is much less demanding than that of conventional spectroscopy, as they do not require magnetic fields of high intensity or homogeneity. They are often performed in systems based on permanent magnets and at Larmor frequencies in the range of a few megahertz to a few tens of megahertz. In addition, the measurements are suited for ex situ applications, i.e., situations where the sample is outside the apparatus. There has been rapid progress in the development of such "inside–out" instrumentation. Applications include well logging, i.e., the evaluation of earth formations from a borehole [11], materials testing with one-sided NMR devices [12], and sensitive diffusion measurements with strayfield NMR setups [13, 14].

3.1.1 Relaxation

Relaxation measurements have long been used to infer fluid properties. In their classic work [15], Bloembergen, Pound, and Purcell showed that the relaxation times of simple fluids are controlled by the correlation times of rotational reorientation and that the viscosity is closely correlated with the measured relaxation time at low Larmor frequencies [16]. The initial development of NMR well logging was motivated by the prospect to distinguish fluids in earth formation by their differences in

M.D. Hürlimann (✉)
Schlumberger-Doll Research, Cambridge, MA 02141, USA

F. Casanova et al. (eds.), *Single-Sided NMR*,
DOI 10.1007/978-3-642-16307-4_3, © Springer-Verlag Berlin Heidelberg 2011

bulk relaxation rates. It was soon discovered that relaxation of the wetting phase in porous media is dominated by surface relaxation [17]. In this case, relaxation gives information on the geometry of the pore space filled by the fluid, rather than the viscosity of the fluid. In the typical case of weak surface relaxivity, the relaxation rate of a fluid diffusing in a pore with surface to volume ratio S/V_p is given by [5]

$$\frac{1}{T_{1,2}} = \rho_{1,2}\frac{S}{V_p}. \tag{3.1}$$

Here $\rho_{1,2}$ is the surface relaxivity for the longitudinal or transverse magnetization, respectively. This expression describes the relaxation behavior in a wide range of porous media, including biological samples [18, 19], sedimentary rocks [6], wood [20], and cement[21]. In heterogeneous samples, the resulting relaxation decay is non-exponential and the distribution of relaxation rates can be used to infer the heterogeneity of the pore geometry.

3.1.2 Diffusion

The translational displacements of molecules undergoing Brownian motion can be measured directly when a magnetic field gradient is applied across the sample [3, 22]. The mean squared displacement is related to a diffusion coefficient, a sensitive probe of molecular size [23, 24]. In complex fluids consisting of mixtures of molecules, the determination of the distribution of diffusion coefficients can be used to infer the composition and distribution of molecular size [25]. Bloembergen suggested already in 1954 to implement diffusion measurements in well logging to distinguish different fluids in earth formations [26].

When a fluid is confined in a porous medium, the translational diffusion becomes restricted [27]. The degree of restriction and its time dependence are powerful tools to obtain geometrical information of the pore space [1, 2, 4].

3.1.3 Diffusion–Relaxation Distribution Functions

A useful extension of relaxation and diffusion measurements is the measurement of two-dimensional distribution functions in inhomogeneous fields [10]. Analogous to the case in conventional multi-dimensional NMR spectroscopy [28], two-dimensional diffusion–relaxation [10] or $T_1 - T_2$ distribution functions [8, 9] contain intrinsically much more information than the corresponding one-dimensional measurements. As an example, from relaxation measurements alone, it is not possible to infer whether the signal originates from the wetting or the non-wetting phase. This makes the interpretation of relaxation measurements ambiguous. Further applications of two-dimensional measurements are illustrated in Sect. 3.5.

3.2 Pulse Sequences and Spin Dynamics in Inhomogeneous Fields

Ex situ NMR measurements on samples placed outside the apparatus are complicated by the inhomogeneities of the magnetic and rf fields and the resulting off-resonance effects in the spin dynamics. With standard NMR equipment, transverse relaxation can be determined from a one-dimensional measurement using the Carr-Purcell-Meiboom-Gill sequence(CPMG) [29, 30], while longitudinal relaxation and diffusion generally require two-dimensional measurements. We review here briefly the extension of these techniques to grossly inhomogeneous fields when the range of Larmor frequencies across the sample exceeds the nutation frequency of the rf pulses.

3.2.1 Relaxation Measurement: Carr-Purcell-Meiboom-Gill Sequence

The CPMG sequence plays a central role in many measurements in grossly inhomogeneous fields for two distinct reasons. First, it can be used to measure relaxation times over a wide range in one-dimensional experiments, even in inhomogeneous fields [31–33]. In well-logging applications, the distribution of relaxation times often spans from sub-milliseconds to several seconds and requires CPMG measurements with thousands of 180° pulses [11]. Second, the CPMG sequence can be used to increase the signal-to-noise ratio by averaging adjacent echoes. For this purpose, it is best to operate the sequence with the shortest possible echo spacing, t_E, to generate the maximum number of echoes in a given time window. CPMG detection with the minimal echo spacing is therefore a common element in many applications in inhomogeneous fields.

In an inhomogeneous field, complications arise in calculating the exact spin dynamics for a sequence with thousands of pulses that are all slice selective. In the limit when t_E is much shorter than the relaxation times, the evolution of the magnetization from one echo to the next for a spin at a given offset frequency ω_0 and rf field strength ω_1 can be well described by an overall rotation around an axis $\hat{n}(\omega_0, \omega_1)$ with an angle $\alpha(\omega_0, \omega_1)$ [32, 34]. At short enough observation, times before relaxation and diffusion become important, the total magnetization $\mathbf{M}$ for the kth echo is then given by

$$\frac{\mathbf{M}(t = kt_E)}{M_0} = \left[\int d\omega_0 f(\omega_0) \left(\mathbf{M}(0^+) \cdot \hat{n}\right) \hat{n}\right] + \\ + \left[\int d\omega_0 f(\omega_0) \left(\mathbf{M}(0^+) - \left(\mathbf{M}(0^+) \cdot \hat{n}\right) \hat{n}\right) \cos(k\alpha)\right] + \\ + \left[\int d\omega_0 f(\omega_0) \left(\mathbf{M}(0^+) \times \hat{n}\right) \sin(k\alpha)\right]. \quad (3.2)$$

Here $f(\omega_0)$ describes the distribution of Larmor frequencies across the sample.

The second and third terms depend on the echo number k and the angle $\alpha(\omega_0, \omega_1)$. Since $\alpha(\omega_0, \omega_1)$ has in general a wide distribution in grossly inhomogeneous fields, these terms quickly average out with increasing echo number k and only generate a detectable contribution for the first few echoes. These terms give rise to the initial transient in echo amplitude and variation in echo shape, as was discussed in the previous chapter. After the first few echoes, the signal is completely dominated by the first term in Eq. (3.2) that is independent of echo number, k.[1]

3.2.1.1 Relaxation of Echoes

After the transient of the first few echoes, the CPMG signal can then be well approximated by:

$$\frac{\mathbf{M}(t = kt_{\mathrm{E}})}{M_0} \simeq \left[\int \mathrm{d}\omega_0 f(\omega_0) \left(\mathbf{M}(0^+) \cdot \hat{n}\right) \hat{n}\right] \exp\left\{-\frac{kt_{\mathrm{E}}}{T_{2,\mathrm{eff}}}\right\}. \tag{3.3}$$

In this expression, the square bracket determines the echo shape, whereas the exponential term indicates the decay of the amplitudes. The time constant $T_{2,\mathrm{eff}}$ both depends on T_2 and T_1 [31, 32] since the effective axis $\hat{n}$ has both longitudinal and transverse components [32]:

$$\frac{1}{T_{2,\mathrm{eff}}} = \left\langle n_\perp^2 \right\rangle \frac{1}{T_2} + \left\langle n_z^2 \right\rangle \frac{1}{T_1}. \tag{3.4}$$

Here the angle brackets indicate a weighted average over all Larmor frequencies that is affected by the filtering used. To maximize the signal-to-noise ratio, it is optimal to use the asymptotic echo shape as matched filter to extract the echo amplitudes. After correcting the amplitudes of the first few echoes for the transient effect, the decay of the amplitudes can then be analyzed in conventional ways used for data acquired in homogeneous fields.

For samples with $T_1/T_2 = 1$, the decay time $T_{2,\mathrm{eff}}$ equals T_2 and is independent of filtering. In contrast, when $T_1/T_2 > 1$, the CPMG decay rate becomes somewhat slower than $1/T_2$ and depends on the details of the filter used [31, 32]. This effect is noticeable in the experimental results shown in Fig. 3.1b on a sample with $T_1/T_2 = 4$. With typical values for $\langle n_z^2 \rangle$ of 0.15, Eq. (3.4) shows that the deviations of the decay rate $1/T_{2,\mathrm{eff}}$ from $1/T_2$ are less than 15%, even for large values of T_1/T_2.

[1] In fields of higher homogeneity, the second and third terms typically contribute to many echoes and often interfere with the relaxation measurement [35]. In this rare instance, large-field inhomogeneities improve the lives of NMR practitioners!

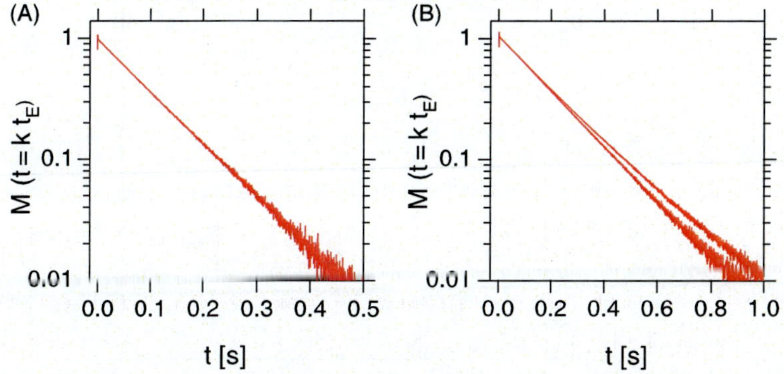

Fig. 3.1 Experimental results for the decay of the CPMG amplitudes for (**a**) sample of doped water with $T_1/T_2 = 1$ and (**b**) milk with $T_1/T_2 = 4$. For each sample, echo amplitudes were extracted both by matched filtering and by echo integration. For the water sample with $T_1 = T_2$, the two decays overlap and the extracted decay rates are independent of filtering. In contrast, the decay rates extracted for the milk sample with the higher T_1/T_2 ratio depend on the filtering used. The decay of the amplitudes obtained by echo integration is mainly affected by T_2 and is faster than the decay of the amplitudes extracted by matched filtering, where there is a noticeable component due to T_1

For this reason, the distinction between T_2 and $T_{2,\mathrm{eff}}$ is often ignored, especially in complex samples where the distributions of relaxation times extend over several decades.

3.2.1.2 Diffusion Effect

Diffusion in inhomogeneous fields leads to additional attenuation of the CPMG echo amplitudes. This effect is mitigated by minimizing the echo spacing t_E. For the CPMG sequence with on-resonance pulses, the magnetization is repeatedly refocused and unrestricted diffusion contributes an additional decay term with a rate of T_{diff}^{-1}, given by

$$\frac{1}{T_{\mathrm{diff}}} = \frac{1}{12}\gamma^2 G_0^2 D t_E^2. \tag{3.5}$$

Here G_0 is the gradient of the static magnetic field and D is the diffusion coefficient. In inhomogeneous fields, numerous additional coherence pathways contribute to the signal. For the higher echoes, these additional coherence pathways have higher sensitivity to diffusion than the CPMG-type coherence pathways considered in Eq. (3.5) [36, 37]. As shown in Fig. 3.2, the result in inhomogeneous fields is a non-exponential decay that is somewhat faster than indicated by Eq. (3.5). However, the decay still scales with $\gamma^2 G_0^2 D t_E^2$ [36]. As mentioned above, the echo spacing t_E is often set to the minimal value to maximize the signal-to-noise ratio. This minimizes diffusion effects during the CPMG decay. In many cases when moderate gradients are used, the decay due to diffusion becomes negligible compared to relaxation.

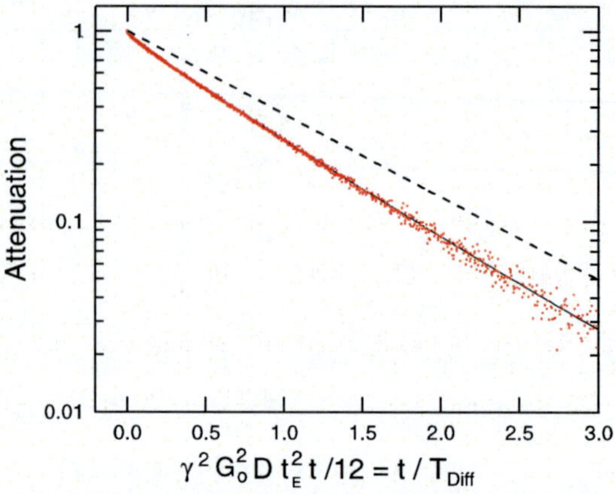

Fig. 3.2 Diffusion effect for CPMG sequence in grossly inhomogeneous fields [36]. Diffusion-induced attenuation is non-exponential and depends on the ratio of the echo time t to the diffusion time T_{diff} given by Eq. (3.5)

3.2.2 Diffusion Measurements with Static Gradients

Conventional NMR diffusion measurements are commonly performed by applying pulsed gradients of variable amplitudes [38, 39]. In many inside-out applications, it is more practical to take advantage of the existing static gradients in the applied magnetic field. In this case, diffusion is best measured with a STRAFI technique [14, 40], where pulse spacings are varied systematically to change the diffusion sensitivity. Such an approach leads to a diffusion-dependent overall attenuation of the signal amplitudes. This is in general superior compared to the CPMG-based approach discussed above that leads to a diffusion-dependent enhancement of the relaxation rate.

Examples of pulse sequences suitable for diffusion measurements in ex situ operations are shown in Fig. 3.3 [10, 13, 41]. The sequences consist of an initial encoding sequence followed by CPMG detection. In all these cases, diffusion attenuates the CPMG echoes by a factor of the form $\exp\{-bD\}$. The diffusion sensitivity b is specific for each coherence pathway, depends on the pulse intervals, and is proportional to the square of the gradient, G_0^2 [36, 42].

Using the intrinsic static field inhomogeneities of the magnet system, these diffusion measurements are performed as two-dimensional measurements by varying systematically b through a change in the pulse spacings in the encoding sequence. Following the diffusion encoding sequence, it is advantageous to use the CPMG detection as shown in Fig. 3.3. This generates a series of echoes that can be averaged to increase the signal-to-noise ratio. Alternatively, the full data set can be used to extract two-dimensional distribution functions. This is discussed in more detail below.

The most basic diffusion encoding sequence is based on the single spin echo, shown in Fig. 3.3a. This is the best choice if the diffusion encoding should be accom-

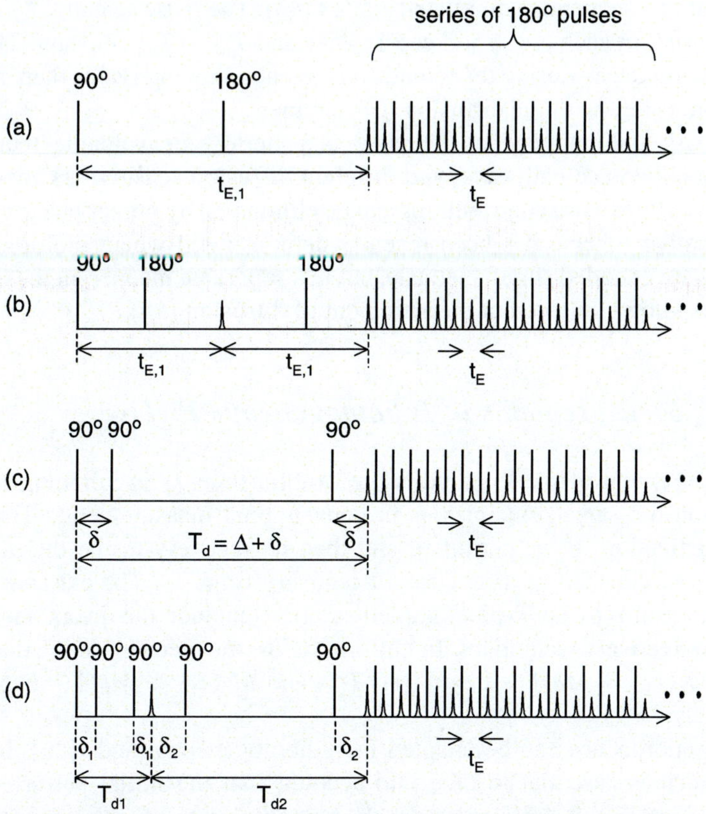

Fig. 3.3 Examples of pulse sequences to measure diffusion and diffusion–relaxation distribution functions in a static gradient. The sequences consist of an initial diffusion encoding sequence followed by CPMG detection. See text for a discussion of the different sequences

plished in the shortest possible time. The single spin echo coherence pathway has the highest diffusion sensitivity per unit time of all possible coherence pathways that refocus the magnetization at time $2t_{E,1}$ and is given by: $b = \gamma^2 G_0^2 t_{E,1}^2/12$.

In samples of low-viscosity fluids, slow convection currents can interfere with the diffusion measurement. Under such circumstances, the sequence shown in Fig. 3.3b is preferable. Here the diffusion encoding is achieved using the second echo, which compensates for convection or flow effects to first order [29].

Diffusion encoding in cases (a) and (b) is achieved by varying the pulse spacing $t_{E,1}$. For samples with large T_1/T_2 ratios or for the study of time-dependent diffusion processes or restricted diffusion, it is useful to select a sequence based on the stimulated echo, as shown in Fig. 3.3c. In this case, the diffusion time T_d is held constant while the pulse intervals δ are varied.

When the duration of the diffusion encoding sequences in Fig. 3.3a–c approaches the relaxation times, it is important to compensate for relaxation effects during the diffusion encoding period. This can be difficult for samples with $T_1 \neq T_2$. This complication can be overcome with the encoding sequence shown in Fig. 3.3d that consists of two sequential stimulated echoes [43]. The pulse spacings $\delta_1, \delta_2, T_{d1}$, and T_{d2} are varied in such a way to keep $\delta_1 + \delta_2$ and $T_{d1} + T_{d2}$ constant. The resulting net relaxation decay during the total diffusion encoding interval is then constant and does not interfere with the diffusion measurement.

In grossly inhomogeneous fields, it is generally unavoidable that additional undesired coherence pathways with different diffusion sensitivities get excited off-resonance. These extra contributions can be eliminated by proper phase cycling [44]. An alternative scheme has been recently demonstrated where multiple coherence pathways are encoded simultaneously, but in a separable manner, in the CPMG echo train. This allows single-shot measurement of diffusion [45].

3.2.3 T_1 Measurements in Inhomogeneous Fields

The measurement of the longitudinal relaxation time T_1 in inhomogeneous fields is based on the same principles as in conventional measurements. The longitudinal magnetization M_z is perturbed and then its recovery to the thermal magnetization M_0 is detected as a function of recovery time, τ_1. The extracted time constant corresponds to T_1. Typical implementations include the inversion–recovery or saturation–recovery techniques. In uniform fields, the recovery of the magnetization follows $M_z(\tau_1) = M_0(1 - 2\exp\{-\tau_1/T_1\})$ and $M_z(\tau_1) = M_0(1 - \exp\{-\tau_1/T_1\})$, respectively.

These approaches can be adapted for inhomogeneous fields [36]. In its implementation, it is essential to take into account that the initial perturbation of the magnetization is non-uniform across the sample due to off-resonance effects. The recovery can always be expressed as a sum of a decaying and recovering signal [36]:

$$S(\omega_0, \tau_1) = s_1(\omega_0)\exp\{-\tau_1/T_1\} + s_2(\omega_0)\left(1 - \exp\{-\tau_1/T_1\}\right). \tag{3.6}$$

Since the decaying and recovering contributions have different spectra, $s_1(\omega_0)$ and $s_2(\omega_0)$, the resulting echo shapes will vary with recovery time τ_1 and make the analysis sensitive to the details of filtering. As shown in [36], this complication can be circumvented by first acquiring data with fully recovered magnetization, $S(\omega_0, \tau_1 \to \infty) = s_2(\omega_0)$. Since the difference signal $\Delta S(\omega_0, \tau_1) \equiv S(\omega_0, \tau_1) - S(\omega_0, \tau_1 \to \infty)$ has a uniform spectrum $s_1(\omega_0) - s_2(\omega_0)$ and decays exponentially as $\exp\{-\tau_1/T_1\}$, it can be analyzed by conventional means. When the basic T_1 sequences are followed by CPMG detection, it is possible to extract two-dimensional $T_1 - T_{2,\text{eff}}$ distribution functions from the data [10].

3.3 One-Dimensional Distribution Functions

Many systems that are studied by one-sided NMR systems are intrinsically heterogeneous and are not characterized by a single relaxation time or a unique diffusion coefficient. A typical example is a water saturated rock. The geometry of the pore space of sedimentary rocks can be very complicated and is in general characterized by a distribution of length scales. Water molecules in small pores will relax faster than molecules in larger pores. Consequently, the relaxation of the overall measured echo amplitudes consists of a superposition of exponential decays and is in general non-exponential. In such cases, it is useful to analyze the measurements in terms of a distribution of relaxation times, $f(T_2)$, and fit the data M_i to an expression of the form:

$$F(t_i) = \int f(T_2)\, e^{-t_i/T_2}\, \mathrm{d}T_2, \tag{3.7}$$

where the distribution $f(T_2)$ fulfills the non-negativity constraint $f(T_2) \geq 0$. In the case of sedimentary rocks, the extracted distribution function $f(T_2)$ reflects the pore size distribution of the sample through the expression given in Eq. (3.1) [46].

A similar situation arises for diffusion measurements in heterogeneous or complex systems. The simple exponential fitting function $F(b_i) = \exp\{-b_i D\}$ has to be replaced by an expression that includes a distribution function of diffusion coefficients, $f(D)$, to describe the data

$$F(b_i) = \int f(D)\, e^{-b_i D}\, dD, \tag{3.8}$$

where $f(D) \geq 0$. The diffusion behavior of a complex fluid, such as a hydrocarbon oil sample, is an example for Eq. (3.8). In the case of mixtures of alkanes, it has been shown that the distribution of diffusion coefficients, $f(D)$, is directly related to the distribution of chain lengths in the fluid [25].

Analysis of experimental data in terms of relaxation or diffusion distributions based on Eqs. (3.7) and (3.8) is particularly well suited for complex or heterogeneous systems. It provides a model-free approach which is often the most appropriate approach because in such systems there are typically no detailed models available with only a limited number of free parameters that can be fit to the data.

3.3.1 Data Inversion

The extraction of the distribution functions $f(T_2)$ or $f(D)$ from the experimental data is based on mathematical routines that incorporate the inverse Laplace transformation [47–51]. The application of these routines requires some care due to the ill-conditioned nature of the inverse Laplace transformation that underlies this analysis. A consequence is that this inversion of experimental data is in general not unique: there are typically many different distribution functions that fit the data within experimental error, even with data of high signal-to-noise ratio. This is illustrated in Fig. 3.4. The top panel shows relaxation data of water filling the pore space of a carbonate rock, measured with the CPMG sequence at a frequency of 1.76 MHz in the fringe field of a superconducting magnet. It is evident that the

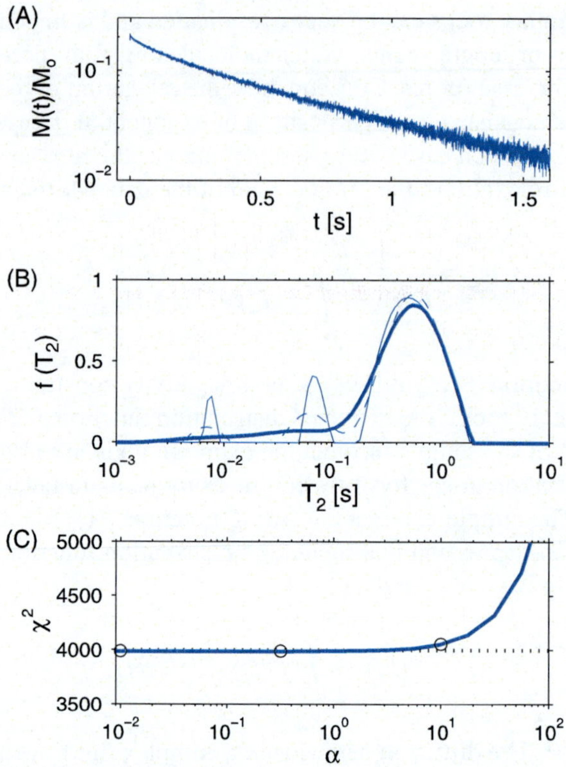

Fig. 3.4 Extraction of distribution function for relaxation data. (**a**) Experimental results for relaxation of water filling the pore space of a carbonate rock. The measurements were performed at 1.764 MHz in the fringe field of a superconducting magnet using the CPMG sequence with 4,000 echoes. The rock has a distribution of pore size, which leads to a non-exponential decay. (**b**) Three different distribution functions $f(T_2)$ that all fit the data of (**a**) well, based on Eq. (3.7). The three distributions correspond to the inversion results with regularization, corresponding to $\alpha = 0.01$, 0.3, and 10, respectively. (**c**) Normalized fitting error, $\chi^2 = \sum_{i=1}^{N} [M_i - F(t_i)]^2 / \sigma^2$, versus the regularization parameter α. For $\alpha \leq 10$, the normalized fitting error is close to $N = 4,000$, the number of echoes, shown as *dotted line*. This indicates a good fit. The values of α for the three distributions shown in (**b**) are indicated by *circles*

decay of the echo amplitudes deviates from a single exponential decay which reflects the distribution of pore size in this sample. The second panel shows three possible distributions $f(T_2)$ that all fit the data well within experimental error.

Even though the three distributions in Fig. 3.4(B) are quite different, the resulting fits $F(t_i)$ based on Eq. (3.7) are nearly identical and differ by less than $5 \times 10^{-4} M_0$ over the whole range of the data. These differences are about five times smaller than the experimental noise associated with the amplitude of each echo, σ. In addition, the three different distributions in Fig. 3.4 have nearly identical total areas, $\int \mathrm{d}T_2 f(T_2) = (0.208 \pm 0.002) M_0$, and mean life times, $\langle T_2 \rangle = \int \mathrm{d}T_2 f(T_2) T_2 / \int \mathrm{d}T_2 f(T_2) = 563.0 \pm 4.2$ ms. These quantities correspond to the magnetization extrapolated to the time of the initial 90° pulse, $M(t = 0)$, related to the porosity of the rock and to the normalized area $\int_0^\infty \mathrm{d}t M(t)/M(t=0)$, respectively.

The ambiguities in the solutions demonstrate the intrinsic challenge of the inverse Laplace transform. Certain features of the solutions are well constrained, whereas other aspects, such as the number of distinct components, are difficult to determine [52–54]. In the current example, the data can be equally well described by a distribution consisting of three separate components centered at 8, 80, and 600 ms (thin solid line), or by a single continuous distribution (solid line). In this case, it is therefore not possible to reliably infer the number of discrete components from the experimental data in Fig. 3.4a.

These issues have to be taken into account in the interpretation of results. Conventional least-squared fitting finds the particular distribution that minimizes the deviation between fit and the data with its random noise. This solution likely contains some sharp, delta-function like features. When the measurement is repeated, a different realization of the random noise is present and the least-squared fitting procedure selects a different solution. The solution obtained by standard least-squared fitting has therefore two undesirable features. First, the solution is not robust, i.e., the detailed shape of the extracted distribution function is strongly affected by the actual noise in the data and is in general not repeatable. Second, distribution functions with sharp, delta-function like features are in most cases un-physical for complex and heterogeneous systems and are not a good representation of the system.

3.3.2 Regularization

It has long been recognized that a more robust and representative solution can be obtained by using regularization in the inversion procedure [48–51]. In the simplest implementation, the cost function to be minimized includes a second term proportional to α, the so-called regularization parameter, that favors extended, smooth distribution functions:

$$C(\alpha) = \sum_{i=1}^{N} \left[\frac{M_i - F(t_i)}{\sigma} \right]^2 + \alpha \int f(T_2)^2 \mathrm{d}T_2, \tag{3.9}$$

Here σ is the experimental uncertainty in the individual amplitudes. The regularization parameter α controls the relative size of the two terms and determines the trade-off between stability and bias. For very small values of α, the inversion reduces to conventional least-squared fitting and the selected distribution function likely exhibits sharp features that are unstable and not reproducible. With very large values of α, the cost function is dominated by the second term in Eq. (3.9) that does not depend on the measured data and it is minimized by solutions that are extended in T_2. While these smooth solutions are stable and reproducible, they become significantly biased, i.e., the fit to the data becomes poor. It is essential to adjust α to an optimal value that gives the best compromise between instability and bias. Inversion procedures are therefore set up to adjust α such that it yields the smoothest solution that still fits the data well. Among all possible solutions that fit the data, it is the distribution function with the fewest features.

The three solutions shown in Fig. 3.4b correspond to the solutions of such an optimization procedure using the cost function of Eq. (3.9) with three different values of α. The fitting error of the solution, i.e., the first term of Eq. (3.9), versus the value of α is shown in Fig. 3.4c. For α up to about 10, the resulting χ^2 is independent of α and close to the number of echoes, N, indicating an excellent fit to the data. For larger values of α, the misfit increases significantly. For this reason, the optimal regularization parameter is $\alpha_{\text{opt}} = 10$ for this case. The corresponding distribution function, shown as solid line in Fig. 3.4b, is therefore chosen as the optimal representative solution for this inversion problem: it is the solution with the minimum amount of structure that still fits the data well.

3.3.3 Systematic Errors

After finding a solution to the inversion problem, it is essential to analyze the residuals between the fit and the data, $F(t_i) - M_i$. The mathematical inversion algorithms are based on the assumption that the experimental data M_i are given by $M_i = F(t_i) + \epsilon_i$, where $F(t_i)$ is given by Eq. (3.7) and ϵ_i is random noise with a standard deviation of σ. If the residuals show a systematic trend, then the kernel in Eq. (3.7) does not adequately describe the data or there are systematic errors in the measurement. In that case, the inversion results should not be trusted.

As an example, relaxation measurements in moderately homogeneous static fields, but with inhomogeneous rf fields or incorrectly set pulse durations often exhibit oscillations in the detected echo amplitudes [35]. Under this experimental condition, the second and third terms in Eq. (3.2) give significant contributions to a large number of echoes. In that case, the data cannot be described by an expression of the form in Eq. (3.7) and the inversion results will be erroneous.

3.3.4 Uncertainties

If there are no apparent systematic errors present, it is important to estimate the uncertainties of the inversion results. Parker and Song [54] have pointed out that it is mathematically not possible to assign conventional error bars to the distribution functions directly. However, most quantities of interest are related to the distribution function by linear functionals of the form $\int_a^b \mathrm{d}T_2\, w(T_2)\, f(T_2)$, to which error bars can be assigned. Examples of such functionals are the porosity $\phi = \int_0^\infty \mathrm{d}T_2 f(T_2)$ and the mean life time $\langle T_2 \rangle = \int \mathrm{d}T_2 f(T_2) T_2 / \int \mathrm{d}T_2 f(T_2)$.

There are two aspects to uncertainty of such derived quantities, related to the stability and to the bias of the solution $f(T_2)$. The stability depends on the sensitivity of the solution to the noise in the experimental data. Ideally, repeated measurements under identical conditions but with different noise realizations yield identical solutions. Stability can be tested by analyzing a simulated set of repeat measurements, generated from the fit to the original data, $F(t_i)$, with the addition of random noise ϵ_i that has the same magnitude σ as in the original experimental data. For the data shown in Fig. 3.4 with the observed noise level of $\sigma = 0.0023\, M_0$ per echo, we find that the stability is $\delta\phi = \pm 0.0011 M_0$ for the porosity and $\delta\, \langle T_2 \rangle = \pm 3.1$ ms for the mean life time, using a regularization parameter of $\alpha = 10$. In general, an increase of the regularization parameter will generate solutions with higher stability, but also with larger bias. Figure 3.4c shows that the solutions for $\alpha > 10$ have a fitting error that is larger than expected from the random noise present in the data. These solutions are therefore biased.

Results should also be tested for the dependence on α for values smaller than α_{opt}. For the example shown in Fig. 3.4, the variations in the extracted values for ϕ and $\langle T_2 \rangle$ with α are listed above and are comparable to the values of stability. This indicates that the porosity and average life time are not very sensitive to the smoothness of the solution. However, the situation might be different for other quantities of interest and has to be checked. A more complete discussion of these issues and a systematic procedure to assign uncertainties can be found in the paper by Parker and Song [54]. Their approach does not rely on regularization, but considers all possible solutions to the inversion problem.

3.4 Two-Dimensional Diffusion–Relaxation Distribution Functions

A powerful extension of the measurement of relaxation and diffusion properties in grossly inhomogeneous fields is the measurement of multi-dimensional distribution functions involving diffusion, transverse, or longitudinal relaxation [9, 10]. Analogous to conventional multi-dimensional NMR spectroscopy [28], two-dimensional diffusion–relaxation or $T_1 - T_2$ distribution functions contain inherently more information than the corresponding one-dimensional distribution functions. We illustrate here the concept with measurements of diffusion–relaxation distribution functions. The approach of two-dimensional NMR measurements is rather general and applies to both homogeneous and inhomogeneous fields, and can be extended to other distribution functions, such as $T_1 - T_2$ [8, 10], flow – T_1 [55] and flow – T_2 distribution functions [56], and to exchange measurements including diffusion–diffusion [57] and $T_2 - T_2$ [58] correlation functions.

3.4.1 Two-Dimensional Diffusion-Relaxation Measurements

The measurement of diffusion–relaxation distribution functions in inhomogeneous fields is based on sequences such as shown in Fig. 3.3. They consist of an initial sequence that encodes diffusion, followed by a long train of refocusing pulses that

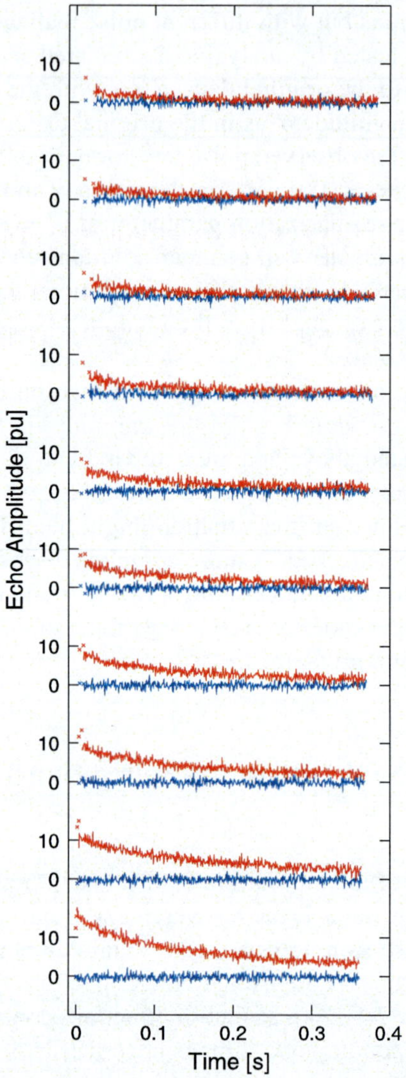

Fig. 3.5 An early example of a diffusion–relaxation measurement performed with an NMR logging tool at a Larmor frequency of 1.2 MHz in an oil well 885 m below surface [59]. The figure shows the in-phase and out-of-phase amplitudes of the signal originating 4 cm into the formation. From the *bottom* to the *top*, the first two echo spacings are systematically increased to encode the diffusion information in the applied static gradient. In this case, the logging tool was stationary and the measurements were averaged 12 times

encodes relaxation. There are two independent variables, the diffusion sensitivity b, controlled by the pulse spacing in the diffusion sequence, and the decay time kt_E during the CPMG part of the sequence. A typical two-dimensional set of data is shown in Fig. 3.5. The data were acquired with a well-logging NMR tool [59] using the sequence of Fig. 3.3b in an oil well almost 1 km underground. As the first two echo spacings are increased for the diffusion encoding, the overall amplitudes of the echo trains decreases systematically.

To analyze the two-dimensional data quantitatively, a spin dynamics correction has to be applied to the first few echoes, in analogy to the case discussed earlier for the CPMG sequence. The expression in Eq. (3.2) still applies for this sequence as long as $\mathbf{M}(0^+)$ is reinterpreted as the magnetization at the end of the diffusion encoding sequence [44]. Therefore, the same type of diffusion-independent correction can be applied to the first few echoes after the diffusion encoding sequence.

Following this correction, the two-dimensional set of experimental data, $M_{i,j}$, is then fit to an expression that for the current pulse sequence is of the form:

$$F(b_i, t_j) = \iint \mathrm{d}D\,\mathrm{d}T_2\, f(D, T_2)\, e^{-b_i D}\, e^{-(2t_{\mathrm{E},1}+t_j)/T_2}. \tag{3.10}$$

Here $f(D, T_2)$ is the diffusion–relaxation distribution function with $f(D, T_2) \geq 0$, and $t_j = jt_\mathrm{E}$ $(j = 1, 2, \ldots)$ are the echo times in the CPMG trains that follow the diffusion encoding sequences. Strictly speaking, the relaxation times T_2 should be replaced by $T_{2,\mathrm{eff}}$, as given in Eq. (3.4).

3.4.2 Data Analysis

The measured amplitudes $M_{i,j}$ are related to the two-dimensional distribution functions by a two-dimensional Laplace transformation, as given by Eq. (3.10). The extraction of the distribution function from the experimental data therefore requires a two-dimensional inverse Laplace transformation [60]. The complications discussed in the previous section for one-dimensional inversions also apply to two-dimensional inversions. The ill-conditioned nature of the two-dimensional inverse Laplace transforms means that there exist a number of distribution functions $f(D, T_2)$ that fit the experimental data within the measurement uncertainties. Many of these distribution functions contain sharp, delta-function like features with slightly different positions and amplitudes. Conventional fitting of the data that minimizes the fitting error will pick one of these possible distribution functions. When the measurement is repeated, the noise will differ and a different distribution function will be chosen among all possible distribution functions. Therefore, the conventional fitting procedure leads to results that contain details that are unstable and not reproducible. In many complex systems of interest here, the distributions are expected to be continuous without any such sharp features.

In analogy to the one-dimensional case, a regularization term is introduced into the cost function to be minimized:

$$C(\alpha) = \sum_i \sum_j \left[\frac{M_{i,j} - F(b_i, t_j)}{\sigma}\right]^2 + \alpha \iint \mathrm{d}D\,\mathrm{d}T_2\, f(D, T_2)^2. \tag{3.11}$$

The regularization parameter α again controls the trade-off between stability and bias. For small values of α, the selected distribution function will fit the data well, but details of the results are unstable and depend on the exact noise realization in the data set. For very large values of α, the selected distribution functions are very stable and reproducible. However, the fitting error becomes larger than expected from the experimental noise and the solution is therefore biased.

The implementation of this approach to the inversion of two-dimensional data sets is closely related to the approach for the one-dimensional case. However, an additional problem arises: for typical two-dimensional sets, the data sets become very large and the inversion becomes unpractically slow. This problem has been overcome in the algorithm developed by Venkataramanan et al. [60] that is well suited to analyze the multi-dimensional data sets of interest here. In this algorithm, the data is first filtered using singular value decomposition. In this reduced data space, the inversion problem is solved using a regularization term and an additional non-negativity constraint, based on the algorithm by Butler et al. [61].

Results of this inversion procedure applied to the downhole data of Fig. 3.5 are shown in Fig. 3.6. For regularization parameters α of 100 or less, the distribution

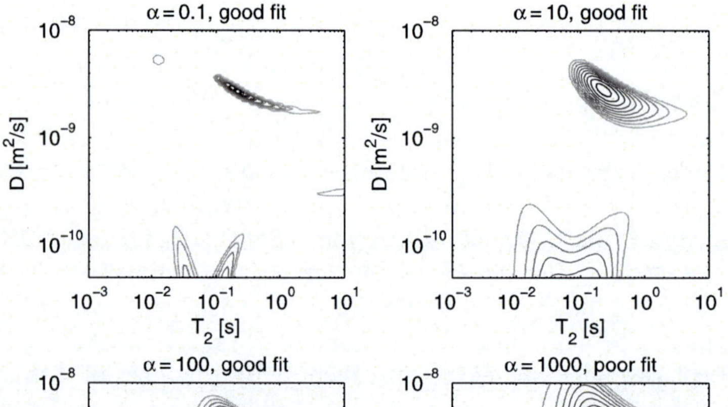

Fig. 3.6 Diffusion–relaxation distribution functions extracted from borehole data shown in Fig. 3.5. The contributions with diffusion coefficients larger than 10^{-9} m^2/s correspond to water and they are well separated from the oil contributions. The four panels show inversion results for different regularization parameters α. As α is increased, the distributions broaden and spurious contributions disappear. For values of α larger than 100, the fit becomes poor. The optimal value of α is therefore between 10 and 100

functions fit the data well and clearly separate into two main components that are associated with the oil and water phase, respectively. A comparison of the four panels illustrates the influence of α on the inferred distribution functions. For the smallest value of α, extra features appear. Increasing α to a value of 10 and 100 removes these spurious features but broadens the two main contributions somewhat. The distribution functions for $\alpha \leq 100$ all fit the data almost equally well, i.e., the fitting errors are almost identical. As the regularization parameter is further increased, the fitting error noticeably increases and the extracted distribution function is excessively broadened and shows a bias. In this case, $\alpha_{\text{opt}} \simeq 100$.

3.4.3 Interpretation of Distribution Functions

When diffusion–relaxation distribution functions are analyzed and interpreted, it is essential to consider the intrinsic limitations of the data inversion due to the Laplace transformation. The results presented in Fig. 3.6 for $\alpha = 0.1, 10, 100$ show three different examples of distribution functions that fit the data almost equally well. This implies that some of the finer features in $f(D, T_2)$ for the smallest values of α might not correspond to real features in the sample. Unlike for Fourier transformations, it is not straightforward to infer the resolution of Laplace transformations. The scale of features that can be resolved in the distribution functions depends on the ratio of signal to noise and the details of the data acquisition [52–54].

To avoid the misinterpretation of spurious fine features in the extracted distribution function, it is generally advisable to present the solution among all solutions $f(D, T_2)$ that is the smoothest but still fits the data well. This corresponds to the distribution function associated with the largest regularization parameter and small fit error. This optimal regularization parameter α [60] has to be determined for each experimental condition. In the example shown in Fig. 3.6, the optimal value for α is around 100.

Similar to the one-dimensional distribution functions, it is not possible to assign conventional error bars to the values of $f(D, T_2)$ for a given value of D and T_2. However, many quantities of interest are integrals or moments of $f(D, T_2)$ over a some limited regions in the $D - T_2$ plane. As an example, in the results of Fig. 3.6, the fraction of contributions with a diffusion coefficient larger than $0.7 \times 10^{-9}\text{m}^2/\text{s}$ corresponds to the water saturation. Despite the intrinsic uncertainties in the inversion, this quantity is well constrained and has small bounds: $53.0 \pm 0.7\%$.

Similarly, the average diffusion coefficient for a given relaxation time

$$\langle D(T_2) \rangle = \frac{\int \mathrm{d}D D f(D, T_2)}{\int \mathrm{d}D f(D, T_2)} \tag{3.12}$$

is well defined. Examples of this approach are shown in Sect. 3.5 for restricted diffusion.

It is important that the experimental parameters are well adapted to optimize the sensitivity to the parameters of interest [53], including the gradient strength, $t_{E,1}$, t_E and the number of echoes, N. There is limited sensitivity to relaxation times less than t_E or diffusion coefficients above b_{max}^{-1}. Contributions with diffusion coefficients less than b_{min}^{-1} or with relaxation times longer than the total echo time are detected, but it is not possible to determine accurately its diffusion or relaxation properties, respectively. In general, it is desirable to have the echo spacing t_E as short as possible. At the same time, it is important to ensure that the duration of the total echo train, Nt_E, covers the whole relaxation decay. In practical applications, limitations in total power consumptions, acquisition memory, or data transmission might require a compromise of these conditions.

3.5 Applications of Two-Dimensional Distribution Functions

In this section, we illustrate the utility of $D - T_2$ and $T_1 - T_2$ distribution functions with some recent applications. Such two-dimensional diffusion and relaxation measurements are particularly useful in heterogeneous systems when the sample consists of a complex fluid, a mixture of fluids, or a fluid filled porous medium. Such systems are ubiquitous and important examples include mixtures of crude oil and brine in geological formations, water content in plants and wood, fat and water distribution in biological or food products, and water content in construction materials such as concrete. The characterization of heterogeneous systems is challenging for approaches based on standard NMR spectroscopy due to susceptibility-induced inhomogeneities that might require magic angle spinning [62]. In contrast, the technique of $D - T_2$ and $T_1 - T_2$ distribution functions does not require high magnetic fields and is well suited to be performed at low-field strength with "inside–out" configurations.

Two-dimensional distribution functions have already found wide spread use in well logging of hydrocarbon bearing earth formations [59]. Such measurements are an essential tool to identify and quantify the fluids in the formation and to understand the geometrical configurations of these fluids in the pore space. In commercial applications, measurements are typically performed continuously while the NMR apparatus is moved up the well [63]. In addition, measurements are performed at different depths of investigation in order to monitor the invasion profile of fluids that infiltrated the formations from the borehole. These measurements have also been used in medical and biological applications [64] and in the characterization of food products [65, 66].

To illustrate the wealth of information that can be extracted from two-dimensional distribution functions, a number of further examples are presented in this section. These examples were measured at different positions in the fringe field of a superconducting magnet at Larmor frequencies of 1.75 or 5 MHz, with corresponding gradients of 132 or 545 mT/m, respectively. For each measurement, the optimal regularization parameter has been determined as outlined above and the displayed distribution has been determined with this value of α_{opt}.

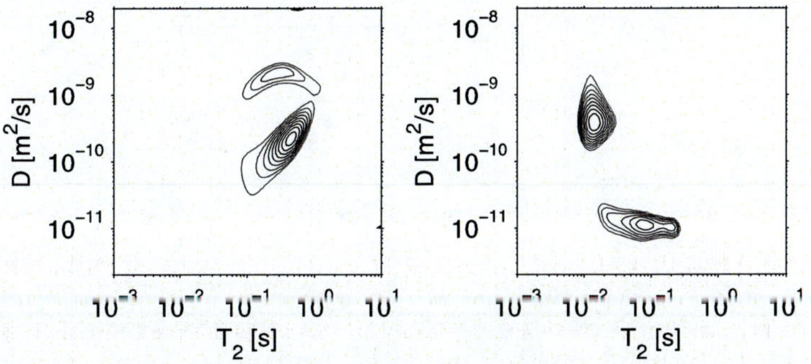

Fig. 3.7 Identification and quantification of different components by diffusion–relaxation distribution functions. *Left*: Results for a core of carbonate rock that was saturated with a mixture of brine and crude oil [67]. The *upper* component is due to brine with a diffusion coefficient close to the molecular diffusion coefficient of water and relaxation times reflecting the pore size distribution. The large extent of the crude oil signal reflects the wide distribution of molecular sizes in the crude oil. *Right*: $D - T_2$ distribution function for a piece of Gruyère cheese [66] showing clear separation of fat and water phase. The fat is contained in small globules that reduces the apparent diffusion coefficient to a low value

3.5.1 Two-Component Systems

In many multi-component systems, diffusion – T_2 distribution functions allow an identification and clear separation of the different components. Two representative examples are shown in Fig. 3.7 that stem from very different applications. The left panel shows measurements on a core of carbonate rock that was saturated with a mixture of brine and crude oil. The two fluid phases can be clearly separated and quantified. The contribution from the aqueous phase can be identified by its diffusion coefficient that is close to the molecular diffusion coefficient of water, $2 \times 10^{-9}\,\mathrm{m}^2/\mathrm{s}$. The relaxation times of this contributions are distributed over a decade and indicate the distribution of local surface-to-volume ratios of the local pore space. The composition of the crude oil consists of a wide distribution of molecules, leading to a wide distribution of T_2 and D values. This explains the observed shape of the oil contribution with its strong correlation between T_2 and D values. The second example shown on the right panel in Fig. 3.7 illustrates the application of this technique to food products, in this case a dairy product. For the sample of Gruyère cheese [66], the two components correspond to the continuous aqueous phase (higher diffusion coefficients) and the fat globules (lower diffusion coefficients).

In both examples, the $D - T_2$ measurements allow a quantitative determination of the content of the two components. In addition, the position and shape of the contributions are related to important chemical, physical properties or, the physical arrangements of the different components. This will be more evident with the additional examples presented below.

In systems where different components exhibit contrasting T_1/T_2 ratios, the measurement of $T_1 - T_2$ distribution functions can also be used to separate different

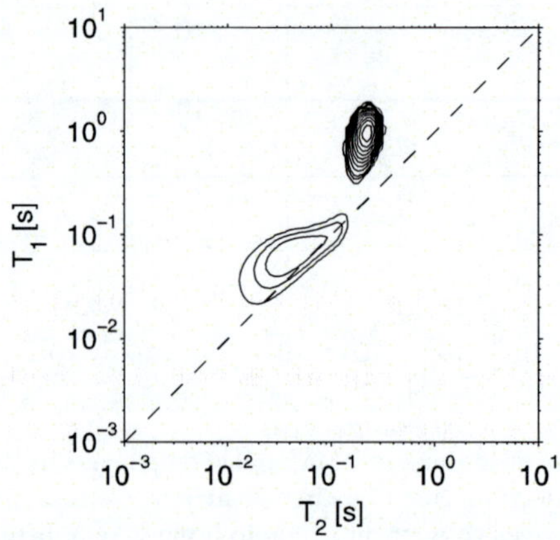

Fig. 3.8 $T_1 - T_2$ distribution function for heavy cream. The fat and water components can be identified based on their different T_1/T_2 ratio. For fat, the T_1/T_2 ratio is close to 1, shown as *dashed line*, whereas for the aqueous phase, $T_1/T_2 \approx 4$ due to the presence of proteins [66]

components. In many systems, the T_1/T_2 ratio at low fields is restricted to a narrow range. However, there are interesting exceptions when the spin dynamics of at least one of the components is affected by motion slow compared to the Larmor frequency. In this case, a significant separation of T_1 and T_2 occurs. As an example, in dairy products interactions between water molecules and proteins result in a T_1/T_2 ratio that is much higher for the aqueous phase than the fat component at a Larmor frequency of 5 MHz [66]. This effect can be used to separate and identify the two components by the T_1/T_2 ratio in a $T_1 - T_2$ measurement, as is illustrated in Fig. 3.8.

3.5.2 Wettability

Diffusion–relaxation distribution functions have also been used to infer wettability and fluid configurations in fluid saturated porous media. This is illustrated in the examples shown in Fig. 3.9. Two fine-grained rock cores were selected from the water zone (A) and oil zone (B) of an oil well, respectively. The top row shows the results when the cores are water saturated. The relaxation times of the two cores are essentially identical, indicating that the pore size distributions for the two cores are very similar. The measured diffusion coefficients are reduced from the molecular diffusion coefficient of water. This demonstrates that diffusion is restricted and confirms that the pore size is smaller than the diffusion length. The second row in Fig. 3.9 shows measurements on the same cores saturated with dodecane. In this case, the measured diffusion coefficients are reduced from the oil molecular diffusion coefficient by the same amount as for the water measurements, as the fluid configurations are nearly identical. However, there is a distinct difference in the relaxation behavior caused by the contrasting wetting behaviors of these two cores. The core from the water zone is predominately water wet. After imbibition

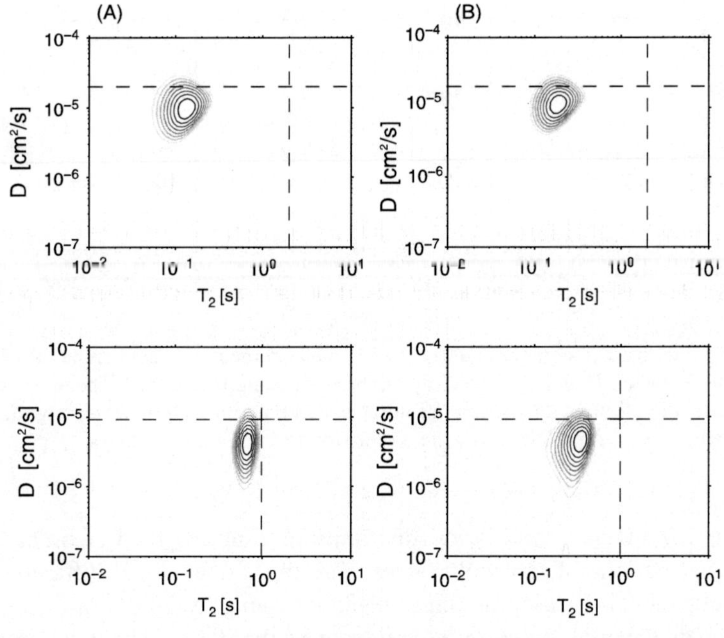

Fig. 3.9 Effect of wettability. $D - T_2$ distribution functions for two fine-grained carbonate rock cores with similar pore size and permeability. Core (**A**) is from the water zone and core (**B**) from the oil zone in this well. The *top row* shows measurements on the cores after initial saturation with brine, the *bottom row* is after subsequent imbibition with dodecane. The *dashed lines* indicate the relaxation times and diffusion coefficients for bulk water and dodecane, respectively. In all cases, the measured diffusion coefficients are reduced from the corresponding bulk values due to restricted diffusion by the same factor. This reflects the fact that the geometrical configurations of the fluid phase are mainly determined by the geometry of the pore spaces and are nearly identical. The relaxation measurements are sensitive to the surface properties and wetting conditions of the cores. Core (**B**) has been in contact with crude oil and has become partially oil wet, which is reflected in significant surface relaxation for the oil. These two cores have similar permeabilities, but the contrasting surface properties lead to large difference in relative permeabilities and related two-phase flow properties

of dodecane, most surfaces are still covered with a thin film of residual water which greatly reduces the surface relaxation for the oil molecules. In contrast, the surfaces of core (B) originating from the oil zone have been modified by previous exposure to crude oil and its surface active molecules and the core has become partially oil wet. As a consequence, the residual water film after imbibition of dodecane covers a smaller fraction of the pore surfaces and the oil molecules can reach the mineral surfaces more easily and relax more effectively.

3.5.3 Complex Miscible Fluid

Many naturally occurring fluids are complex mixtures of molecules with a wide range of size and chemical properties. In such fluids, the relaxation times and

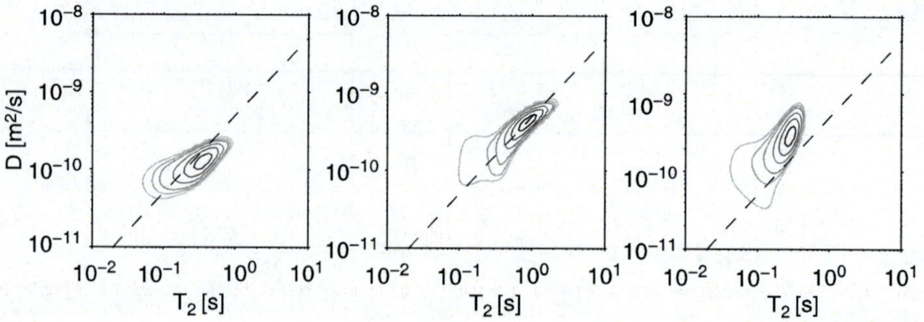

Fig. 3.10 Fingerprint of complex fluids: $D - T_2$ measurements on three crude oils show different correlations between D and T_2 depending on the fluid composition [68]. The oil on the *left* is rich in aromatics, the oil in *middle* is rich in saturates, and the oil on the *right* has significant amounts of asphaltene. The *dashed line* shows the correlation found for pure alkanes

diffusion coefficients show wide distributions that are to first order determined by the distributions of molecular sizes. The two-dimensional diffusion–relaxation measurements can be used as finger-prints of complex fluids [68]. The diffusion–relaxation distribution functions are affected by the shapes, chemical properties, and interactions of the molecules since the relaxation times are sensitive to the rotational diffusion properties of the molecules. The diffusion–relaxation distribution functions are therefore affected by the correlation between translational and rotational diffusion coefficient. The results in Fig. 3.10 show $D - T_2$ measurements for three different crude oils. Oils of different chemical composition show systematically different correlations between the diffusion coefficient and relaxation time.

3.5.4 Structured Fluid

Structured fluids, including gels, emulsions, and liquid crystals, are another important class of materials suitable to be studied with this new technique. In these systems, the fluid becomes inhomogeneous on the micron level or lower and the structure generally modifies the diffusion or relaxation properties of the fluid. Diffusion–relaxation distribution functions can then be used as a sensitive probe of the fluid properties. Examples are shown in Fig. 3.11. A comparison of the diffusion–relaxation distribution functions of a homogeneous fluid, a gel, and an emulsion shows obvious differences that reflect the underlying structure of the three fluids [68]. The first sample shown in Fig. 3.11 is a crude oil similar to those discussed in the previous section. In this homogeneous fluid, the diffusion coefficient and relaxation time are strongly correlated. The second sample is a waxy oil that has been cooled below the wax transition temperature and has become a gel. In this case, the network is formed by only a small fraction of molecules. The rotational properties of the remaining molecules are only weakly affected by the presence of

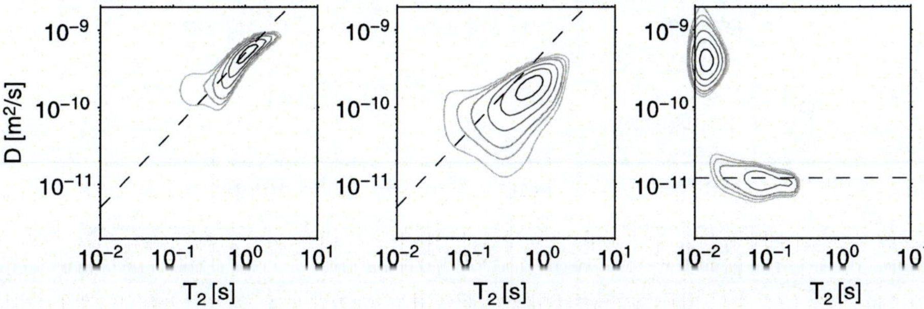

Fig. 3.11 Relaxation–diffusion distribution functions for fluids reflecting its structure. *Left*: Crude oil with a high content of alkanes and other saturates shows a nearly linear correlation between D and T_2 [68]. *Middle*: Waxy crude oil below the wax appearance temperature [68]. The formation of a gel in this fluid leads to a reduction of D without affecting T_2 significantly. *Right*: Example of oil in water emulsion. In this sample of cheese, the measured diffusion coefficient of the fat component is determined by the bubble size and independent of T_2 [66]

the network, but the translational diffusion is noticeably restricted. This leads to the significant broadening along the diffusion dimension in the $D - T_2$ distribution function. Finally, the third sample corresponds to an emulsion. The fat molecules in this sample of cheese have a wide distribution of size, leading to a spread in relaxation times. However, the molecules are all confined in small globules that restrict the diffusion to the bubble size. This leads to the characteristic horizontal signature of the fat contribution in the $D - T_2$ distribution function.

3.5.5 Pore Geometry of Porous Media

The pore geometry of fluid saturated porous media is routinely studied by relaxation measurements, as outlined already in Sect. 3.1.1. In addition, diffusion measurements can be used to infer geometrical length scales by monitoring the degree of restrictions encountered by the diffusing spins [4]. Since collisions with the pore walls are controlling both the extra relaxation and the reduction in apparent diffusion coefficient, we expect a strong correlation between these two effects. This is generally confirmed by the results shown in Fig. 3.12. This figure shows measurements of the average diffusion coefficient $\langle D(T_2)\rangle$ versus relaxation times (see Eq. 3.12) on a series of brine-saturated carbonate cores from an oil well in the Middle East, using the pulse sequence of Fig. 3.3c with a diffusion time $T_d = 20$ ms. These rock cores are characterized by a wide distribution of pore sizes. In large pores, the collision rate of the water molecules with the grain surfaces is low, which leads to relaxation times close to the bulk relaxation time of water and an apparent diffusion coefficient close to the molecular diffusion coefficient, D_0. In smaller pores, both the relaxation times and the measured average diffusion coefficients decrease.

The cores in this well have a fairly uniform porosity of around 29%, but the rock textures and average pore sizes vary widely. This leads to the wide range of perme-

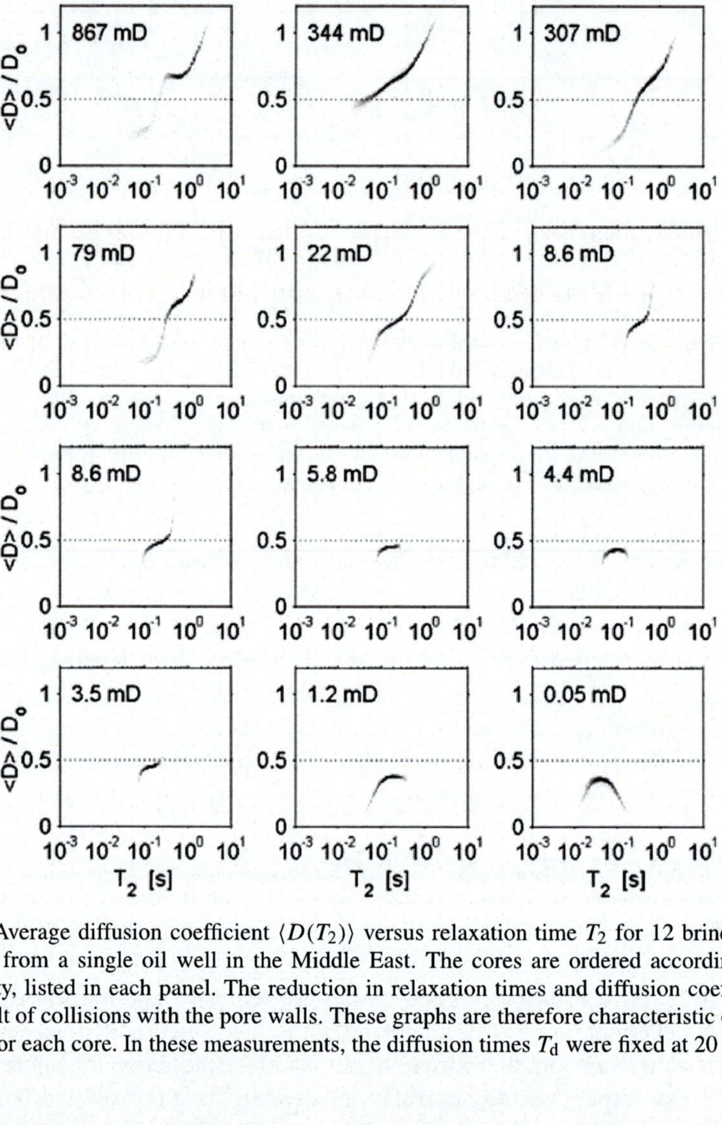

Fig. 3.12 Average diffusion coefficient $\langle D(T_2)\rangle$ versus relaxation time T_2 for 12 brine saturated rock cores from a single oil well in the Middle East. The cores are ordered according to their permeability, listed in each panel. The reduction in relaxation times and diffusion coefficient are both a result of collisions with the pore walls. These graphs are therefore characteristic of the pore geometry for each core. In these measurements, the diffusion times T_d were fixed at 20 ms

abilities in these cores that vary by more than four orders of magnitude, as indicated in the figure. Figure 3.13 shows that both the average relaxation time and the average diffusion coefficient are correlated with the permeability, but it is interesting to note that the average diffusion coefficient of the brine-saturated samples shows a tighter correlation with permeability than the relaxation time. The geometrical parameters controlling diffusion are more directly related to the length scales that control permeability than the corresponding parameters for relaxation. The reduction of the measured diffusion coefficient is controlled by the pore size and pore connectivity on a length scale of the diffusion length, $2\sqrt{D_0 T_d} \simeq 14\,\mu\text{m}$, whereas the relaxation rate is determined by the local surface-to-volume ratio, surface roughness, and the surface relaxivity.

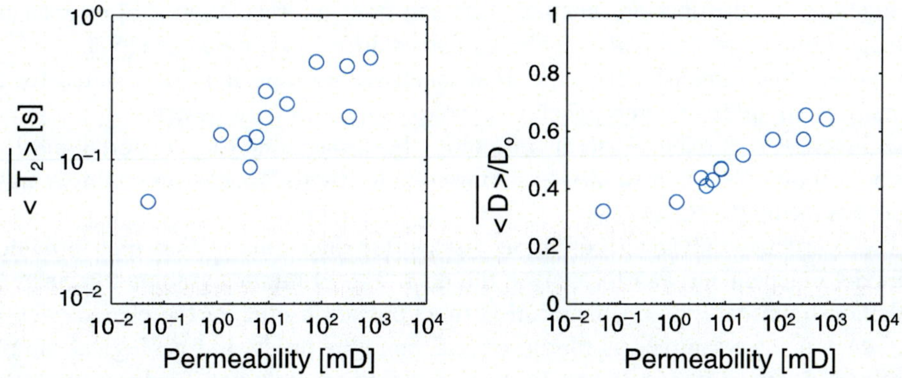

Fig. 3.13 Correlation of average relaxation time (*left*) and average diffusion coefficient (*right*) versus permeability for the brine-saturated rock cores shown in Fig. 3.12. The diffusion data show a tight correlation with permeability

3.6 Conclusion

There has been much recent progress in the measurement of one- and two-dimensional distribution functions for ex situ NMR applications. These developments have been made possible by a number of separate advances. First, they are based on the thorough understanding of the spin dynamics in inhomogeneous fields for sequences consisting of thousands of slice selective pulses, which include dissipative effects of relaxation and diffusion. This has made it possible to design optimized pulse sequences for the measurement of specific properties of interest. Second, efficient inversion algorithms have been developed to extract the distribution functions from the experimental data. These routines have to deal with the mathematical difficulties created by the ill-conditioned nature of the inverse Laplace transformation. Another challenge to overcome has been posed by the often large size of typical sets of experimental data to be analyzed.

The acquisition of one-dimensional T_2 distribution functions is based on the analysis of one-dimensional CPMG measurements, whereas the determination of diffusion or T_1 distribution functions requires the acquisition of two-dimensional measurements. For these measurements in inhomogeneous fields, it is generally advantageous to follow the encoding sequence with a long train of 180° pulses to multiplex the signal. This so-called CPMG detection can then be used to increase the signal-to-noise ratio significantly.

An important extension is the measurement of two-dimensional distribution functions, such as diffusion-T_2 and $T_1 - T_2$ distribution functions. In analogy to conventional NMR spectroscopy, such multi-dimensional measurements contain intrinsically much more information than the corresponding one-dimensional measurements. While the initial developments have focused on the measurements of $D - T_2$ and $T_1 - T_2$ distribution functions, the approach is rather general. The same general strategy can be used to design ex situ measurements for other related quantities, including the distribution functions between flow and relaxation [56] or exchange measurements between different relaxation or diffusion components [57].

These two-dimensional distribution functions are especially well suited for the study of complex or heterogeneous samples. Such complex systems are ubiquitous and include soft matter systems in biology [69], medicine [64], material and food sciences [65, 66, 70], naturally occurring complex fluids [10, 68], and fluid-saturated porous media [9, 10, 44, 59].

A particular advantage of these distribution functions is that they provide a model-independent way to analyze the data. The examples presented in Sect. 3.5 demonstrate that $D - T_2$ distribution functions are useful for the characterization of such diverse samples as food products, complex fluids, and geological samples saturated with different fluids. In these cases, the distribution functions provide a wealth of unique new information that can be obtained without the need to handle or contact the sample. This includes the identification of fluids in geological samples, the quantification and separation of multiple miscible fluid phases, the characterization of the molecular composition of a complex fluid, the study of wetting properties and surface interaction in porous media, and the characterization of the geometrical arrangement of fluids inside porous media.

These NMR techniques have another critical advantage compared to conventional NMR spectroscopy: the requirements on the strength, homogeneity, and stability of the static magnetic field are much less stringent. This has made it possible to design and build instrumentations that greatly enhance the applicability of these measurements beyond the standard laboratory setting. As an example, the new techniques are already routinely used in commercial NMR well-logging measurements [63]. In this application, a measuring device is lowered into a borehole that can be several kilometers deep with temperatures in excess of 170°C and pressures in excess of 150 MPa. While the NMR apparatus is moving along the borehole, two-dimensional distribution functions are continuously acquired that are used to infer detailed information on the fluids in the earth formation adjacent to the borehole. This illustrates that ex situ NMR diffusion and relaxation measurements can be implemented successfully even in very harsh environments. It is expected that these techniques will find numerous other applications in many diverse fields.

Acknowledgments I would like to thank all my colleagues at Schlumberger–Doll Research who have contributed to the development of this field, in particular Lalitha Venkataramanan and Yi-Qiao Song.

References

1. Cory DG, Garroway AN (1990) Measurement of translational displacement probabilities by NMR: an indicator of compartmentation. Magn Reson Med 14:435–444
2. Callaghan PT, Coy A, MacGowan D, Packer KJ, Zelaya FO (1991) Diffraction-like effects in NMR diffusion studies of fluids in porous solids. Nature 351:467–469
3. Callaghan PT (1991) Principles of nuclear magnetic resonance microscopy. Clarendon Press, Oxford

4. Mitra PP, Sen PN, Schwartz LM, and Le Doussal P (1992) Diffusion propagator as a probe of the structure of porous media. Phys Rev Lett 68:3555–3558
5. Brownstein KR, Tarr CE (1979) Importance of classical diffusion in NMR studies of water in biological cells. Phys Rev A 19:2446
6. Kenyon WE, Day PI, Straley C, Willemsen JF (1988) A three-part study of NMR longitudinal relaxation properties of water-saturated sandstones. Soc Petrol Eng Form Eval 3:622–636; Erratum: Soc Petrol Eng Form Eval 4:8 (1989)
7. D'Orazio F, Tarczon JC, Halperin WP, Eguchi K, Mizusaki T (1989) Application of nuclear magnetic resonance pore structure analysis to porous silica glass. J Appl Phys 65:742–751
8. English AE, Whittal KP, Joy MLG, Henkelman RM (1991) Quantitative two-dimensional time correlation relaxometry. Magn Reson Med 22:425–434
9. Song Y-Q, Venkatarmanan L, Hürlimann MD, Flaum M, Frulla P, Straley C (2002) $T_1 - T_2$ correlation spectra obtained using a fast two-dimensional Laplace inversion. J Magn Reson 154:261–268
10. Hürlimann MD, Venkataramanan L (2002) Quantitative measurement of two dimensional distribution functions of diffusion and relaxation in grossly inhomogeneous fields. J Magn Reson 157:31–42
11. Kleinberg RL (1996) Well logging. In: Encyclopedia of nuclear magnetic resonance, vol 8. John Wiley & Sons, Chichester, pp 4960–4969
12. Eidmann G, Savelsberg R, Blümler P, Blümich B (1996) The NMR MOUSE, a mobile universal surface explorer. J Magn Reson A 122:104–109
13. Kimmich R, Fischer E (1994) One- and two-dimensional pulse sequences for diffusion experiments in the fringe field of superconducting magnets. J Magn Reson A 106:229–235
14. McDonald PJ (1997) Stray field magnetic resonance imaging. Prog Nucl Magn Reson Spect 30:69–99
15. Bloembergen N, Purcell EM, Pound RV (1948) Relaxation effects in nuclear magnetic resonance absorption. Phys Rev 73:679–712
16. Zega A, House WV, Kobayshi R (1989) A corresponding-states correlation of spin relaxation in normal alkanes. Physica A 156:277–293
17. Brown RJS (2001) The Earth's-field NML development at Chevron. Concepts Magn Reson 13:344–366
18. Belton PS, Jackson RR, Packer KJ (1972) Pulsed NMR studies of water in striated muscle. Transverse nuclear spin relaxation times and freezing effects. Biochim Biophys Acta 286: 16–25
19. Hazelwood CF, Chang DC, Nichols BL, Woessner DE (1974) Nuclear magnetic resonance transverse relaxation times of water proton in skeletal muscle. Biophys J 14:583–606
20. Araujo CD, MacKay AL, Whittall KP, Hailey JRT (1993) A diffusion model for spin-spin relaxation of compartmentalized water in wood. J Magn Reson B 101:248–261
21. Halperin WP, Jehng JY, Song YQ (1994) Application of spin-spin relaxation to measurement of surface area and pore size distributions in a hydrating cement paste. Magn Reson Imaging 12:169–173
22. Hahn EL (1950) Spin echoes. Phys Rev 80:580–594
23. Einstein A (1906) Eine neue Bestimmung der Moleküldimensionen. Annalen der Physik 19:289–306
24. Douglass DC, McCall DW (1958) Diffusion in paraffin hydrocarbons. J Phys Chem 62:1102–1107
25. Freed DE, Burcaw L, Song Y-Q (2005) Scaling laws for diffusion coefficients in mixtures of alkanes. Phys Rev Lett 94:067602
26. Bloembergen N (1966) Paramagnetic resonance precession method and apparatus for well logging. United States Patent No. 3,242,422A. Filed 1954, issued 1966.
27. Woessner DE (1963) NMR spin-echo self-diffusion measurements on fluids undergoing restricted diffusion. J Phys Chem 67:1365–1367
28. Ernst RR, Bodenhausen G, Wokaun A (1987) Principles of nuclear magnetic resonance in one and two dimensions. Clarendon Press, Oxford

29. Carr HY, Purcell EM (1954) Effects of diffusion on free precession in nuclear magnetic resonance experiments. Phys Rev 94:630–638
30. Meiboom S, Gill D (1958) Modified spin-echo method for measuring nuclear relaxation times. Rev Sci Instrum 29:688–691
31. Goelman G, Prammer MG (1995) The CPMG pulse sequence in strong magnetic field gradients with applications to oil-well logging. J Magn Reson A 113:11–18
32. Hürlimann MD, Griffin DD (2000) Spin dynamics of Carr – Purcell – Meiboom – Gill – like sequences in grossly inhomogeneous B_o and B_1 fields and application to NMR well logging. J Magn Reson 143:120–135
33. Bãlibanu F, Hailu K, Eymael R, Demco DE, Blümich B. (2000) Nuclear magnetic resonance in inhomogeneous magnetic fields. J Magn Reson 145:246–258
34. Jaynes ET (1955) Matrix treatment of nuclear induction. Phys Rev 98:1099–1105
35. Bull TE (1974) Effect of RF field inhomogeneities on spin-echo measurements. Rev Sci Instrum 45:232–242
36. Hürlimann MD (2001) Diffusion and relaxation effects in general stray field NMR experiments. J Magn Reson 148:367–378
37. Song Y-Q (2002) Categories of coherence pathways for the CPMG sequence. J Magn Reson 157:82–91
38. Stejskal EO, Tanner JE (1965) Spin diffusion measurements: spin echoes in the presence of a time-dependent field gradient. J Chem Phys 42:288–292
39. Cotts RM, Hoch MJR, Sun T, Markert JT (1989) Pulsed field gradient stimulated echo methods for improved NMR diffusion measurements in heterogeneous systems. J Magn Reson 83:252–266
40. Kimmich R, Unrath W, Schnur G, Rommel E (1991) NMR measurement of small self-diffusion coefficients in the fringe field of superconducting magnets. J Magn Reson 91: 136–140
41. Rata DG, Casanova F, Perlo J, Demco DE, Blümich B (2006) Self-diffusion measurements by a mobile single-sided NMR sensor with improved magnetic field gradient. J Magn Reson 180:229–235
42. Woessner DE (1961) Effects of diffusion in nuclear magnetic resonance spin-echo experiments. J Chem Phys 34:2057–2061
43. Fischer E, Kimmich R (2004) Constant time steady gradient NMR diffusometry using the secondary stimulated echo. J Magn Reson 166:273–279
44. Hürlimann MD, Venkataramanan L, Flaum C (2002) The diffusion – spin relaxation time distribution function as an experimental probe to characterize fluid mixtures in porous media. J Chem Phys 117:10223–10232
45. Hürlimann MD (2007) Encoding of diffusion and T_1 in the CPMG echo shape: single-shot D and T_1 measurements in grossly inhomogeneous fields. J Magn Reson 184:114–129
46. Kenyon WE (1992) Nuclear magnetic resonance as a petrophysical measurement. Nucl Geophys 6:153
47. Provencher SW (1982) A constrained regularization method for inverting data represented by linear algebraic or integral equations. Comput Phys Commun 27:213–227
48. Kroeker RM, Henkelman RM (1986) Analysis of biological NMR relaxation data with continuous distributions of relaxation times. J Magn Reson 69:218–235
49. Whittall KP, MacKay AL (1989) Quantitative interpretation of NMR relaxation data. J Magn Reson, 84:134–152
50. Fordham EJ, Sezginer A, Hall LD (1995) Imaging multiexponential relaxation in the (y, $\log_e$ T_1) plane, with application to clay filtration in rock cores. J Magn Reson A 113: 139–150
51. Borgia GC, Brown RJS, Fantazzini P (1998) Uniform-penalty inversion of multiexponential decay data. J Magn Reson 132:65–77
52. Brown RJS (1989) Information available and unavailable from multiexponential relaxation data. J Magn Reson 82:539–561

53. Borgia GC, Brown RJS, Fantazzini P (2000) Uniform-penalty inversion of multiexponential decay data II. Data spacing, T_2 data, systematic data errors, and diagnostics. J Magn Reson 147:273–285
54. Parker RL, Song YQ (2005) Assigning uncertainties in the inversion of NMR relaxation data. J Magn Reson 174:314–324
55. Britton MM, Graham RG, Packer KJ (2001) Relationships between flow and NMR relaxation of fluids in porous solids. Magn Reson Imaging 19:325–331
56. Scheven UM (2005) Stray field measurements of flow displacement distributions without pulsed field gradients. J Magn Reson 174:338–342
57. Callaghan PT, Furó I (2004) Diffusion–diffusion correlation and exchange as a signature for local order and dynamics. J Chem Phys 120:4032–4038
58. McDonald PJ, Korb JP, Mitchell J, Monteilhet L (2005) Surface relaxation and chemical exchange in hydrating cement pastes: a two-dimensional NMR relaxation study. Phys Rev E 72:011409
59. Hürlimann MD, Venkataramanan L, Flaum C, Speier P, Karmonik C, Freedman R, Heaton N (2002) Diffusion-editing: new NMR measurement of saturation and pore geometry. In: Transactions of the SPWLA 43rd Annual Logging Symposium, Oiso, Japan, Paper FFF
60. Venkataramanan L, Song Y-Q, Hürlimann MD (2002) Solving Fredholm integrals of the first kind with tensor product structure in 2 and 2.5 dimensions. IEEE Trans. Signal Process 50:1017–1026
61. Butler JP, Reeds JA, V.Dawson S (1981) Estimating solutions of first kind integral equations with nonnegative constraints and optimal smoothing. SIAM J Numer Anal 18:381–397
62. de Swiet TM, Tomaselli M, Hürlimann MD, Pines A (1998) In situ NMR analysis of fluids contained in sedimentary rock. J Magn Reson 133:385–387
63. Freedman R, Heaton N (2004) Fluid characterization using nuclear magnetic resonance logging. Petrophysics 45:241–250
64. Seland J, Bruvold M, Anthonsen H, Brurok H, Nordhøy W, Jynge P, Krane J (2005) Determination of water compartments in rat myocardium using combined $D - T_1$ and $T_1 - T_2$ experiments. Magn Reson Imaging 23:353–354
65. Godefroy S, Creamer LK, Watkinson PJ, Callaghan PT (2003) The use of 2d Laplace inversion in food materials. In: Webb GA, Belton PS, Gil AM, Delgadillo I (eds) Magnetic resonance in food science: a view to the future. Royal Society of Chemistry, Cambridge
66. Hürlimann MD, Burcaw L, Song YQ (2006) Quantitative characterization of food products by two-dimensional $D - T_2$ and $T_1 - T_2$ distribution functions in a static gradient. J Colloid Interface Sci 297:303–311
67. Hürlimann MD, Flaum M, Venkataramanan L, Flaum C, Freedman R, Hirasaki GJ (2003) Diffusion–relaxation distribution functions of sedimentary rocks in different saturation states. Magn Reson Imaging 21:305–310
68. Mutina AR, Hürlimann MD (2008) Correlation of transverse and rotational diffusion coefficient: a probe of chemical composition in hydrocarbon oils. J Phys Chem A112:3291–3301
69. Windt CW, Vergeldt FJ, Van As H (2007) Correlated displacement – T_2 MRI by means of a pulsed field gradient-multi spin echo method. J Magn Reson 185:230–239
70. Hills B, Benamira S, Marigheto N, Wright K (2004) $T_1 - T_2$ correlation analysis of complex foods. Appl Magn Reson 26:543–560

Chapter 4
Magnets and Coils for Single-Sided NMR

Juan Perlo

The concept of single-sided NMR was introduced with inside-out NMR in the 1980s by the oil industry breaking the myth that NMR experiments can only be performed in the homogeneous fields located inside the magnet [1]. Although nowadays there are already several different NMR well-logging tools, all of them are based on the same basic concept: to position the sample to be studied outside the sensor in the stray field of both the magnet and the rf coil. An unavoidable consequence is that the magnetic field is low and inhomogeneous, leading to extremely low SNR. For slightly inhomogeneous magnetic fields the sensitivity loss can be quantified simply by the frequency bandwidth, while for grossly inhomogeneous fields not only the bandwidth but also the maximum sensitive volume that can be excited must be taken into account. The sensitive volume size and with it the number of spins excited is determined by the gradient strength and by the excitation bandwidth of rf pulses. For example, in the presence of a gradient of 0.1 T/m rf pulses 100 μs long excite a slice 2.4 mm thick, while the same pulses in the presence of a 1 T/m excite only 0.24 mm.

Keeping the static gradient strength as low as possible was the concept followed in the construction of the first NMR well-logging tool [2]. The Jackson geometry uses two cylindrical bar magnets with opposite polarization to produce a toroidal region of radial and nearly uniform B_0 field (Fig. 4.1a). The sensor provides a very good volume selection; by moving 1 cm in any direction the signal is lost. This fact has also the consequence that just a small volume can be excited leading to very low SNR. Another tool (NUMAR) uses a long cylindrical magnet with polarization transverse to its axis to generate a "2D" dipolar field [2] (Fig. 4.1b). Although the sensitive volume is only 1 mm thick along the radial direction due to the strong gradient, the sensitive volume can be as long as half a meter along the cylinder axis. Therefore a sensitivity increment of about two orders of magnitude has been obtained with respect to the Jackson geometry.

J. Perlo (✉)
Institut für Technische Chemie und Makromolekulare Chemie, RWTH Aachen University, D-52074 Aachen, Germany

F. Casanova et al. (eds.), *Single-Sided NMR*, 87
DOI 10.1007/978-3-642-16307-4_4, © Springer-Verlag Berlin Heidelberg 2011

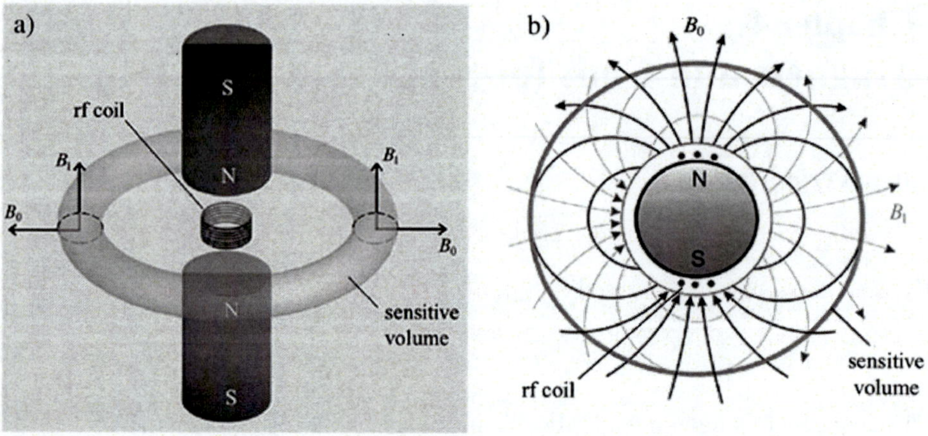

Fig. 4.1 (**a**) Jackson geometry. (**b**) NUMAR tool

As discussed in Chap. 1, the concept of inside-out NMR has been largely expanded by the developers of well-logging NMR and some engineers at the Southwest Research Institute in San Antonio, Texas, to be of use also for other purposes, like, for example, the detection of moisture in soil, and studies of concrete bridge decks [1, 3–7]. Such open magnets were large and heavy, and the electronic at these times was bulky (Fig. 1.3). The change from transportable to mobile NMR happened in 1995 with the development of the NMR-MOUSE (Mobile Universal Surface Explorer) [8]. This small hand-held sensor was designed to scan the surface of objects in a similar way as when moving a computer MOUSE on a pad. The instrument, built from permanent magnets, could be operated with a computer-size spectrometer (desktop spectrometer).

During the last decade sensors like the NMR-MOUSE have evolved into instruments capable of running profiles with microscopic resolution [9], 3D imaging [10], velocity measurements [11], and even ^{1}H spectroscopy in the stray field of the magnet [12]. In the following an overview of non well-logging hardware is presented in three sections, devoted to magnets, rf coils, and gradient coils.

4.1 Magnets

The first NMR-MOUSE design employed a magnet in the shape of a "U" or a horseshoe (Fig. 4.2b) [4, 8]. This shape is obtained by opening up a conventional c-shaped magnet with a homogeneous field between its poles (4.2.a). While in a conventional c-shaped magnet, the sample rests inside the rf coil and between the poles to achieve maximum strength and homogeneity, in unilateral NMR the sample rests near the sensor in the inhomogeneous stray fields of the magnet and the rf coil. This lifts the constraints on the sample size, but introduces variations in magnitude and orientation of the static and rf fields B_0 and B_1, respectively, across an extended sample. Once accepting inhomogeneous fields for the NMR measurement, the most

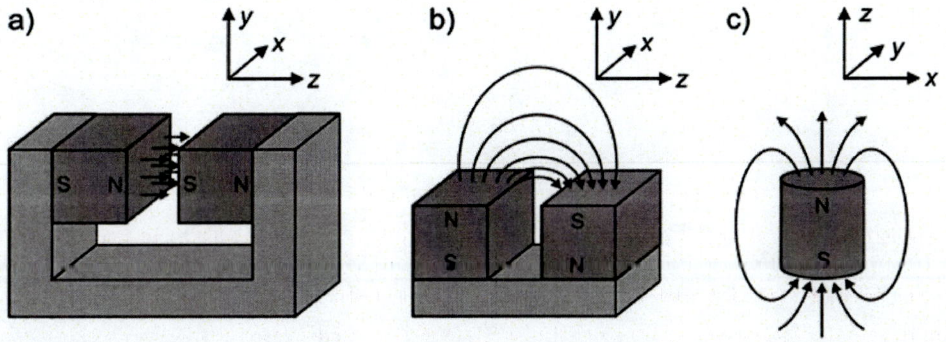

Fig. 4.2 (**a**) Conventional c-shaped geometry where the sample is positioned in the homogeneous field inside the gap. (**b**) U-shaped open magnet. The sample is positioned above the gap in the stray fields of the magnet and the rf coil. The static magnetic field is parallel to the sensor surface. (**c**) Single bar magnet generating a magnetic field perpendicular to the magnet surface. The sample is placed in the stray field of the magnet as in (**b**)

simple NMR sensor is obtained as the simple bar magnet (Fig. 4.2c) [13]. If the rf coil is placed on one of the faces and the orientation of the B_0 field is perpendicular to the sensor surface an rf coil which produces a field substantially parallel to the surface needs to be employed. Such a coil is the figure-8 coil, where the current flow follows the path of the figure-8.

The two mentioned geometries, the U-shaped and the single bar magnets, are the bases of the many different magnet designs for single-sided NMR reported in the last decade. A number of U-shaped magnets have been built either with different aspect ratios and additional magnetic material placed at some strategic positions to generate stronger fields or combining concentric U-shaped magnets to reduce the gradient strength; however, all have a common feature, the magnetic field points parallel to the sensor surface. The case of the single bar magnet is not less diverse. Although magnets with cylindrical or square cross sections, single bars with a hole like the barrel magnet, barrel magnets with an additional single bar placed inside the hole, etc., are available, they share as common feature the generation of a static magnetic field perpendicular to the sensor surface. The direction of the field imposes strong restrictions to the rf coil and gradient coil designs, therefore, magnets with B_0 perpendicular and parallel to the sensor surface are discriminated.

4.1.1 B_0 Perpendicular to the Sensor Surface, the Bar Magnet Geometry

One of the simplest single-sided sensors uses a single bar magnet block to generate the static field (Fig. 4.2c) [13]. A cylindrical block magnet polarized along its axis generates a B_0 field along the depth direction. This is combined with a figure-8 rf coil placed on one of the pole faces. The magnetic field generated by the cylindrical block magnet is the same as the one of a solenoid with the same size (Fig. 4.3a).

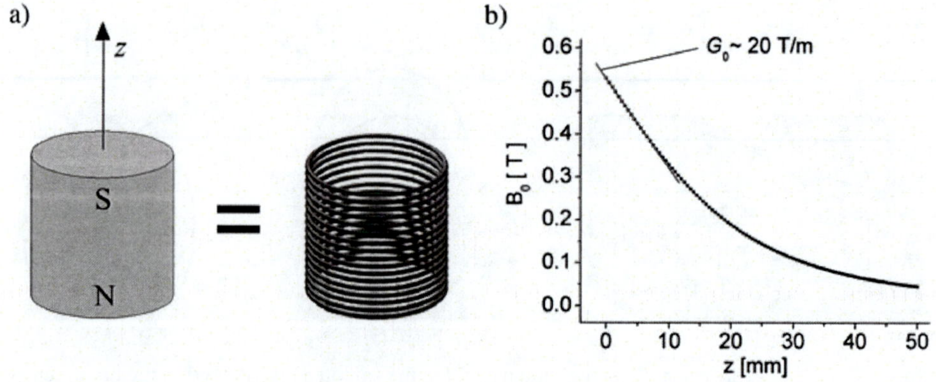

Fig. 4.3 (**a**) Cylindrical bar magnet and equivalent solenoidal coil generating the same magnetic field as the magnet block. (**b**) Dependence of the magnetic field strength B_0 on depth z

Hence, it is similar to a conventional superconducting magnet, which generates a very uniform field in the central region with an extremely strong gradient in the fringe field region. Thus, the situation is completely equivalent to the one of the STRAFI setup.

Figure 4.3b shows the dependence on the depth z of the magnitude of the static field for a cylinder 50 mm in diameter and 50 mm high, the one used in [13]. At 3 mm from the magnet surface, the field magnitude is $B_0 \sim 0.5\,\mathrm{T}$ (20 MHz) while the gradient strength is $G_0 \sim 20\,\mathrm{T/m}$. According to Eq. (2.42) these two values are enough to evaluate, for instance, the sensitivity of the magnet (B_0^2/G_0). However, taking advantage of the simplicity of the calculations the aim of this section is mainly to perform a more general study of the SNR as a function of the magnet dimensions.

An important trend to study is the dependence of the field strength on the size of the cylinder. The natural intuition would suggest that in order to increment B_0 one should increase the size of the magnet; the larger the amount of magnetic material the stronger the field. Interestingly, what can be observed from Fig. 4.4a is that B_0 diminishes with the magnet diameter. This surprising effect can be explained in two ways: First, consider again the equivalence shown in Fig. 4.3 and then note that if the magnet diameter is increased the only effect is to increase the diameter of the current loop, yet the current magnitude remains constant. Considering that the magnetic field produced by a circular coil at its center is inversely proportional to the diameter, the behavior of Fig. 4.4a is obvious.

A second explanation can be given in terms of the magnetic field lines depicted in Fig. 4.4b; the use of square magnets instead of cylindrical ones is simply to make the explanation more clear. The field lines are produced by a block magnet curl approximately as shown in Fig. 4.4b. They emanate from one pole face, then turn in the free space to finally come inside the block again through the other pole face. The direction of the magnetic field is always tangential to the lines; therefore, the magnetic field at the sides of the block has a sign opposite to that of the polarization. Let us consider the case of the big block B built from nine identical blocks A. The

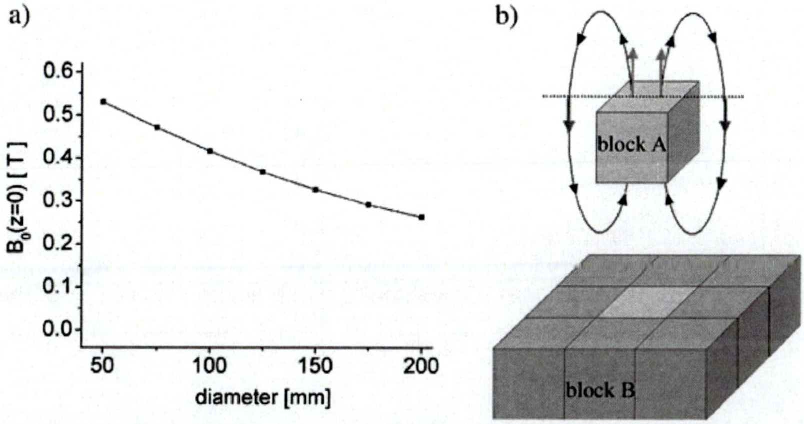

Fig. 4.4 (**a**) Magnitude of the static field at the surface of a cylindrical bar magnet plotted as a function of the magnet diameter. (**b**) The field lines produced by the block magnet A curl approximately as depicted in the drawing. At the surface and on the axis, the field has the same direction as the magnet polarization; however, at the side of the magnet the field inverts its sign. The magnet block B is built from nine block magnets identical to block A. The field generated at the surface by block B is lower than the one generated by A due to the fact that the contribution of the central block has a sign different from the other blocks (see text)

magnetic field is the addition of the contribution from all nine blocks, but only the central one produces a positive field, all other eight blocks have a negative contribution at the center of the magnet. Hence, we can conclude that A produces a stronger magnetic field than B, which is in agreement with the results of Fig. 4.4a.

Another possibility for increasing the size of the sensor is to vary the height of the cylinder. In such a case the increment in the amount of magnetic material leads to an increment in the field as shown in Fig. 4.5a. However, the increment is not proportional to the amount of magnetic material, after a height of about 100 mm B_0 reaches a plateau. Considering again the magnet as a solenoid, an increment in the height can only be produced by adding new turns at the bottom side because the top side defines the magnet surface which is fixed. Hence, although the height can be increased, after a certain threshold the turns are too far away from the surface and their contribution to the total field becomes negligible.

It is worth noting that the height where the plateau is reached depends on the magnet diameter; thus, the larger the diameter the larger the maximum height that can efficiently be used. But, by considering again the results of Fig. 4.4a one can also say that the larger the diameter the lower the field. These two facts tell us that there is a critical relation between the diameter and the height.

Interestingly, it is found that the magnitude of the static field reached in the plateau is independent of the diameter and the height. Furthermore, if the results of Fig. 4.5a are plotted as a function of the height but measured in diameter units the curve does not depend any longer on the diameter (Fig. 4.5b). This scaling law is even more general; for a given aspect ratio h/d the 3D spatial dependence of the B_0 does not depend on the magnet size if the space coordinates are measured in

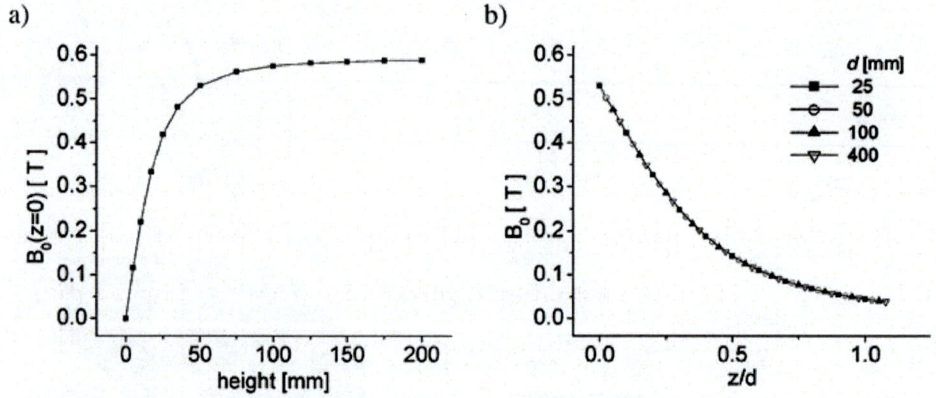

Fig. 4.5 (**a**) Magnitude of the static field at the surface of a cylindrical bar magnet with 50 mm diameter as a function of magnet height. (**b**) Dependence of the static field plotted as a function of the normalized depth coordinate. Independent of the magnet diameter, all curves overlap

diameter units. For example, the field produced by a magnet with $d = 50$ mm and $h = 20$ mm at 10 mm from the surface is the same as the one produced by a magnet $d = 500$ mm and $h = 200$ mm at 100 mm from the surface. The scaling law can also be used to obtain the relationship between the gradient strength and the size of the sensor. Let us introduce a set of variables $\mathbf{r}'$ defined as

$$\mathbf{r}' = \mathbf{r}/d = (x/d, y/d, z/d) \text{ and } h' = h/d. \tag{4.1}$$

The scaling law be can then written as

$$B_0(\mathbf{r}', h', d_1) = B_0(\mathbf{r}', h', d_2) \tag{4.2}$$

for any value of the diameter d. By taking derivative with respect to $\mathbf{r}'$ we obtain

$$\nabla' B_0(\mathbf{r}', h', d_1) = \nabla' B_0(\mathbf{r}', h', d_2) \tag{4.3}$$

from Eq. (4.1) $\nabla' = d^{-1}\nabla$; therefore,

$$d_1^{-1}\nabla' B_0(\mathbf{r}', h', d_1) = d_2^{-1}\nabla' B_0(\mathbf{r}', h', d_2). \tag{4.4}$$

This result shows that the gradient scales exactly as $1/d$. For higher order derivatives the scaling is $1/d^n$ for nth order. By combining Eqs. (4.2) and (4.4) the term B_0^2/G_0 can be easily evaluated to compare the sensitivity of two magnets with diameters d_1 and d_2

$$\frac{\Psi(\mathbf{r}', h', d_1)}{\Psi(\mathbf{r}', h', d_1)} = \frac{d_1}{d_2}. \tag{4.5}$$

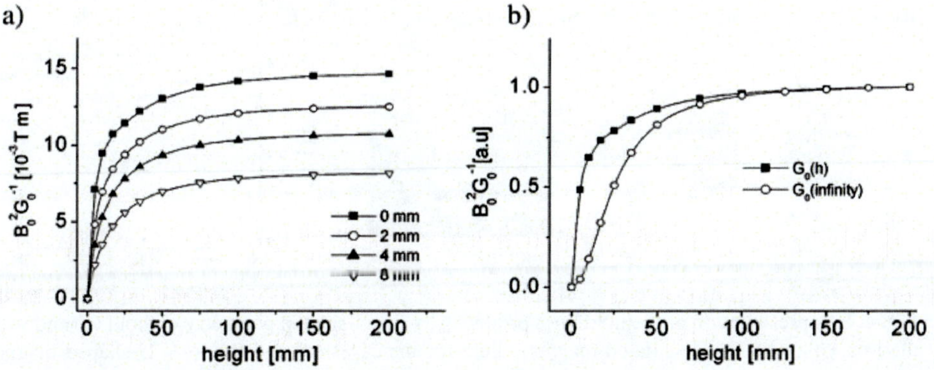

Fig. 4.6 (**a**) Dependence of the magnet sensitivity B_0^2/G_0 as a function of the height of a bar magnet for different depths. (**b**) The case depth = 0 of (**a**), but taking into account the variations of the gradient with h, denoted as $G_0(h)$ and neglecting the variation of the gradient, denoted as $G_0(\infty)$

The sensitivity scales in proportion to the diameter of the magnet. Notice that the comparison is done at the same point $\mathbf{r}'$ but not at the same real position $\mathbf{r}$. Equation (4.5) must be carefully interpreted. For example, by increasing the size of the sensor by a factor of 2 the magnet sensitivity is doubled, even though the working depth is also increased by a factor 2.

Finally, the dependence of the term B_0^2/G_0 can be analyzed as a function of the height. Figure 4.6a shows the results obtained at different depths for a 50 mm diameter magnet. It is observed that the sensitivity is zero for height zero (no magnet), then increases quickly to finally reach a plateau. The values calculated at different depths show the expected trend; the higher the depth the lower the sensitivity. Interestingly, when working at the surface, a sensitivity loss of only a factor 2 is obtained when reducing the height of the sensor from $h = 200$ to 5 mm. This behavior is due to the fact that the gradient strength is strongly reduced at the surface when $h \ll d$.

Figure 4.6b compares the results obtained by taking into account the variations of the gradient with h, denoted as $G_0(h)$, and by neglecting them, denoted as $G_0(\infty)$. For $h \gg d$ both cases give the same value, but when $h \ll d$ the gradient variation becomes dominant. For $h = 5$ mm the gradient strength is only 2 T/m compared to 24 T/m obtained for $h = 200$ mm, leading to a sensitivity gain that almost compensates the sensitivity loss due to lower field strength when going from 25 to 4.8 MHz.

4.1.2 B_0 Parallel to the Sensor Surface, the U-Shaped Geometry

The U-shape is one of the most common geometries used to generate a static field parallel to the sensor surface. Two magnet blocks with opposite polarization are placed on an iron yoke leaving a gap between them as shown in Fig. 4.7b [4, 8]. The iron plate works as a guide for field lines due to its high magnetic permeability leading to an increment in the field strength. Figure 4.7c shows the dependence of the field magnitude as a function of the depth for the cases with and without the iron

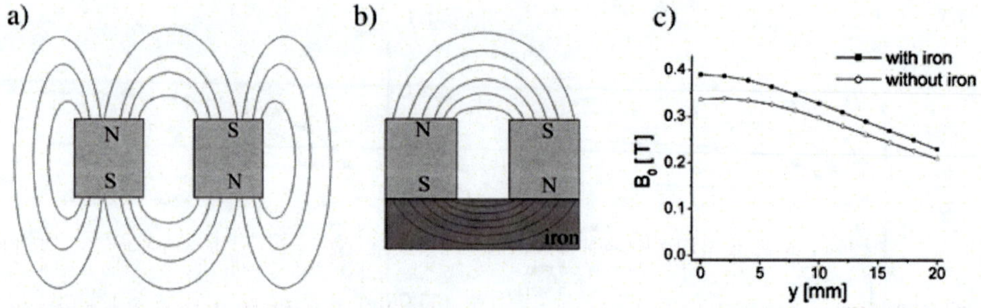

Fig. 4.7 Representation of the field lines produced by the U-shaped geometry without (**a**) and with the iron yoke (**b**). (**c**) Calculated magnetic field for the cases (**a**) and (**b**) for a U-shaped magnet with dimensions: $Lx = 100\,\text{mm}$, $Ly = 50\,\text{mm}$, $Lz = 50\,\text{mm}$, and a gap of 30 mm

yoke (Fig. 4.7a, b, respectively). Both cases show the same trend, a relatively weak gradient at the surface, but then for depths comparable to the gap size the gradient becomes stronger and also constant, i.e., the field varies linearly with the depth. In the following we describe the dependence of the field strength with the size of the sensor, as was previously done for the bar magnet.

Figure 4.8a shows the magnitude of the magnetic field at the sensor surface as a function of the gap while keeping the magnet blocks, dimensions fixed ($Lx/2 = Ly = Lz = 50\,\text{mm}$). The dependence can be explained by first considering again the magnet blocks as solenoidal coils (Fig.4.3a), square ones in this case, and then taking into account that the field is mostly the one generated by the two wires delimiting the gap. The current through the two wires has the same direction; therefore, when the gap is zero both become one with double the current. The divergent behavior of the field for a gap tending to zero is due to the fact that two wires become closer and closer to the point where the magnetic field is calculated. Such behavior is not an artifact of the calculation. This can be experimentally corroborated: magnetic fields up to 1.4 T can be measured. The important drawback of working in this limit is the spatial dependence of the field, the gradients are enormous, up to 400 T/m, and also the direction of the field changes dramatically in small distances.

To study the dependence of the field with the lateral dimensions, the gap and the height were fixed to 15 and 50 mm, respectively. Figure 4.8b shows that the field strength is zero for $Lx = 0$ (no magnets), then increases almost proportional to the size, and finally reaches a plateau. Again by considering the two currents, the plateau is in perfect agreement with the fact that the field produced by infinitely long wire is finite, in other words does not diverge for $Lx \to \infty$. Figure 4.8c shows the dependence of the field with Lz. For $Lz = 0$ (no magnet) the positive (inner wires) and negative current (outer wires) are in the same position; thus they cancel, resulting in zero field. When Lz is increased, keeping the gap constant, the inner wires remain at the same position but the outer ones move away from the gap. What is observed is how the negative contribution coming from the outer wires vanishes when $Lz \to \infty$. Hence, the value reached at the plateau corresponds to field generated just by the two inner wires.

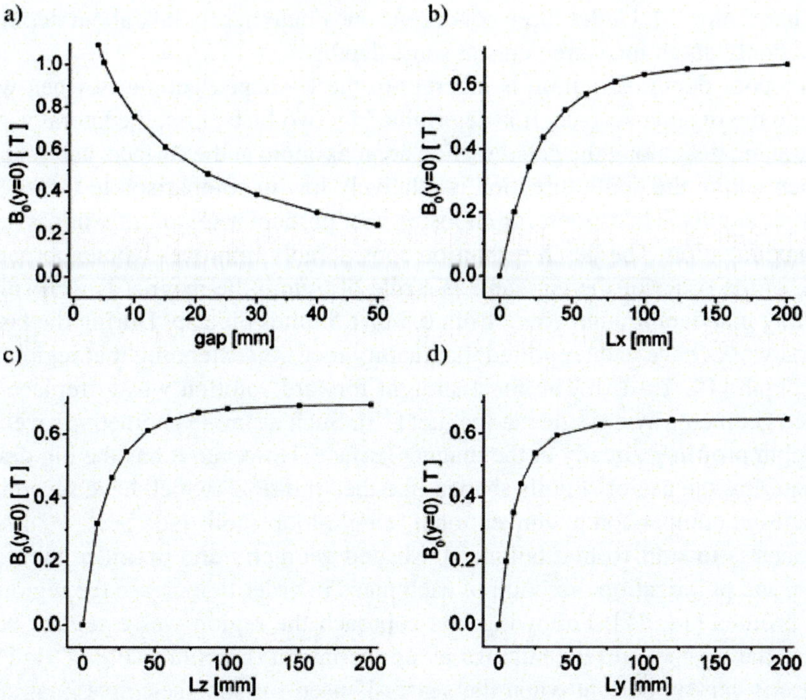

Fig. 4.8 Magnitude of the magnetic field generated by a U-shaped magnet at the surface as function of the gap (**a**), the lateral dimensions Lx (**b**) and Lz (**c**), and the magnet height Ly (**d**). For all cases the magnet dimensions were fix $Lx/2 = Ly = Lz = 50$ mm and gap $= 15$ mm, except the variable one

Finally, Fig. 4.8d shows the dependence of the field magnitude with the magnet height Ly. The trend is the same as in the case of the bar magnet, and it can be explained in the same way: new turns added to increase the height are always attached to the bottom side; therefore, its contribution becomes zero for larger values of Ly. The variation of B_0 as a function of Lx, Ly, and Lz tells us that once a certain limit has been reached (plateau region) it is useless to further increase the size of the sensor.

4.1.3 Magnets for Depth Profiling

The two magnet geometries described above have a strong gradient along the depth direction which can be used to obtain spatial localization into the object simply by varying the excitation frequency. However, the one-to-one relationship between resonance frequency and depth holds only if the lateral field variations are zero. In the general case such variations are not negligible and impose a limit to the depth resolution that can be achieved. The early design of the NMR-MOUSE offered a typical resolution of about a millimeter while the maximum penetration depth was less than 5 mm [8]. Under these conditions one could hardly talk about depth profiling but only about measurements at some depths.

The poor depth resolution is inherent to the U-shaped geometry when working close to the magnet surface. It is determined by two facts, first, the lateral variations of the static field along the gap direction are maximum at the surface, and second, the gradient along the depth direction is relatively low in comparison to the one found at larger depths. The reason for choosing that particular region to work is sensitivity maximization. The depth resolution considerably improves when approaching a depth of the order of the gap, and as a rule of thumb the magnet is well suited for profiling in a depth range going from 0.5 to 1.5 times the gap. During the last years several works have been reported that mainly focus on extending that region toward zero depth [10, 13–15]. The most straight forward solution was to replace the U-shaped geometry by a single bar magnet [13]. Such a simple geometry is well suited for depth profiling already at the magnet surface. However, it has the big drawback of requiring the use of rf coils shaped like the "figure-8" which have several disadvantages in comparison to conventional "single-loop" coils (see Sect. 4.2). Another approach is to start from a tailored U-shaped geometry and optimize the position, shape, and polarization direction of each piece in order to increase the region of flat field profiles [14, 15]. Following this approach the region was extended not only toward the magnet surface but also along the lateral directions (Fig. 4.9). This last feature is very important when the sensor is used for slice-selective imaging [10]. The typical depth resolution achieved was between 0.1 and 0.5 mm for a depth range

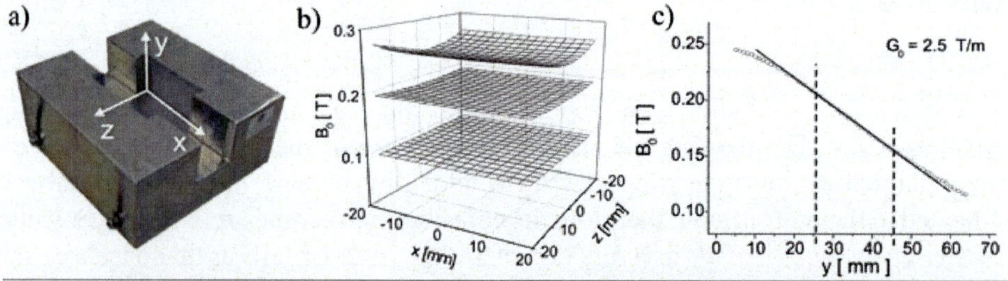

Fig. 4.9 (**a**) Modified U-shaped geometry designed for slice-selective imaging [15]. (**b**) Magnetic field magnitude scanned at 0, 30, and 75 mm from the magnet surface. The field profile is curved at the surface and reaches a flat shape at 30 mm. (**c**) Dependence of the field strength along the depth direction. The gradient can be considered to be constant (2.5 T/m) between 25 and 45. (**d**) Magnet optimized for obtaining flat field profiles over a large depth region (**e**)

going from 0 to 20 mm. In the same range the field strength drops from 14 to 6 MHz requiring the retuning, either manually or electronically, of the rf circuit [14].

An alternative way of measuring a large depth range is to design a sensor with reduced gradient; thus short rf pulses can excite larger regions in a single experiment [16, 17]. Although a sensor working according to this principle would be the ideal solution from a measurement procedure point of view (FT of the echo signal gives already the profile), the sensor design is the most challenging one. First, a reduction of the main gradient magnitude can only be achieved at the expense of field strength. Second, the curvature of the sensitive slice or in other words the best attainable spatial resolution is proportional to the inverse of the main gradient; therefore, a reduction of G_0 must be accompanied by a similar reduction of the lateral field variations in order to keep the spatial resolution. Third, the requirements to the rf coil are very high, good lateral selection, low B_1 gradient, and high efficiency (short rf pulses)/low Q (broadband) at the same time.

In spite of that there are realizations of sensors working according to this principle. A hollow bar magnet was used in a region (deeper than the sweet spot) of reduced gradient (5.7 T/m). In this work it was possible to excite a region of a 1 mm while keeping a resolution of about 0.1 mm [17]. In comparison to the basic geometry, a bar magnet of the same size, only a modest reduction of less than 40% of main gradient was achieved. Another approach using high-permeability materials, like iron, to shape the surface of bar magnet (Fig. 4.10a) has shown to be much more efficient in generating a gradient of only 0.3 T/m with a sensor of the size of the one described above [16]. In the presence of such a gradient the frequency spread within an object 10 mm thick is only 130 kHz; thus rf pulses of "typical" length ($\sim 5\,\mu$s) can excite the whole object. The penalty for achieving such small gradient is the low magnetic field strength (4 MHz) and the quite unfavorable lateral field profile. The region of moderate field curvature (some hundreds of micrometer) is extended over 8 mm depth but it is only 5 mm wide. Outside that region the resolution rapidly deteriorates becoming worse than a millimeter.

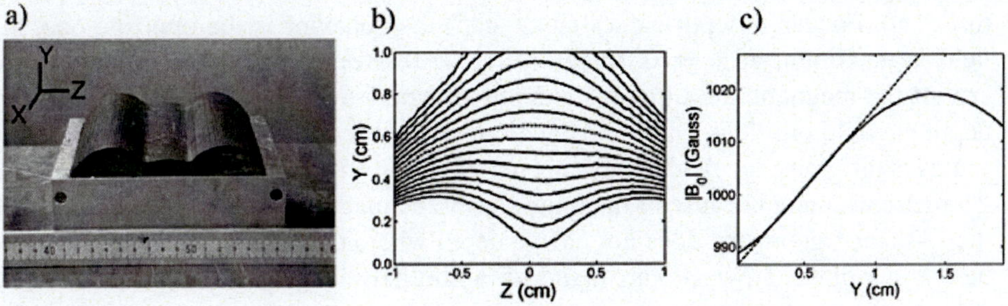

Fig. 4.10 (**a**) Magnet designed using scalar potential model [16]. (**b**) Measured magnetic field in a region 2 cm wide and centered 0.5 cm above the face of the magnet.The *dashed line* indicates $B_0 = 1,006$ G and the contours are at 2 G intervals. (**c**) Measured field profile at $z = 0$ through the sensitive volume (*solid line*). A least-squares fit to the linear region (*dashed line*) gives a gradient of 0.30 T/m

The magnets for depth profiling described up to here do not require either the sample or the sensor to be moved in order to obtain the profile. The sample remains on top of the sensor and the profile is acquired by retuning the rf circuit or by exciting the whole region of interest in a single experiment. Although the procedure is simple and fast, there are some important deficiencies in it. Density profiles are contrasted by relaxation times or self-diffusion coefficients in order to improve the discrimination of heterogeneities in the material, but when the depth is changed the values of these parameters vary, introducing systematic changes in the contrast. For example, considering that the static field has a gradient of some T/m the resonance frequency changes by several megahertz in a few millimeters. This imposes a restriction to the use of T_1 contrast in samples with frequency-dependent T_1. Moreover, the transverse relaxation time ($T_{2\mathrm{eff}}$) measured by a CPMG sequence in inhomogeneous fields, which is a mixture of T_1 and T_2, changes with the depth due to the variation of the B_0 and B_1 field profiles. Even the contrast by diffusion is distorted as a consequence of the variation of the static gradient as a function of the depth [14].

A different way of obtaining depth profiles is by changing the relative position of the sample with respect to the sensitive slice keeping the excitation frequency constant as in STRAFI [18]. The main advantage of this approach is being a truly distortion-free procedure; the whole profile is measured under exactly the same conditions. Furthermore, it has also important consequences on the sensor design. Both magnet and rf coil can be optimized to work only at a fix depth instead of in the whole scanning range; therefore, better performance and/or simpler geometry are expected.

Let us consider again the basic U-shaped geometry and use the following field expansion:

$$|B_0(\mathbf{r})| = B_0(y) + \alpha_z(y)z^2 + \alpha_x(y)x^2, \qquad (4.6)$$

where $B_0(y)$ takes into account the main spatial variation along the depth direction, and α_z and α_x account for the lateral deviations at each depth y. For small depth values ($y \ll$ gap) the field strongly increases when approaching the magnet blocks reflecting the inverse dependence of the field on the distance to the magnetic source ($\alpha_z > 0$). For large depth values ($y \gg$ gap) the behavior is the opposite one, the field is maximum at $z = 0$; therefore $\alpha_z < 0$. Because of the continuity property of the magnetic field there is a depth y_0 where $\alpha_z(y_0) = 0$. At this particular depth the field profile is said to be flat along the z-direction. The behavior along the x-axis is different; the field always decreases ($\alpha_x < 0$) when moving away from the geometrical center because of the finite size of the magnet along x. Strictly speaking the U-shaped geometry does not have a depth where the field can be considered to be flat in a plane. However, at y_0 and for a infinitely long magnet along x the field can be considered to be flat. Such a solution leads to very large and heavy magnets. Another solution is to split the magnet along x introducing a second gap d_S as shown in Fig. 4.11a [9]. To illustrate the effect of splitting, a set of B_0 profiles for different values of d_S is shown in Fig. 4.11b. For $d_S = 0$ (basic U-shape) a strong variation of the field can be observed, while for $d_S = 2$ mm the magnitude of the field is almost

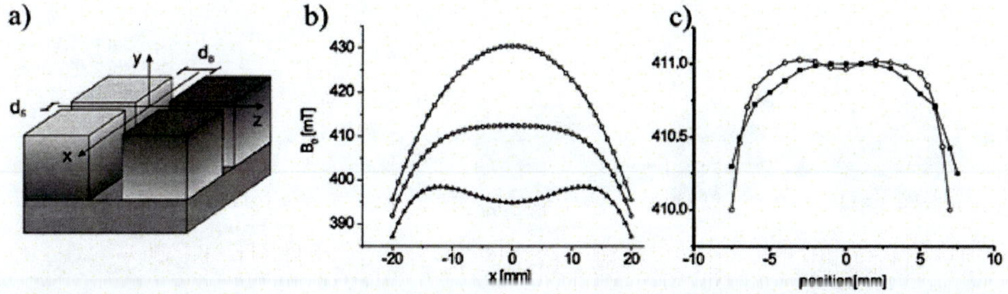

Fig. 4.11 (**a**) Modified U-shaped geometry. The U-shaped geometry is split into two introducing a small gap d_S. The magnet blocks are cubes with 5 cm sides, the main gap d_B is 14 mm, and the small gap $d_S = 2$ mm. At 10 mm depth the magnetic field has a magnitude of about 0.4 T and the gradient is about 20 T/m. (**b**) Magnitude of the magnetic field along x for $d_S = 0$, 2, and 4 mm, at 10 mm from the magnet surface, and for $z = 0$. (**c**) Magnetic field magnitude measured at $y = 10$ mm via the resonance frequency. The field variation in a region of 10 mm along both directions is smaller than 0.2 mT, which corresponds to a depth deviation smaller than 10 μm

constant in a region of 20 mm. If the gap is further increased to 4 mm the profile has the opposite curvature, showing that there is an optimum d_S for which $\alpha_x = 0$. Sensors based on this approach [9] can archive the micrometer resolution range while keeping simplicity and high sensitivity. Typical characteristic for a sensor $10 \times 10\,\text{cm}^2$ large are $y_0 = 10$ mm, $B_0(y_0) = 0.5$ T, $G_0 = 20$ T/m, depth resolution better than 5 μm over a plane $1 \times 1\,\text{cm}^2$.

4.1.4 Sweet-Spot Magnets

In the case discussed above the main motivation is to discriminate NMR information as a function of the depth; however, there are cases like, for example, spatially homogenous samples, where such information is simply not needed. Therefore the requirements to the sensor are different; basically, the maximization of the detection sensitivity is the target. Magnets designed for such applications use the so-called sweet-spot approach in order to increase the size of the excited volume. A sweet spot is a point in space where all first-order derivatives of the magnetic field magnitude are zero while the field strength is different from zero. In general, because of symmetry reasons, the first-order derivatives along the lateral directions are zero while the only coefficient that must be canceled is G_0 (gradient along the depth direction).

For a given magnet geometry such cancelation can be obtained by combining two units of the same type with opposite polarization and different sizes. Let us consider the simple case of the bar magnet geometry. Two concentric bar magnets with different diameters and opposite polarization generate at a certain distance from the sensor surface a point where the total gradient is zero. Notice that two bar magnets cannot be really placed one inside the other; however, the same effect can be produced by drilling a hole in the center of the bar (Fig. 4.12a). A hole can always be considered as a negatively polarized material inside a positively polarized

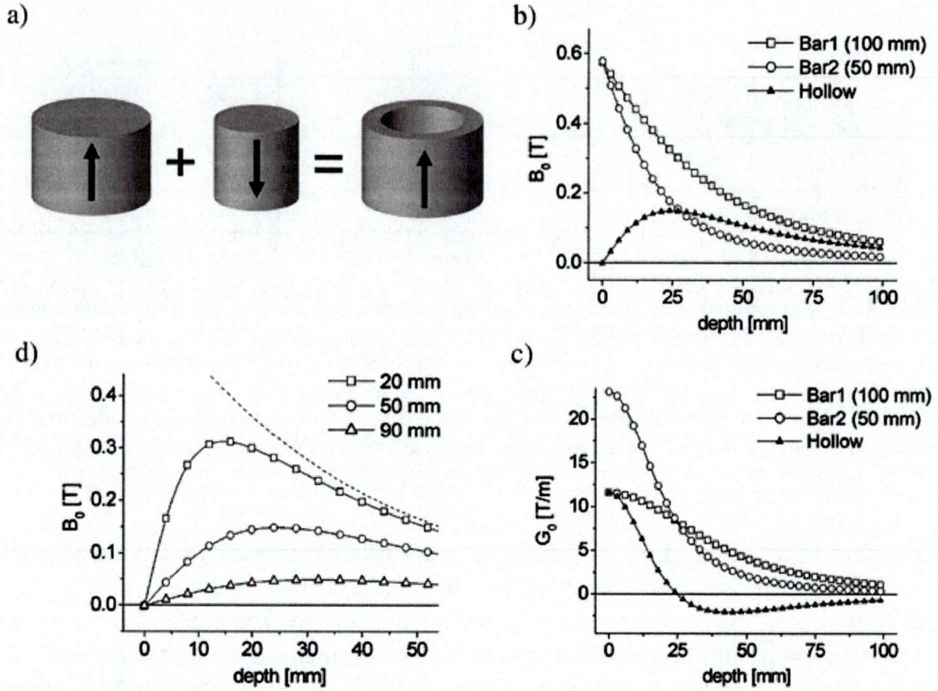

Fig. 4.12 (**a**) Cylindrical bar magnet with a hollow (barrel magnet [19, 20]). The geometry is obtained by combining two concentric bar magnets with opposite polarization (see text). In this calculation the diameters are 50 and 100 mm. (**b**) Field magnitude as a function of the depth. (**c**) Gradient strength as a function of the depth. At a depth of about 25 mm the gradients of the two bar magnets (100 and 50 diameter) have the same value; therefore, the total gradient is zero (Sweet-spot condition). (**d**) Field profile for different inner diameters. The *dash line* is the limit value corresponding to inner diameter equal to zero (single bar magnet)

one. This is the basic idea behind the barrel magnet [19, 20]. Figure 4.12b shows the dependence of the field strength for two bar magnets with different diameters (50 and 100 mm). At a depth equal to zero both magnets generate the same field strength while one gradient value doubles the other one (assuming the magnet height to be much larger than the diameter). These two facts are in agreement with the scaling law presented in Sect. 4.1.1; in particular, they can be predicted from Eqs. (4.2) and (4.4). Another consequence of the scaling law is that the gradient of the smaller magnet must go to zero faster than the one of the larger magnet does; therefore, at a certain depth both magnets must have the same gradient value. Since both magnets have opposite polarization the gradient of the total field (subtraction) is zero. Interestingly, that particular depth z_{sp} has the value of the inner radius of the hollow magnet. Furthermore, the remanent field strength at the sweet spot B_{0sp} is half on the one generated by the larger bar magnet alone.

The depth where the sweet spot is generated depends on the inner diameter. Figure 4.12d shows the total field for three different values of the inner diameter. As the inner diameter approaches the outer one the field strength tends to zero as

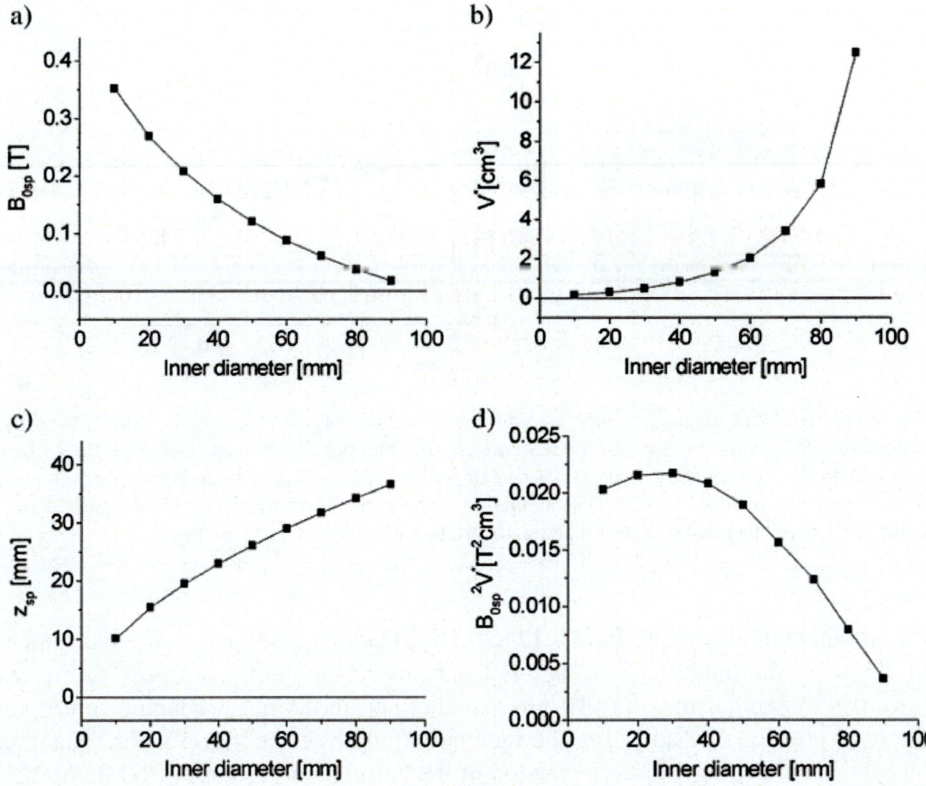

Fig. 4.13 Dependence of the remanent magnetic field at the sweet spot (**a**), its size over a 50 kHz range (**b**), and the depth where it is formed (**c**) as a function of the inner diameter for an outer diameter of 100 mm. There are two limiting cases of the abscise, 0 and 100 mm, corresponding to a simple bar magnet (no sweet-spot formation) and to no magnet, respectively. **d** Magnet sensitivity as a function of the inner diameter obtained from results (**a**) and (**b**)

expected. Notice that the case inner diameter equal to outer diameter means no magnet; thus, the field is zero. Figure 4.13 shows the main results of a more detailed analysis where the remanent field (a), the depth (b), and the size (c) of the sweet spot were calculated as a function of the inner diameter. Based on the results of Fig. 4.13a, b one can evaluate the sensitivity of the magnet as $B_0^2 V^*$ (Eq. 2.42). Figure 4.13d shows that in the trade-off field strength vs size of the sweet-spot there is an optimum combination that maximizes the magnet sensitivity.

A possibility to introduce certain degree of control over the field profile is to introduce a second magnet (central magnet) in the hole of the barrel and vary its position and size [19, 20]. Figure 4.14 shows field profiles as a function of the position of the central magnet. There is a position about −10 mm that maximizes the size of the homogeneous field region. The main two advantages of that modification are the increments of the sweet-spot size and the field strength. Both changes contribute to an increment of the magnet sensitivity. For comparison with simple hollow magnet, the size of the sensitive volume was incremented from 1.3 to 1.7 cm^3 and

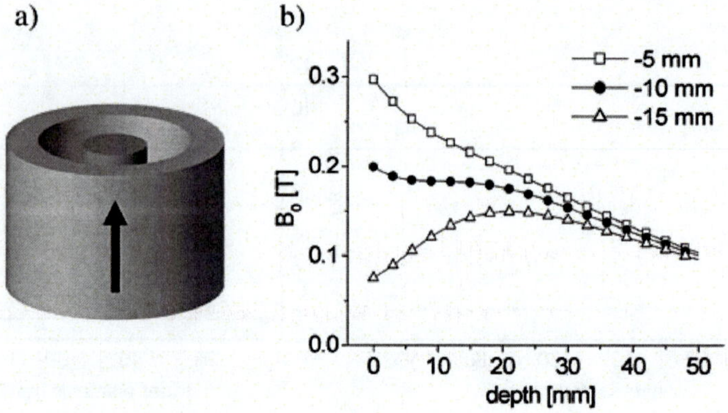

Fig. 4.14 (**a**) Barrel magnet set with an additional central magnet [19, 20]. The position of the central magnet is varied to modify the field profile (**b**). The position is measured from the surface of barrel magnet. For a central magnet position of −10 mm the homogenous field region is extended over depth range of about 15 mm. The calculations were done for a barrel magnet (100 mm/50 mm) outer/inner diameters and a central magnet 40 mm in diameter and a 25 mm high

the remanent field strength from 0.12 to 0.18 T, leading to a sensitivity enhancement of about 3. The main disadvantage is the fact that the sweet-spot position moves toward zero depth, from 35 to 10 mm. Another possibility of introducing control on the field profile is by adjusting the inclination angle of the barrel walls. That can be achieved by using a discrete version of the barrel as in the case of NMR-MOLE (MObile Lateral Explorer) [21].

4.2 RF Coils

A variety of different radio-frequency coils have been explored for single-sided NMR. Similar to the magnets that provide the polarization and detection field B_0, rf coils are discriminated that produce B_1 fields mainly perpendicular and mainly parallel to the surface of the coil. Those with a field perpendicular to the face of the coil are based on a single current loop in the plane (Fig. 4.15a), while those with a field parallel are build from at least two opposed current loops (Fig. 4.15b). The simplest coil of this type is the figure-8 coil. In general, single current loop coils are combined with U-shaped magnets and figure-8 ones with bar magnets. Notice that figure-8 coils can also be used with U-shaped magnets [22] but a single current loop coil is not well suited to be combined with a bar magnet because B_0 and B_1 are mostly parallel. In addition to the basic requirement to generate a B_1 field perpendicular to B_0 there are several other general requirements like, for example, high sensitivity, and some more specific ones like the need of special spatial dependencies, for example, to achieve lateral selection of a defined area or to implement $B_0 - B_1$ matching techniques for spectroscopy in inhomogeneous fields [23, 24] (See Chap. 6).

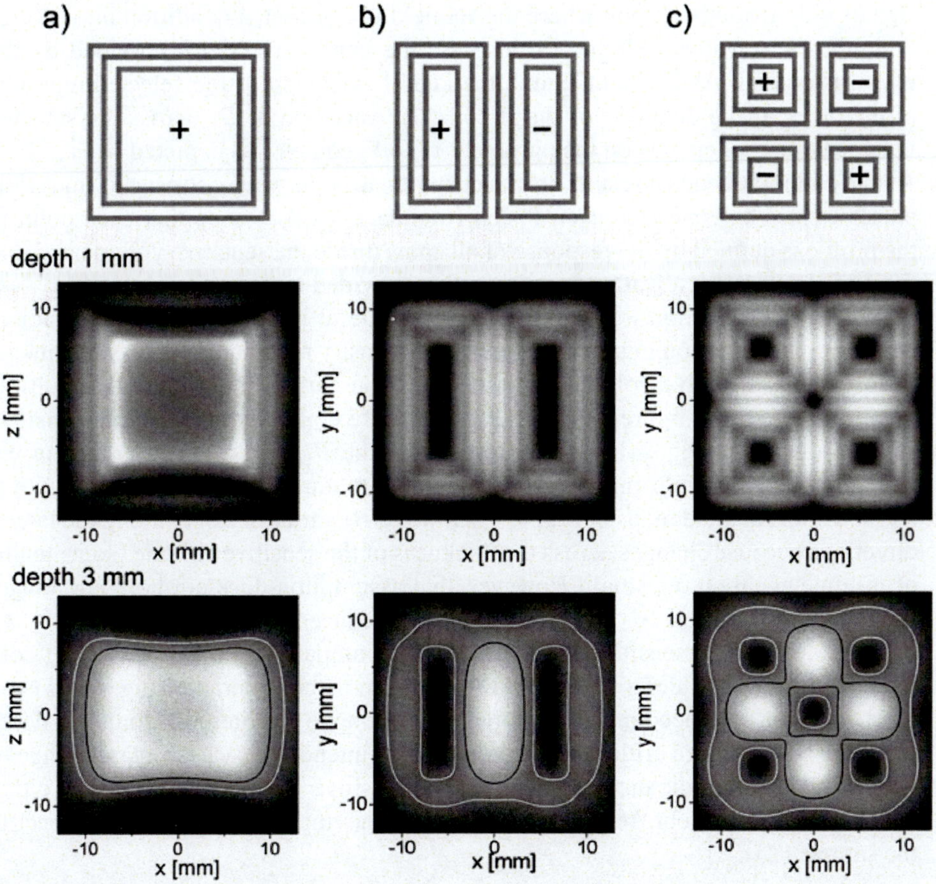

Fig. 4.15 Radio-frequency coils used in single-sided NMR. (**a**) Single-loop coil generating a B_1 field perpendicular to the coil surface. The coil is best to be combined with a U-shaped magnet. (**b**) Figure-8 coil producing a B_1 field essentially parallel to the coil surface. The coil is built from two single-loop coils with opposite currents. (**c**) Butterfly coil: Basically two figure-8 coils one next to the other with opposite currents. In the *lower rows* 2D B_1 maps are shown for two different depths. At 1 mm depth the B_1 profile closely reproduces the current paths while at 3 mm the fine structure already disappears. The contour lines 0.7 and 0.5 of the maximum value are shown in *black* and *white*, respectively

4.2.1 Coils for Depth Profiling

In single-sided NMR the rf coil plays a more important role than in conventional NMR. It is so because the rf coil in combination with the magnet determines the size of the sensitive volume, a concept not applicable to NMR in homogeneous or weakly inhomogeneous fields where the whole sample is excited. In the determination of the sensitive volume the rf coil efficiency B_1/i, coil resistance R, coil inductance L, and the spatial variations of B_1 are important issues. This set of parameters can be optimized for maximum SNR according to Eq. (2.42). In the case of sensors intended for depth profiling the rf coil must be designed to generate and detect signals only from the region where the B_0 field is constant, a condition that assures a highly flat sensitive volume. In this case the lateral selectivity provided by the rf coil is crucial. As a rule of thumb the lateral extension of the selected region is of the order of the diameter of the rf coil. The variation of B_1 across the selected volume strongly depends on the particular rf coil geometry as depicted in Fig. 4.15. Any B_1 inhomogeneity results in a sensitivity loss, as well-defined 90 and 180° pulses cannot be generated even for on-resonance spins. Notice that each point in the profile is defined by the response of all spins inside the sensitive volume and that there is no lateral localization other than that provided by the coil.

When slice selection is combined with 2D lateral imaging by pulsing gradient fields, any B_1 variation manifests itself as an intensity modulation across the image. A continuous sensitive volume is provided by the current loop while the multiple current loops of the figure-8 coil and the butterfly coil generate multiple sensitive spots in the plane (Fig. 4.15). Close to the coil, the B_1 field maps copy the current paths of the coil, while further away, the fine structure in the sensitive volume is washed out. The field maps in Fig. 4.15 show the B_1 variations only along the lateral directions because changes across the thickness of the sensitive volume (some tenths of millimeter) are quite small. However, that assumption does not hold any longer in the case of magnets for depth profiling with reduced main gradient [16, 17]. As the thickness of the sensitive volume becomes comparable to the size of the coil, the variations of B_1 along the depth direction can be considerable (Fig. 4.16a) and must be taken into account. The problem of strong B_1 variations along the depth direction is even more critical when the sensor is intended to access a large range of depths by changing the distance between the sensitive volume and the rf coil either by retuning at different frequencies [14] or by moving the magnet with respect to the sample [25].

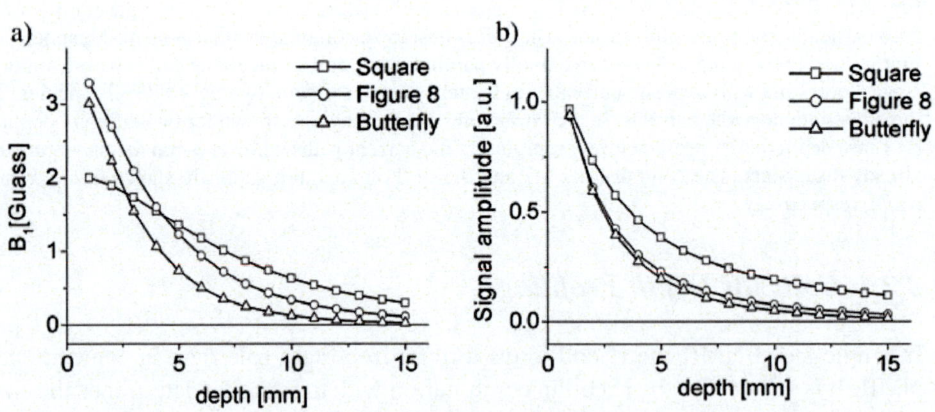

Fig. 4.16 (**a**) B_1 field as a function of the depth for the three coils of Fig. 4.15. The B_1 values were calculated at the center of the coil, for the square and figure-8 ones, and at $x = 5$ mm for the butterfly coil. Notice that for the last case $B_1 = 0$ at the center of the coil. (**b**) Signal amplitude as a function of the depth. It was calculated as the signal response to a CPMG train integrated in the xy plane over a region larger than the coil size. The 90° pulse was calibrated for every depth by maximizing the signal amplitude

4.2.2 Coils for Sweet-Spot Magnets

The requirements for rf coils used in combination with magnets that generate a sweet spot are primarily related to sensitivity maximization. There is no need for coils that provide lateral selection; actually, it is the other way around, the optimum coil is the one that does not impose any restriction on the size of the sensitive volume. In general, rf coils used to excite sweet spots are larger than the size of the sensitive volume which is mainly determined by the 3D spatial variation of the B_0 field. For a given magnet the most sensitive rf coil can be found by means of Eq. (2.42), which requires only the knowledge of the inductance of the coil L and the rf field per unit of current B_1/i at the measuring spot.

The self-inductance of a coil L and magnetic energy U is related by [26]

$$\frac{1}{2}LI^2 = U = \int \frac{B^2}{2\mu_0} dr^3, \tag{4.7}$$

where I is the current through the coil, μ_0 is permeability of the free space, and B is the magnetic field generated by the coil. Hence, a straightforward calculation can be used to determine the inductance of the rf coil. Figure 4.17 shows the dependence of the inductance for a single loop and a figure-8 coil as a function of the coil size d and the number of turns N. If the inductance is normalized to the coil "volume" d^3 and the number of turns by the coil size, all curves overlap in one master curve. The normalized N-axis can be understood as the fraction of turns covering the coil surface. For example, all coils with turns filling the whole coil surface have the same N/d value.

Figure 4.18 shows an example of a coil optimization in terms of SNR. The term $B_1^{\frac{4}{3}}/L^{\frac{2}{3}}$ of Eq. (2.42) is evaluated as a function of the number of turns and the coil size for a single-loop coil (Fig. 4.18a, c) and for a figure-8 coil (Fig. 4.18b, d). As

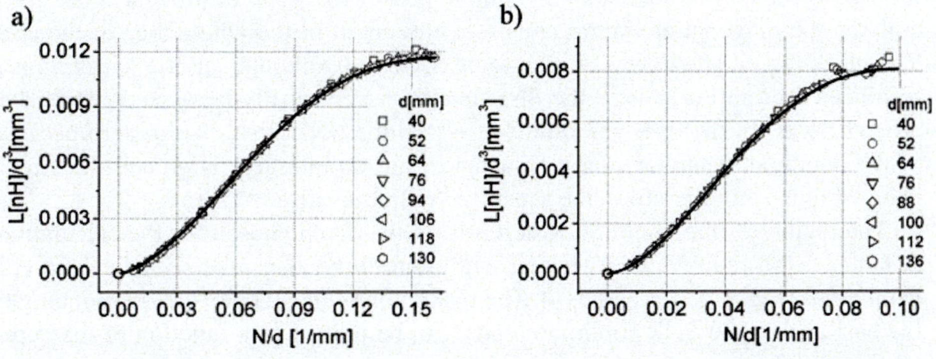

Fig. 4.17 (**a**) Inductance values for a square coil as a function of the number of turns N for different coil sizes d. The inductance is normalized to the coil "volume" d^3 and the number of turns by the coil size. All results can be fit to a master curve (*solid line*). This result corresponds to a wire diameter $a = 2$ mm and distance between wires $b = 3$ mm. (**b**) Same as (**a**) but for a figure-8 coil

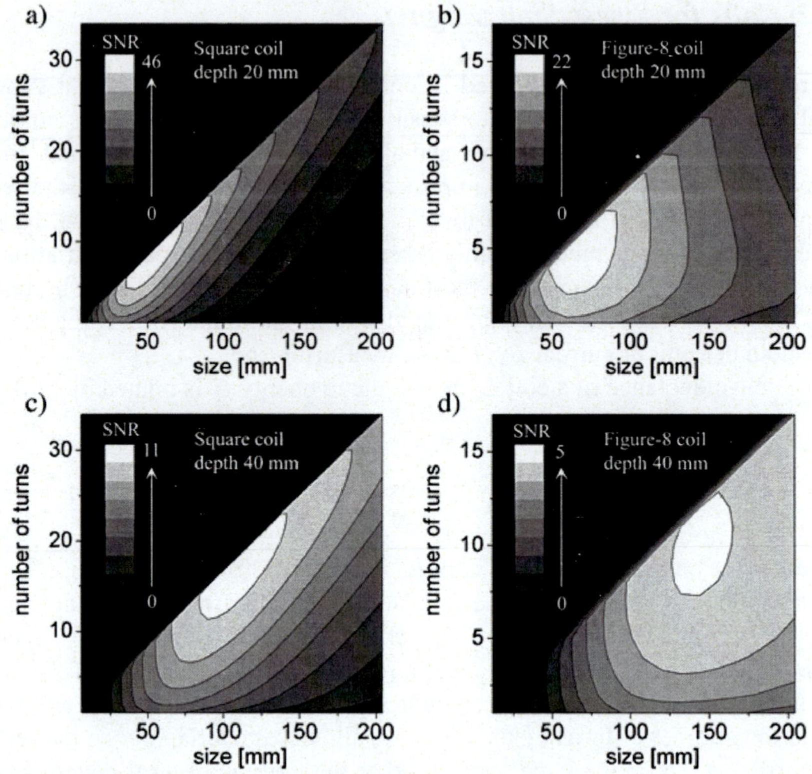

Fig. 4.18 Radio-frequency coil sensitivity ($B_1^{\frac{4}{3}}/L^{\frac{2}{3}}$, Eq. 2.42) as a function of the number of turns and coils size for two coil geometries, single-loop coil (**a** and **c**) and figure-8 (**b** and **d**). Notice that the coil size limits the maximum number of turns, that is, the reason of the *black region* above the diagonal. The SNR units are arbitrary but the same for the four plots

the size of the coil increases the maximum possible number of turns also does. All coils on the diagonal are spiral coils, i.e., the turns fill the whole area of the coil. The black region above was chosen to be SNR $= 0$ although strictly speaking is a forbidden region. For $d = 0$ the SNR is also zero basically because B_1 is zero at a depth $\gg d$. In the opposite limit ($d \rightarrow \infty$) the SNR goes also to zero because of the dominant behavior of the inductance. For an infinitely large coil the circuit bandwidth is zero; therefore, the sensitive volume size is also zero.

The region of maximum SNR depends on the depth chosen for the calculation. A larger depth requires a larger coil with more turns as can be seen for both coil geometries (Fig. 4.18a, c and b, d). The maximum of the 2D plots, corresponding to the best achievable SNR (optimum coil), can be plotted as a function of the depth to quantify the loss of sensitivity with increasing depth (Fig. 4.19). Interestingly, the sensitivity loss is the same for both coils (Fig. 4.19.b). The dashed (single-loop) and dotted (figure-8) lines are linear regressions both with slope -2. meaning that for both coil geometries the SNR decays proportional to the inverse of the squared

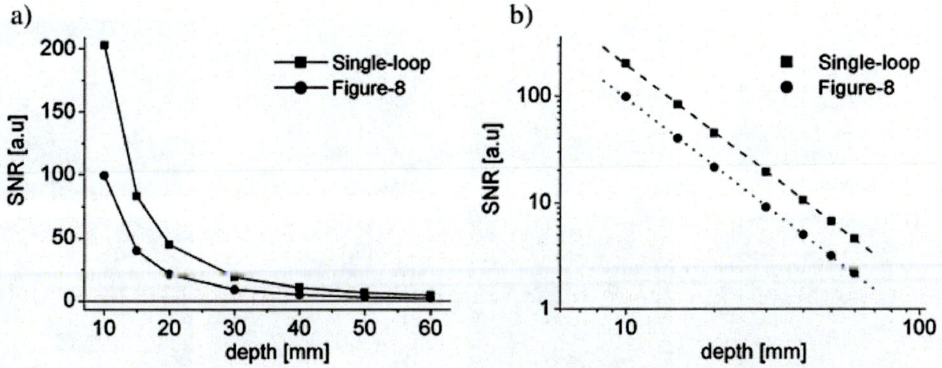

Fig. 4.19 (**a**) The dependence of the best achievable SNR (optimized coil) on the depth for the single-loop and figure-8 coils. (**b**) Same as (**a**) on a log–log scale. The *dashed* (single-loop) and *dotted* (figure-8) *lines* are linear regressions both with slope −2 meaning that for both coil geometries the SNR decays proportional to the inverse of the depth square

depth. In comparison, for every depth the single-loop coil is more sensitive than the figure-8 by a factor of 2.

4.3 Gradient Coils

Pulsed field gradients (PFG) are widely used for space encoding in imaging and flow NMR in homogeneous fields [27–30]. In single-sided NMR typical static gradients can easily be one or two orders of magnitude larger than PFG. In spite of that, pulsed gradients can be used for space and displacement encoding by combining those with suitable NMR techniques (basically pure phase encoding methods) [10, 11, 31–34]. However, the generation of PFG with single-sided coils presents several challenges mostly related to the efficiency of surface or planar gradient coils. For instance, the efficiency is likely to change with the distance to the coil introducing considerable variations of the FoV with the depth. Another important issue is related to signal attenuation observed in liquid samples due to molecular self-diffusion in the presence of strong background gradients. As the minimum encoding times used in imaging or velocity sequences are determined by the maximum strength of the PFG, a weak pulsed gradient leads to strong signal attenuation.

As for the case of rf coils the direction of the polarizing field plays a determining role in the design of gradient coils; only the component of field generated by gradient coils parallel to B_0 has to be considered. For example, the z-component of the field generated by two solenoids arranged as in Fig. 4.20a has a linear variation along x, making them appropriated for imaging along the gap of a U-shaped magnet. In this setup an ill-defined sensitive volume together with changes of the FoV with the depth leads to significant image blurring (Fig. 4.20a). Such a problem is overcome when the sensitive volume is a flat thin slice. In that case the variations of the coil efficiency with the depth can be neglected simply because the thickness of the sensitive volume is few tenths of a millimeter.

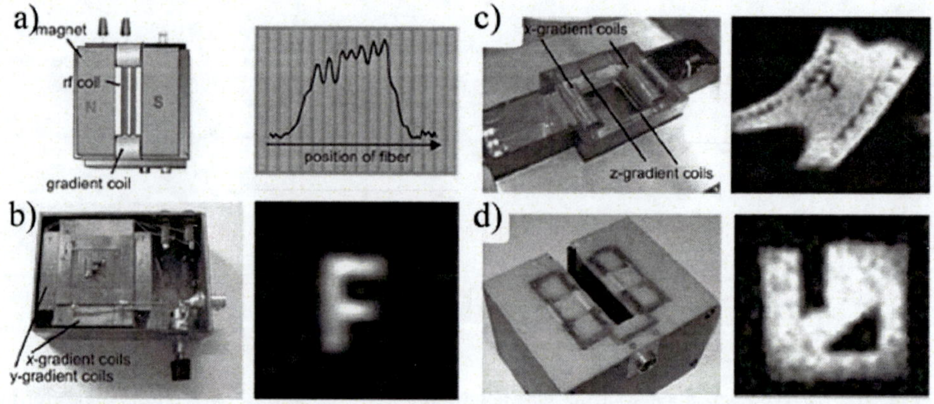

Fig. 4.20 Different approaches to achieve spatial resolution with single-sided sensors. (**a**) NMR-MOUSE for 1D lateral imaging [31] with gradient coils in the magnet gap and an elongated rf coil. Drawing (*left*) and profile of fiber positions in an air-spring bellows (*right*). The fibers are centered on a 1 mm grid. (**b**) Small unilateral MRI scanner with a bar magnet (*left*) and image (*right*) of a letter F cut from rubber and placed on *top* of the rf coil [32]. (**c**) Gradient coil arrangement mounted in the gap of a unilateral MRI scanner and slice-selective NMR image of a defect in a textile fiber reinforced rubber tube (*right*) [35, 36]. The acquisition time for the 50 × 50 pixel image with a field of view of 4 × 4 cm^2 was 2 h. (**d**) Flat gradient coil system built to fit on the *profile* NMR-MOUSE (*left*) and slice-selective NMR image of a silicon rubber phantom

Several setups based on this approach have been built to produce 2D images in planes parallel to the sensor surface (Fig. 4.20b–d) [32, 10]. For example, a planar xy gradient system has been combined with a bar magnet to produce 2D images with sub-millimeter resolution at the near surface (2 mm) of the sensor (Fig. 4.20b) [32]. Notice that B_0 is normal to the sensor surface; therefore, the x and y gradient coils are identical but rotated 90°. Figure 4.20c shows an xz gradient system used in combination with a large U-shaped magnet for 3D imaging [10]. The gradient coils were designed to work in a depth range of about 20 mm. To cover this range the sensor is retuned and the currents through the gradient system are adjusted to keep the FoV constant for all depths. In this case the gradient system is placed in the gap of a U-shaped magnet, however, for similar sensors but with a smaller gap that it is not possible. One solution is to place the gradient coils at the borders of the magnet; however, by doing so the efficiency drops dramatically. A better choice is the planar gradient coil system used with the *profile* NMR-MOUSE (Fig. 4.20d). Based on the same approach as the magnet, the zx gradient system was optimized to work at a fixed depth alleviating constrains on its dimensions.

References

1. Jackson JA, Burnett LJ, Harmon JF (1980) Remote (inside-out) NMR. 3. Detection of nuclear magnetic-resonance in a remotely produced region of homogeneous magnetic-field. J Magn Reson 41(3):411–421
2. Kleinberg RL (1996) Well logging. In: Encyclopedia of NMR. Wiley-Liss, New York, NY

3. Hogan BJ (1985) One-sided NMR sensor system measures soil/concrete moisture. Design News, May 5
4. Matzkanin GA (1989) A review of nondestructive of composites using NMR. In: Nondestructive characterization of materials. Springer, Berlin
5. Paetzold RF, Delossantos A, Matzkanin GA (1987, March) Pulsed nuclear-magnetic-resonance instrument for soil-water content measurement – sensor configurations. Soil Sci Soc Am J 51(2):287–290
6. Paetzold RF, Matzkanin GA, de los Santos A (1985) Surface soil-water content measurement using pulsed nuclear magnetic-resonance techniques. Soil Sci Soc Am J 49(3):537–540
7. Rollwitz WL (1985) Using radiofrequency spectroscopy in agricultural applications. Agric Eng 66(5):12–14
8. Eidmann G, Savelsberg R, Blümler P, Blümich B (1996, Sep) The NMR mouse, a mobile universal surface explorer. J Magn Reson Ser A 122(1):104–109
9. Perlo J, Casanova F, Blümich B (2005, Sep) Profiles with microscopic resolution by single-sided NMR. J Magn Reson 176(1):64–70
10. Perlo J, Casanova F, Blümich B (2004, Feb) 3D imaging with a single-sided sensor: an open tomograph. J Magn Reson 166(2):228–235
11. Perlo J, Casanova F, Blümich B (2005, Apr) Velocity imaging by ex situ NMR. J Magn Reson 173(2):254–258
12. Perlo J, Casanova F, Blümich B (2007, Feb) Ex situ NMR in highly homogeneous fields: H-1 spectroscopy. Science 315(5815):1110–1112
13. Blümich B, Anferov V, Anferova S, Klein M, Fechete R, Adams M, Casanova F (2002, Dec) Simple NMR-mouse with a bar magnet. Concepts Magn Reson 15(4):255–261
14. Prado PJ (2003, Apr) Single sided imaging sensor. Magn Reson Imaging 21(3–4):397–400
15. Popella H, Henneberger G (2001) Design and optimization of the magnetic circuit of a mobile nuclear magnetic resonance device for magnetic resonance imaging. COMPEL 20:269–278
16. Marble AE, Mastikhin IV, Colpitts BG, Balcom BJ (2006) A constant gradient unilateral magnet for near-surface MRI profiling. J Magn Reson 183(2):228–234
17. Rahmatallah S, Li Y, Seton HC, Mackenzie IS, Gregory JS, Aspden RM (2005, Mar) NMR detection and one-dimensional imaging using the inhomogeneous magnetic field of a portable single-sided magnet. J Magn Reson 173(1):23–28
18. McDonald PJ (1997, Mar) Stray field magnetic resonance imaging. Prog Nucl Magn Reson Spectrosc 30:69–99
19. Fukushima E, Jackson JA (2002) Unilateral magnet having a remote uniform field region for nuclear magnetic resonance. US Patent, 6489872
20. Fukushima E, Roeder SB, Assink RA, Gibson AV (1988) Nuclear magnetic resonance apparatus having semitoroidal RF coil for use in topical NMR and NMR imaging. US Patent 4721914
21. Manz B, Coy A, Dykstra R, Eccles CD, Hunter MW, Parkinson BJ, Callaghan PT (2006, Nov) A mobile one-sided NMR sensor with a homogeneous magnetic field: the NMR-mole. J Magn Reson 183(1):25–31
22. Anferova S, Anferov V, Adams M, Blümler P, Routley N, Hailu K, Kupferschlager K, Mallett MJD, Schroeder G, Sharma S, Blümich B (2002, Mar) Construction of a NMR-mouse with short dead time. Concepts Magn Reson 15(1):15–25
23. Meriles CA, Sakellariou D, Heise H, Moule AJ, Pines A (2001, July) Approach to high-resolution ex situ NMR spectroscopy. Science 293(5527):82–85
24. Perlo J, Demas V, Casanova F, Meriles CA, Reimer J, Pines A, Blümich B (2005, May) High-resolution NMR spectroscopy with a portable single-sided sensor. Science 308(5726):1279–1279
25. McDonald PJ, Aptaker PS, Mitchell J, Mulheron M (2007, Mar) A unilateral NMR magnet for sub-structure analysis in the built environment: the surface GARfield. J Magn Reson 185(1):1–11

26. Jackson JD (1998) Classical electrodynamics. Wiley, New York, NY
27. Fukushima E, Roeder SBW (1986)Experimental pulse NMR: a nuts and bolts approach. Addison Wesley, New York, NY
28. Callaghan PT (1991) Principles of nuclear magnetic resonance microscopy. Clarendon Press, Oxford
29. Kimmich R (1997) NMR: tomography, diffusometry, relaxometry. Springer, Berlin
30. Blümich B (2000) NMR imaging of materials. Clarendon Press, Oxford
31. Prado PJ, Blümich B, Udo Schmitz U (2000) One-dimensional imaging with a palm-size probe. J Magn Reson 144:200–206
32. Casanova F, Blümich B (2003, July) Two-dimensional imaging with a single-sided NMR probe. J Magn Reson 163(1):38–45
33. Casanova F, Perlo J, Blümich B (2004, Nov) Velocity distributions remotely measured with a single-sided NMR sensor. J Magn Reson 171(1):124–130
34. Casanova F, Perlo J, Blümich B, Kremer K (2004, Jan) Multi-echo imaging in highly inhomogeneous magnetic fields. J Magn Reson 166(1):76–81
35. Kolz J, Goga N, Casanova F, Mang T, Blümich B (2007) Spatial localization with single-sided NMR sensors. Appl Magn Reson 32(1–2):171–184
36. Blümich B, Perlo J, Casanova F (2008, May) Mobile single-sided NMR. Prog Nucl Magn Reson Spectrosc 52(4):197–269

Chapter 5
Single-Sided Tomography

Federico Casanova

MRI is one of the most powerful non-invasive techniques to assess material heterogeneities [1, 2]. It allows the generation of images contrasted by a variety of parameters that enhance the discrimination of regions with different molecular structure and dynamics for a wide range of materials. In recent years special effort has been made to achieve spatial localization in the presence of the strongly inhomogeneous magnetic fields generated by open magnet geometries, a capability that converts single-sided NMR probes into truly open tomographs. These sensors were originally intended to scan the surface of large objects by simply repositioning the probe across a region of interest. This procedure provides a crude lateral spatial resolution of the order of the centimeter with a pixel size defined by the size of the sensitive volume, which is mainly determined by the extension of the rf coil. Although the resolution can be improved by decreasing the size of the rf coil, this approach leads to a considerable reduction of the maximum penetration depth of the sensor. To achieve finer resolution inside the sensitive volume, Fourier imaging proved to be the right approach. In this way, by combining pulsed gradients along the two lateral directions with the static gradient of the magnet along the depth full 3D localization has been achieved. The next sections describe the steps needed to implement this imaging techniques at maximum sensitivity.

5.1 Depth Resolution Using the Static Field Gradient

Once the sensor is positioned at a certain spot, depth resolution into the object can be gained by taking advantage of the natural field gradient generated by the open magnet. As described in Chap. 4 a number of magnet geometries have been optimized to generate a uniform gradient along the depth direction. Ideally, in the presence of such a gradient, a depth profile of an object could be obtained as the FT of an echo signal. However, two important technical limitations complicate the

F. Casanova (✉)
Institut für Technische Chemie und Makromolekulare Chemie, RWTH Aachen University, D-52074, Aachen, Germany

F. Casanova et al. (eds.), *Single-Sided NMR*,
DOI 10.1007/978-3-642-16307-4_5,

implementation of this technique in a straightforward way. First, the strong gradients generated in practice by real sensors reduce the excited depth range to a very thin slice. For example, in the presence of a gradient of 1 T/m (a relatively weak gradient for open magnets) an rf pulse of 10 μs (typical value needed when working at a few millimeters from the surface of the rf coil) excites a slice of about 1 mm. Second, the efficiency of the rf coil strongly decays for large depths, complicating the uniform excitation of thick slices. Two approaches have been adopted to scan a large depth range (see Chap. 4). In magnets generating a uniform gradient over a large depth range, retuning of the rf probe has been used to excite slices at variable depths. To improve the gradient uniformity the profile NMR-MOUSE was optimized to work at a fixed depth from the sensor, where high depth resolution can be achieved [3]. This strategy requires varying the relative position of the excited slice inside the sample. It has been achieved by displacing the sensor by means of a high-precision lift that repositions the magnet with respect to the object, which is kept static on a holder (see Fig. 5.1a).

Independent of the approach chosen to reposition the excited slice inside the object, to measure a profile over a large depth range requires concatenation of the information acquired at several depths. A typical way to control the resolution in the profile is by varying the length of the rf pulse. As shown in Chap. 2, the frequency bandwidth excited by a CPMG sequence is $\Delta\nu_0 = 1/(2t_p)$, so, by increasing the pulse length the slice thickness is reduced. However, by doing so, the profile is measured point by point, and as the resolution is improved the measuring time becomes excessively long. It is so because if the slice thickness is decreased a factor two, not only four scans are needed to keep sensitivity constant, but also the number of repositioning steps is doubled. Then, a factor two in resolution costs a factor eight in time. On the other hand, if the desired spatial resolution is smaller than the thickest exited slice, it can be set just by controlling the acquisition time T. This

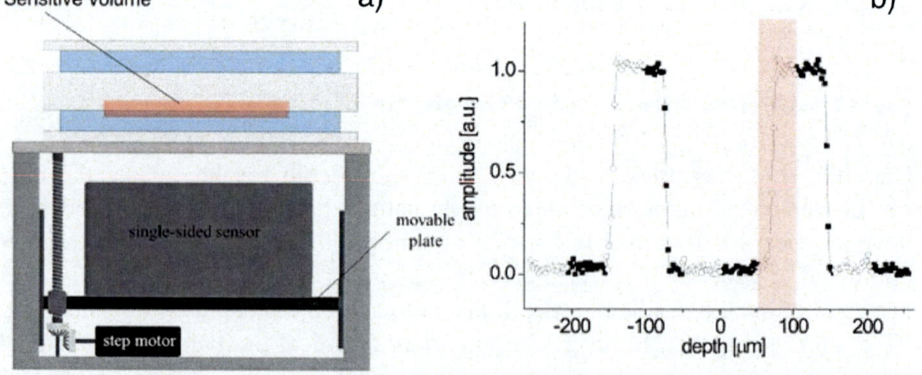

Fig. 5.1 (**a**) Lift used to reposition the sensor with respect to the sample. (**b**) Profile of a sample made of two latex layers 70 μm thick separated by a glass 150 μm thick. The complete profile was obtained as the catenation of 50 μm profiles obtained as the FT of the echo signal measured every 50 μm by moving the sensor

approach requires to set the rf pulse to the shortest duration to maximize the excited region and then to set the acquisition time to achieve the desired resolution inside the excited slice (by acquiring signal during a time T the frequency resolution in the spectrum is $\Delta\nu = 1/T$). Then, at each depth a profile of the excited region is measured. By sweeping the position of the excited slice across the sample the full profile is obtained as the catenation of the individual slices. The implementation of this procedure to measure a high-resolution profile is illustrated in Fig. 5.1b. It shows a depth profile of a simple two-layer rubber sample measured with the profile NMR-MOUSE. The rf pulses were set to 3 μs to excite a bandwidth of about 150 kHz, which under the 20 T/m field gradient generated by this sensor corresponds to 150 μm. Actually, to avoid appreciable variation of the excitation efficiency only a region of 50 μm was extracted from the profile acquired at each position, and the sensor was moved every 50 μm. Setting the acquisition window to 300 μs a nominal resolution of about 4 μm was defined with an echo time $t_E = 345\,\mu s$. Instead, if the rf pulse is set to 300 μs to excite a slice of 4 μm the sensor position has to be stepped every 4 μm, requiring about 12 steps to measure the 50 μm range that otherwise is profiled in a single experiment [3]. Another disadvantage of controlling the resolution by increasing the pulse length is that it leads to an increase of the echo time about twice the pulse length increment. It is so because when a thinner slice is excited the echo becomes broader (the excited frequency range is narrower) and a longer acquisition time must be set in the sequence to maximize the sensitivity. In samples with short T_2 this limitation becomes critical.

5.2 Spatial Encoding by Fourier Imaging

To achieve spatial resolution along the lateral directions of the sensitive slice, Fourier imaging methods based on pulsed gradients can be implemented. It requires to equip the sensor with a flat gradient coil system and an imaging technique working in the presence of a grossly inhomogeneous fields. Getting spatial resolution inside the sensitive volume allows one to look inside the selected spot with a magnifying glass useful to detect a fine structure of the material. The full sample can be scanned with high resolution by combining displacement of the sensor with Fourier imaging (Fig. 5.2).

Although at first sight the mentioned strategy sounds straightforward, it must be noticed that conventional imaging sequences such as spin echo, gradient echo, SPRITE, FLASH, EPI [1, 2] cannot be implemented in the presence of the static magnetic field gradient of single-sided sensors. This large group of sequences is excluded for two reasons. First, no FID is detectable because T_2^* is of the order of a few microseconds and second, although spin echoes can be generated, the signal is acquired in the presence of a strong background gradient preventing the use of read gradients. These limitations leave pure phase encoding as the only alternative. Prado et al. implemented a single-point spin echo imaging method (Fig. 5.3) [4] that combined a Hahn-echo sequence, applied to refocus the phase spreading due to the

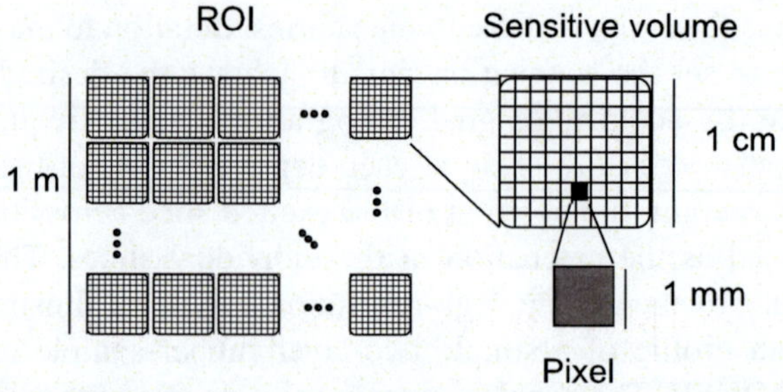

Fig. 5.2 Combination of sensor repositioning with Fourier imaging to map with high resolution a region of interest (ROI) larger than the sensitive volume. The ROI is divided in subregions denoted as (m, n) with the size of the sensitive volume and then each of those is subdivided again in pixels by Fourier imaging

static background gradient, with gradient pulses applied during the free evolution periods to encode the spatial information in the phase of the echo. Single point refers to the fact that only one point of the k-space is acquired per shot, therefore to measure a 1D image with n pixels requires at least n experiments.

In this first attempt the original U-shaped NMR-MOUSE was furnished with a gradient system consisting of two solenoids placed inside the gap to generate a gradient field along the gap direction (Fig. 5.4 a) [4]. By driving the current through the coils with a conventional gradient amplifier a pulsed field gradient was generated. Figure 5.4 shows a phantom built from a multi-turn commercial rubber band (b), the 1D image measured with the NMR-MOUSE (c), and the same result but for a uniform object (d). The last image shows the size and shape of the sensitive region, i.e., the real field-of-view (FOV). Although the sample is uniform its image is modulated by the non-uniform shape of the sensitive volume. As for any pure

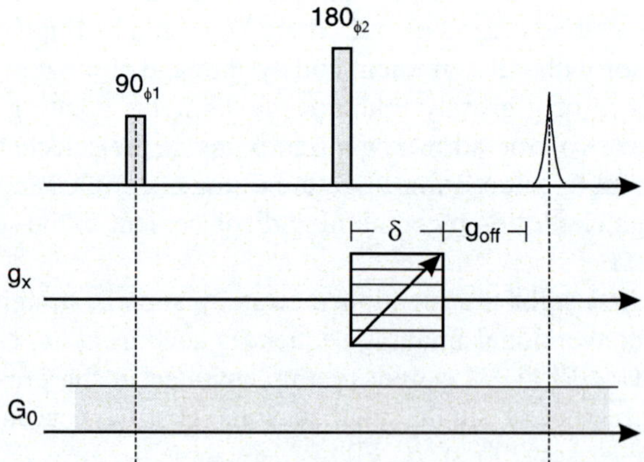

Fig. 5.3 Spin-echo sequence for 1D phase-encoding imaging. In the echo maximum the phase spread produced by the strong static gradient is canceled, while the phase introduced by the pulsed gradient is retained. The amplitude of the pulsed gradient g_x is stepped to sample the k-space from negative to positive values

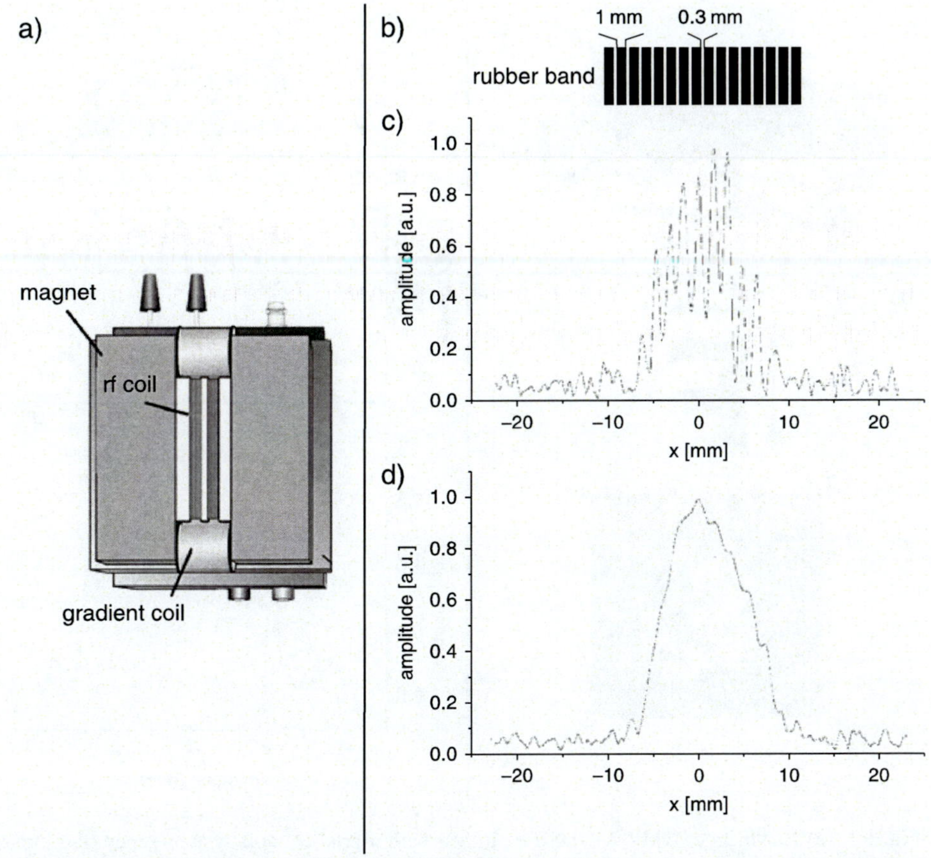

Fig. 5.4 (**a**) U-shaped NMR-MOUSE equipped with a gradient coil system used to generate a pulsed gradient along the gap direction. (**b**) Phantom: Multi-turn commercial rubber band. The 1D images obtained by implementing a single-point acquisition technique for the rubber band sample and for a homogeneous sample are shown in (**c**) and (**d**), respectively

phase-encoding technique the spatial resolution attainable for non-diffusive samples is determined only by the uniformity of the gradient and not by relaxation. In this case a spatial resolution of the order of some hundred of micrometers was reported.

The extension of the technique to obtain 2D images required, in principle, just a second gradient coil to generate a gradient pointing across the gap and the incorporation of a second pulsed gradient in the sequence. However, a more important limitation had to be overcome. The sensor must generate a sensitive volume with a reasonable FOV across the gap, which is the second direction in the image. The elongated sensitive volume generated by the original U-shaped magnet delayed the implementation of a 2D technique demanding the development of a new sensor [5]. In 2002 a new prototype combining a bar magnet polarized along the depth direction and a figure-8 rf coil located on one of the pole faces (Fig. 5.5) was used to generate a sensitive volume symmetrically extended along both lateral directions. The main gradient of the static magnetic field generated by such block is also along

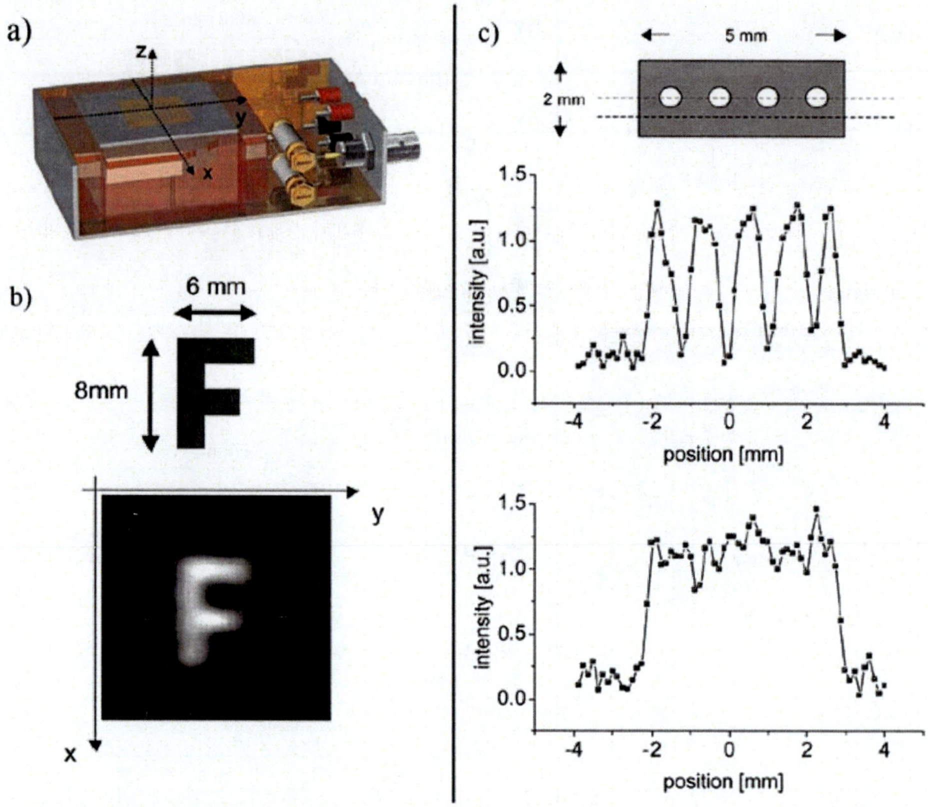

Fig. 5.5 (**a**) Bar-magnet NMR-MOUSE equipped with a gradient coil system generating pulsed gradients along the two lateral directions. (**b**) Letter F cut from a natural rubber sheet 2 mm thick used as phantom to show the performance of the 2D imaging method. (**c**) 1D profiles measured at different depths across a fiber-reinforced rubber sheet. It can be observed that while the profile measured at the center of the fibers clearly reproduces its structure, by shifting the sensitive slice next to the fibers, the slice is flat enough not to intersect any of the fibers in the measured range

the depth direction, defining rather flat slices of constant frequency parallel to the sensor surface. The sensor was equipped with a set of planar gradient coils mounted on the same polar face but at the side of the rf coil (Fig. 5.5a). The 2D k-space was sampled point by point by means of a pure phase-encoding imaging sequence like the one shown in Fig. 5.3, but with the two gradients pulsed simultaneously. As a test phantom, a letter F of about $8 \times 6\,\text{mm}^2$ was cut from a natural rubber sheet. An image measured with an in-plane spatial resolution of $1\,\text{mm}^2$ was measured in an experimental time of the order of 1 h (Fig. 5.5b).

Besides generating a symmetric field of view along the lateral directions, the bar magnet used in this work generates a uniform gradient along the depth direction. In the presence of such field profile a flat sensitive volume can be excited at different depths by retuning the excitation frequency. Figure 5.5c shows 1D images measured at different depths inside a fiber-reinforced rubber layer. While the profile measured at the center of the fiber layer allows for the quantification of position and dimension of the fibers, a profile measured 500 μm off-centered shows already a uniform structure, proving that the slice is flat enough not to intersect the fiber layer. Although this work shows that 3D spatial localization can be achieved by combining a 2D pure phase-encoding method with slice selection in the presence of the background gradient, the poor sensitivity inherent to single-sided sensors extended the experimental times to limits far too long for practical uses. Another important limitation of the sensor, mainly determined by rf coil geometry, was the maximum penetration depth, which was hardly larger than a couple of millimeters. In the next sections we discuss an important modification to the detection scheme that led to a large sensitivity improvement as well as a new open MRI system able to produce 3D images up to depths of at least 1 cm in reasonable short measuring times defined by practical limits (minutes).

5.3 Multi-Echo Acquisition Schemes

The basic idea to speed up data acquisition relies on the fact that the echo time t_E used for the Hahn-echo sequence is usually much shorter than the T_2 of the sample. In this echo time limit a long train of echoes can be generated by applying a CPMG sequence. The complete echo train can be then co-added in order to increase the sensitivity during detection. By doing so a reduction of the experimental time by a factor proportional to the number of echoes acquired in the train is achieved. This concept has been largely exploited in the context of homogeneous static and rf magnetic fields [1, 2]; however, its application in strongly inhomogeneous magnetic fields is not straightforward. Figure 5.6 shows the result of the implementation of the single-point imaging sequence used in [4] followed by a CPMG train (Fig. 5.7a). As it can be seeing, the space information encoded in the phase of the signal by the gradient pulses is clearly distorted as the echo number increases. The results, obtained by computer simulations, show complicated distortion patterns due to the imperfection of the refocusing pulses. The aim of this section is to understand the problematic that leads to a loss of information during the multi-echo train in order to design proper encoding strategies working under these experimental conditions. Considering that real experiments are already distorted by the shape of the sensitive volume (Fig. 5.4), and poor SNR, in the next we prefer to carry this analysis based on numerical simulation. This is one example where the power of numerical calculations based on the rotations described in Chap. 2 can be fully appreciated. Calculating the evolution of the magnetization during the pulse sequence two different encoding approaches offering outstanding performance could be identified.

5.3.1 RARE-Like Imaging Sequence

To understand the evolution of the space information encoded by the gradient pulses as a function of the echo number during the CPMG train, the evolution of the

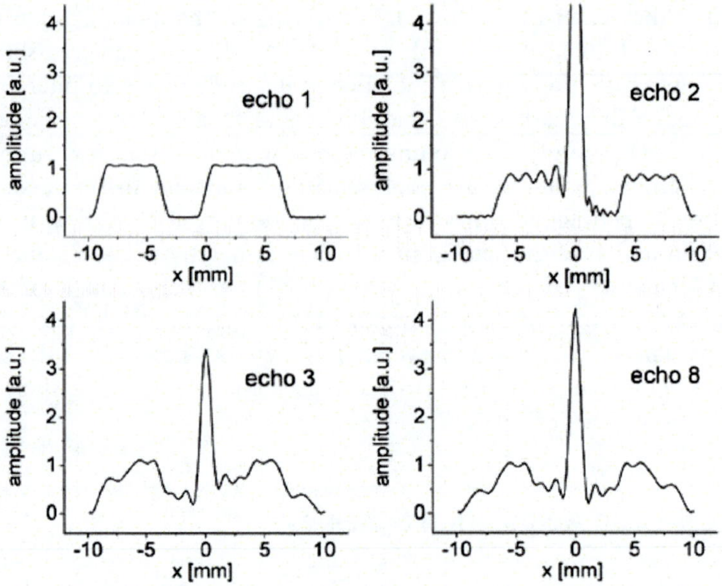

Fig. 5.6 Set of 1D images obtained from different echoes generated by the imaging sequence of Fig. 5.7a. The image obtained from the first echo reproduces well the sample profile. However, for higher echo numbers, the spatial information is completely distorted. This problem is due to the imperfection of the refocusing pulses

magnetization was calculated using the formalism described in Chap. 2. To include the effect of off-resonance in the spatial encoding we assumed for the calculation a static field pointing along the z-axis with a uniform gradient along the y-axis (depth) and a homogeneous rf field pointing along the y-axis. The field generated by the gradient coils, considered also ideal, has a gradient only along the x-axis (imaging direction). An object consisting of a single pixel placed at x_0 is considered in order to simplify the data interpretation. Basically, the idea is to follow the position of the pixel in the image echo by echo. The object is extended along the static gradient direction to introduce a resonance frequency distribution. The size of the object was set in such a way that the frequency distribution is much broader than the excitation bandwidth of the rf pulses ($\omega_1 \ll \Delta\omega_0$).

Figure 5.8a shows the evolution of the images (rows) as a function of the echo number n (columns) for the pulse sequence of Fig. 5.7a. The image obtained from the first echo shows a single peak at the right position x_0, as expected. Although for higher echo numbers the peak remains at the same position, two more contributions appear at position $-x_0$ and zero. The contributions at x_0 and $-x_0$ confirm that the space information is not fully lost, even for the last echo the signal is modulated by the right frequency; however, the sign of the frequency cannot be discriminated. It is a consequence of the loss of the quadrature channel perpendicular to the 180° pulses during the refocusing train. As it is well known, the CPMG sequence preserves the component of the magnetization along the refocusing train, but the component perpendicular to it dies within few echoes [6].

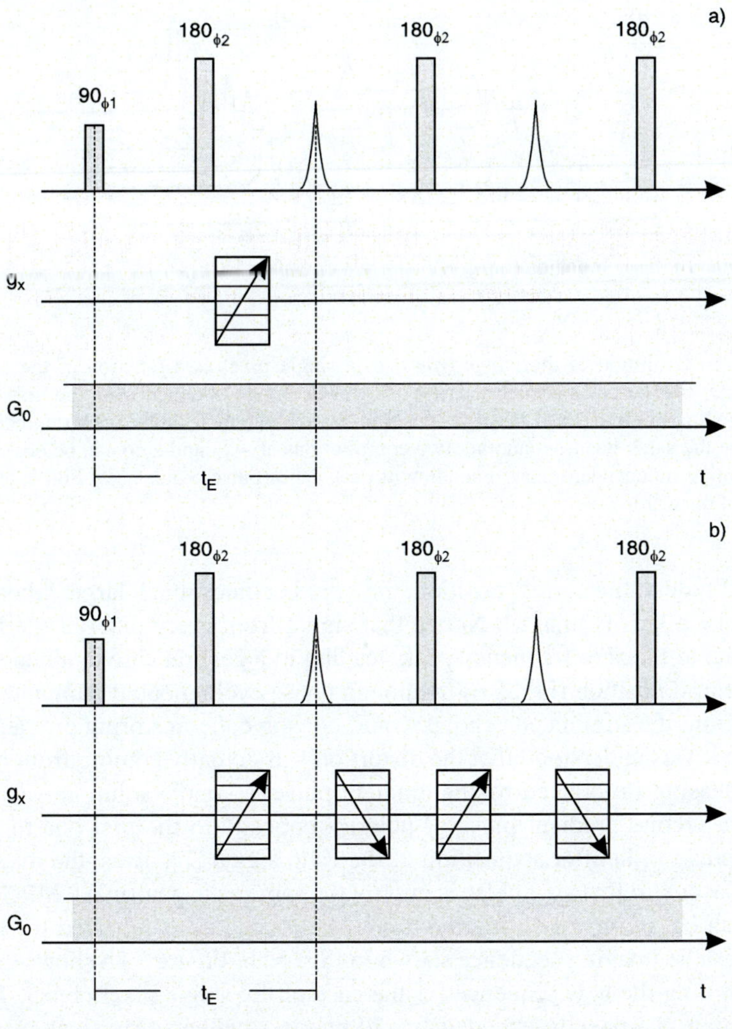

Fig. 5.7 (**a**) Straightforward extension from single- to multi-echo detection of the phase-encoding imaging sequence originally implemented in [4]. (**b**) The phase introduced by the first gradient pulse is canceled by a second one applied with opposite current. Therefore the phase distribution at the end of each evolution period is identical to the one obtained in the CPMG sequence. By doing so no spatial information must be preserved from echo to echo and the high performance of the CPMG sequence to generate a long echo decay is preserved

Although in the sequence of Fig. 5.7a the 180° pulses are applied 90° phase shifted with respect to the first pulse, the gradient pulse rotates the magnetization out of this direction generating a parallel and a perpendicular component. The loss of the perpendicular component (imaginary channel) leads to a mirroring of the image. In the other hand, the contribution at zero frequency ($x = 0$) corresponds to signal that is not affected by the gradient pulse. It comes from the stimulated echo that is stored as longitudinal magnetization during the application of the gradient pulse. Although the amplitude of the distortion at zero frequency is only one-half of

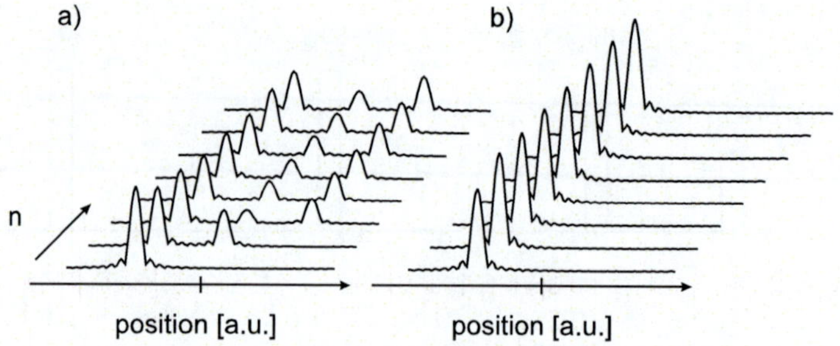

Fig. 5.8 (**a**) Evolution of the image (*rows*) of a single pixel as a function of the echo number n (*columns*) for the pulse sequence of Fig. 5.7a. The image obtained from the first echo shows, as expected, a single peak at the right position x_0. Although for higher echo numbers the peak remains at the same position, another two contributions at $-x_0$ and zero can be observed. (**b**) By implementing independent phase encoding of each echo (pulse sequence of Fig. 5.7b) all images are free of distortions

the amplitude at the correct position, the ratio becomes much larger when an object with many pixels is imaged. Notice that signal from every pixel of the image will contribute to the zero frequency peak, leading in a general case to a huge distortion at the central position (Fig. 5.6). Although phase cycling could eliminate the signal arising from the stimulated echo, it would not solve the mirroring problem.

Once it was understood that the distortion was actually coming from magnetization dephasing introduced by the gradient pulse, a simple solution was presented. It used a second gradient pulse of polarity opposite to the first one to cancel the phase spread right after acquisition of the echo signal [7]. Thus, the magnetization before the next refocusing pulse is exactly the same as in a normal CMPG sequence, and it can be preserved during the train. The procedure is repeated for every echo leading to the imaging sequence shown in Fig. 5.7b. Figure 5.8b shows the result of implementing the new sequence for the case of the single pixel object. The spatial information is perfectly encoded for all echoes; neither distortions at the mirror position nor zero frequency is observed. The amplitude of each gradient pulse can be set equal in all the intervals, sampling the same k-space point for all the echoes of the train or it can be stepped to sample the complete k-space in a convenient way during a CPMG train decay. When the same k-space point is sampled in all the echoes a set of experiments must be repeated, increasing the amplitude of the gradient pulses step by step to sample the complete k-space.

The echoes sampled during the decay can be added to improve the signal-to-noise ratio (Fig. 5.9a). In the case that T_2 changes across the object the addition of all echoes acquired during the train decay results in a T_2-weighted profile, while if a profile is reconstructed from each echo it can be plotted as a function of the time to determine the relaxation rate pixel by pixel. By changing the phase-encoding scheme, each echo can be used to sample a different k-space point scanning the complete image in a single CPMG train. As small values of k determine the amplitude of the image, and the large ones determine the resolution, enhanced T_2 contrast can

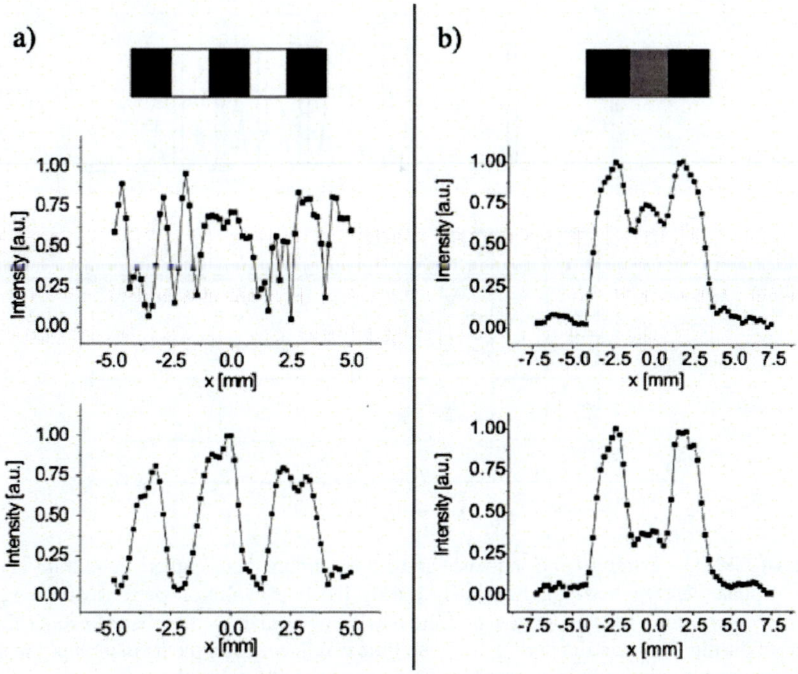

Fig. 5.9 (**a**) 1D images obtained by implementing the multi-echo pulse sequence shown in Fig. 5.7b. On *top* the result obtained from the first echo (single-echo detection) and on the *bottom* from the addition of the complete echo train. (**b**) Taking advantage of the independent phase encoding achieved in each echo, a RARE-type k-space sampling scheme was implemented to obtain T_2-weighted 1D profiles. To compare the contrast the k-space was sampled from zero to k_{max} (*top*), and from k_{max} to zero (*bottom*)

be obtained in a single CMPG experiment by sampling the k values from maximum to zero (Fig. 5.9b). To obtain the maximum number of echoes during the CPMG sequence a short echo time is desired, but it is limited by the rise and fall times of the gradient pulses. The phase defined by the gradient pulse applied before each echo must be completely canceled by the gradient pulse applied following the echo. This condition determines that the echo time must be long enough to assure the second gradient pulse to be zero when the next rf pulse is applied. If this is not fulfilled, a slightly different FOV is obtained from even and odd echoes, a fact that can be used to set the correct echo time. Depending on the T_2 values, use of a current pre-emphasis is needed to shape the current pulse.

5.3.2 CPMG-CP for Quadrature Detection

Shortly after the implementation of the RARE-like sequence a more powerful method was presented [8]. In contrast to the first solution, where the magnetization is spread and refocused before and after each echo without affecting the dynamics

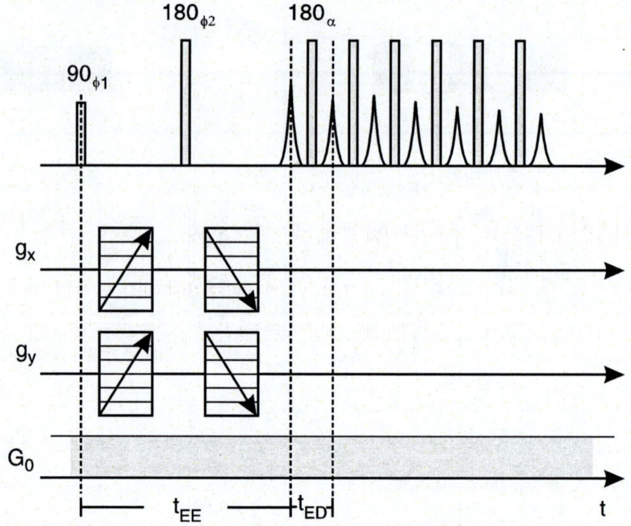

Fig. 5.10 CPMG-CP multi-echo sequence applied to sample the complete echo train decay. Both gradient pulses are applied simultaneously during the free evolution periods of the Hahn-echo sequence determining the encoding echo time t_{EE}. To maximize the number of echoes that can be generated during the detection period, the echo time is reduced after the formation of the first echo. The detection echo time t_{ED} is determined by the dead time of the probe

of the CPMG sequence, the second alternative exploits the fact that a CPMG train at least preserves one component accepting that the second component is lost. It requires the combination of two experiments to acquire both quadrature components, but it increases the sensitivity by maximizing the number of echoes that can be generated during the train and minimizes the power dissipated by the gradient coil system. The new pulse sequence (Fig. 5.10) can be understood as the combination of a first *encoding period* consisting of a Hahn-echo sequence followed by a *detection period* composed of a train of refocusing pulses.

The encoding period generates an echo signal with a phase defined by the pulsed gradients and the position of the spins. Then, for a given phase ϕ measured from the x-axis (here we assume that the Hahn echo is generated along the x-axis in the absence of pulsed gradients, for example, applying the 90° with phase $\pi/2$ and the 180° with phase 0) the components of the magnetization at the echo maxima are $M_x = M_0 \cos(\phi)$ and $M_y = M_0 \sin(\phi)$, with x- and y- directions in the rotating frame (Fig. 5.11a). A first experiment executed with refocusing pulses applied with phase 0 (CPMG sequence) is used to sample M_x (real channel), while M_y is attenuated to zero after a few pulses (Fig. 5.11b). The second experiment is identical but the refocusing pulses are applied with phase $\pi/2$ (CP sequence) to preserve M_y (imaginary channel) (Fig. 5.11c). By combining the real component of the first experiment with the imaginary of the second one the full complex signal $M_0(\cos(\phi)+i\,\sin(\phi))$ is built for each echo. A typical experiment uses add–subtract phase cycling, so a minimum of four phase steps is required (see Table 5.1).

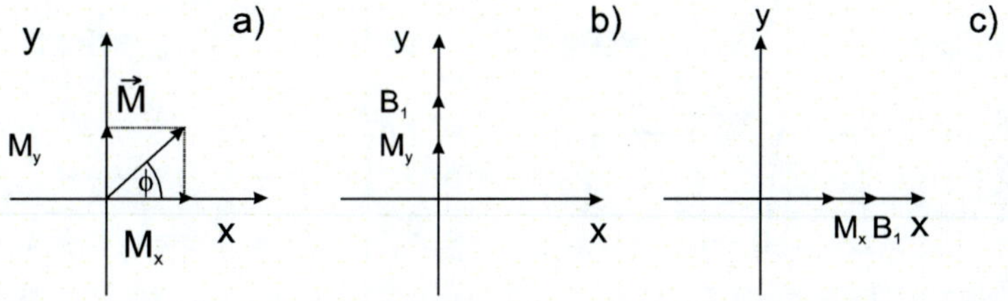

Fig. 5.11 (**a**) Magnetization vector at the time of the first echo maximum. The phase shift ϕ is proportional to position and the gradient strength. (**b**) During the application of the refocusing train with phase $\pi/2$ only M_y is preserved. (**c**) By switching the phase of the refocusing train to 0 it is possible to preserve M_x

Table 5.1 Phase cycle required by the CPMG-CP sequence to sample both channels. It also includes an add/subtract step

ϕ_1	ϕ_2	ϕ_α	ϕ_{rec}	Component
$+\pi/2$	0	0	0	Real
$-\pi/2$	0	0	π	Real
$+\pi/2$	0	$\pi/2$	0	Imaginary
$-\pi/2$	0	$\pi/2$	π	Imaginary

Figure 5.12 shows a set of 1D images (obtained by numerical simulations) plotted as a function of the echo number for the new multi-echo detection scheme. The conditions assumed are the same as in the previous calculations, off-resonance excitation and homogeneous B_1. All images show the structure of the sample proving that the space information is well preserved from echo to echo. As described in Chap. 2, due to the superposition of the Hahn and the stimulated echo stored along z by the second pulse, the CPMG train shows always a second echo larger than the first one. However, the sequence shown in Fig. 5.10 generates a second echo smaller than the first one and a third echo larger than the second one. It is so because as $t_{\mathrm{ED}} \ll t_{\mathrm{EE}}$ the first stimulated echo is lost (the second echo is only a direct echo), and the third echo is the first one where a stimulated echo (magnetization stored along z by the first pulse of the detection train) superimposes with a Hahn echo. Notice that due to the large offset from resonance even with a homogeneous B_1 field, the refocusing pulses introduce a broad distribution of flip angles. Therefore, no significant changes are expected for the case of inhomogeneous B_1 field.

In Table 5.2 the single-echo detection (Hahn echo) and the two multi-echo detection sequences (RARE like and CPMG-CP) are compared in terms of measuring time, average power, and total energy required. In favorable cases, where T_2 is much longer than the echo time, the experimental time reduction obtained when moving from single- to the multi-echo detection can be several orders of magnitude. The improvement from 100 to 15 min achieved when the CPMG-CP sequence is used instead of the RARE-like one is due to the fact that for the first one the echo time in the detection train is limited only by the dead time of the rf probe, while in the

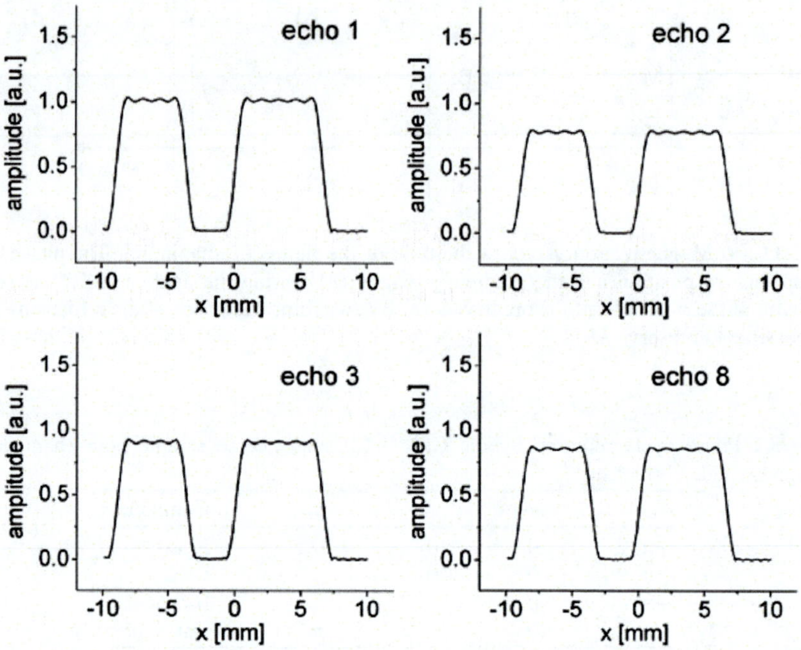

Fig. 5.12 Set of 1D images acquired as a function of the echo number for the CPMG-CP multi-echo detection scheme (Fig. 5.10). All images show the structure of the object assumed for the calculation proving that the space information is well preserved from echo to echo. The oscillation of the amplitude with the echo number is due to splitting of the signal in Hahn and stimulated echoes, a typical behavior observed in CPMG-like sequences

Table 5.2 Comparison of the Hahn echo, RARE like, and CPMG-CP methods

Sequence	Experimental time	RF power (W)	Gradient power (W)	Energy (kJ)
Hahn echo	43 h	0.002	0.4	62
RARE	100 min	0.1	47	282
CPMG-CP	15 min	1.4	0.4	1.6

second one it is limited by the length of the two gradient pulses. The shortening of the echo time increases the number of echoes that can be acquired during the train, maximizing the sensitivity enhancement. The important reduction of the total energy required is because only one instead of $2N$ gradient pulses is required per train of N echoes. This is an important issue determining the autonomy for portable systems where the energy is stored in batteries. Besides these advantages, the most important one is the ability of the CPMG-CP scheme to preserve phases defined in the spin system prior to the application of the detection train, something that the RARE sequence cannot do. It allows that one can attach this acquisition scheme to any encoding sequence like the ones used for flow measurements described in the next section.

5.4 Performance of the Multi-Echo Detection Scheme

5.4.1 Sensitivity Improvement

The time reduction that can be achieved with the new multi-echo technique instead of the Hahn-echo sequence becomes significant for samples having long T_2. To show the improvement obtained with the new sequence a silicone rubber sample with $T_1 = 330$ ms and $T_2 = 90$ ms was imaged (Fig. 5.13a). It was positioned above the sensor centered at 5 mm from the surface and a slice 1 mm thick was selected. The sequence of Fig. 5.10 was implemented using gradient pulses 0.37 ms long and an echo time t_{ED} of 0.11 ms. The gradient amplitude was increased in 32 steps sampling k-space from negative to positive values. The FoV was set to 32 mm along both directions to obtain a spatial resolution of about 1 mm. A single scan was used per gradient amplitude, and 800 echoes were acquired and averaged to improve the signal-to-noise ratio. A recycling delay of 0.45 s was used between experiments resulting in a total experimental time of 15 min to obtain a 2D image.

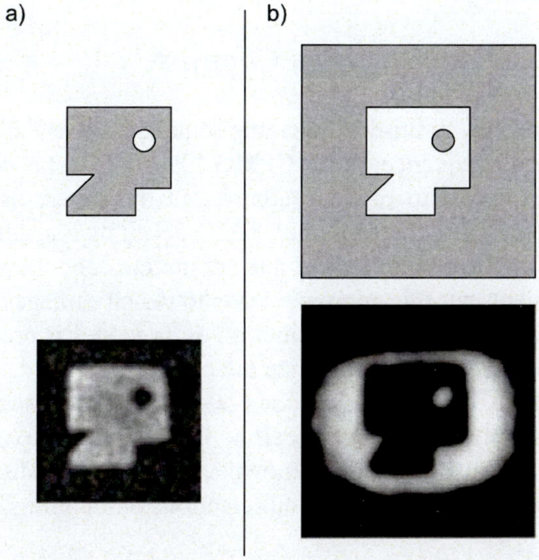

Fig. 5.13 (**a**) (*Top*). Geometry of a test object made of silicon rubber. (*Bottom*) 2D image obtained by applying the new multi-echo acquisition scheme. Thanks to the 800 echoes generated during the echo train decay, a single scan per gradient amplitude was needed to sample each quadrature channel. The echo train was sampled until the amplitude had decayed to 30% of the initial amplitude and before adding the echo train it was apodizated by the same decay function. The total experimental time to obtain the 2D images with a spatial resolution of about 1 mm was 15 min, reducing the acquisition time by a factor of 175 compared to the single-echo method, which would require 43 h. (**b**) Subtraction of the image of (**a**) from the image of a uniform object $50 \times 50\,\text{mm}^2$ used to probe the FoV defined by the RF coil. The dimensions of the selected sensitive volume are 36 mm along x and 26 mm along z

Figure 5.13a shows the cross section obtained using the new pulse sequence. The image reproduces the structure of the object without appreciable distortions demonstrating that 2D spatial localization can be obtained across the selected slice [8]. To obtain a 2D cross section with the same signal-to-noise ratio but using the single-echo sequence a larger number of scans must be used. As the amplitude of the echo train decays exponentially and is sampled up to one-third of its initial amplitude, the sensitivity improvement can be easily calculated. For this rubber sample 175 full Hahn echoes must be averaged to obtain the signal-to-noise ratio provided by the multi-echo sequence, giving an experimental time of about 43 h. The field of view defined by the rf coil was probed by a uniform silicon rubber sample with dimensions $50 \times 50\,\text{mm}^2$ along the x- and z-axes. The fast imaging method was implemented using the parameters of the previous experiment, but a FoV of 42 mm was set. Figure 5.13b shows the result obtained by subtracting the image of Fig. 5.13a from the image of the uniform object. The RF coil defines a FoV of 36 mm along x and 26 mm along z. In case a large object is positioned on the sensor, the FoVs must be set larger than these dimensions to avoid folding of the signal. For the sake of simplicity, objects with limited sizes along x and z will be used from this point on.

5.4.2 Relaxation and Diffusion Contrast

As described above, the multi-echo imaging sequence is basically obtained by combining a Hahn-echo sequence and a CPMG-like multi-echo detection train. This feature can be exploited to produce contrast in the image using the Hahn-echo sequence as a filter or directly from the CPMG decay by considering different parts of the echo train. In the previous section the complete echo train was added for sensitivity improvement, but this approach leads to density images only if T_2 does not vary across the sample. In case a distribution of T_2 values is present, a T_2-weighted image is obtained. The way to maximize the contrast is to add different parts of the train. If, for example, domains with known T_2 values are present in the sample, the integration limits in the decay curve can be easily set to maximize the difference between the two signals, but if an unknown sample is to be studied, different parts of the train can be chosen with a judicious elections of the limits based on the overall signal decay.

An application of this method is shown in Fig. 5.14a, where it was used to discriminate three different rubber types present in the object. A 1 mm thick slice at 5 mm from the surface and inside the object was selected. The multi-echo sequence was applied setting $t_{\text{ED}} = 0.11$ ms and using 1,000 echoes to correctly sample the longest decay present in the object. The gradient amplitude was increased in 24 steps and with the FoV set to 32 mm a spatial resolution of about 1.3 mm was obtained. The repetition time was set to 0.3 s, three times the longest T_1 to avoid any T_1 weighting in the image. Eight scans were averaged for sensitivity improvement. Figure 5.14a shows different T_2-weighted images reconstructed from the addition

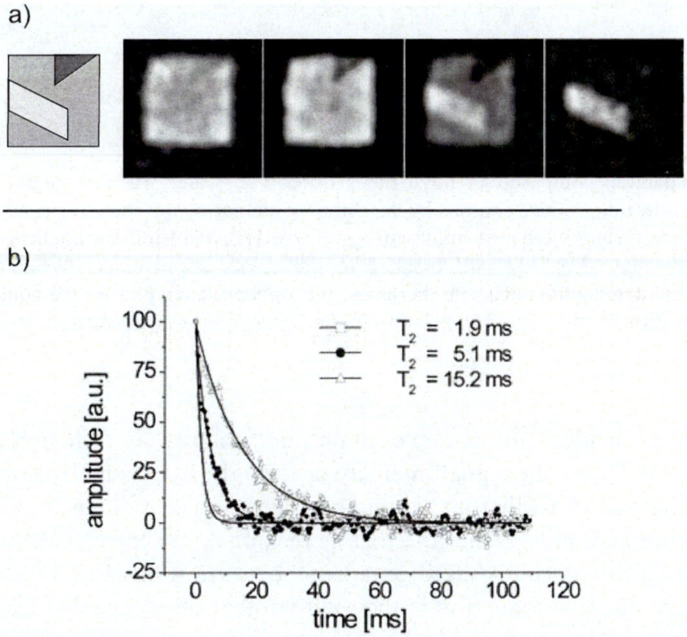

Fig. 5.14 (**a**) Object geometry composed of three different kinds of rubber presenting different relaxation times. To produce different T_2 contrast different train intervals were averaged before image reconstruction. The set of images (from *left* to *right*) was obtained by adding the echoes 1–8, 8–32, 100–200, and 200–400. (**b**) Amplitude decays corresponding to the three characteristic regions inside the object. Fitting the decaying signal in each pixel the local relaxation time was obtained. The values shown in the picture are in agreement with the relaxation times of each rubber sample

of the echoes (from left to right) 1–8, 8–32, 32–200, and 200–400. Normalizing the highest intensity of each image to one, the first image shows the rectangular object with almost uniform intensity, the second one shows the shortest T_2 region with half of the maximum intensity, while the other two regions cannot be differentiated, the third one shows the shortest T_2 with zero intensity, the middle one with the half of the intensity, and the longest with full intensity, and finally in the last image only the region with the longest T_2 is visible. Instead of adding parts of the echo train to produce contrast in the images the local relaxation time can be obtained by fitting the amplitude decay of each pixel. In contrast with the single-echo method the complete decay is sampled in a single imaging experiment, reducing the dimensionality of the experiment by one. Figure 5.14b shows the signal decays corresponding to each region in the object together with the respective fit. From dark to light gray the obtained relaxation times are 1.9, 5, and 15 ms, in agreement with the relaxation times measured from each single rubber sample using a CPMG sequence.

In cases where different regions in the sample possess similar CPMG decay times, but different signal attenuation during the Hahn-echo sequence, the encoding echo time (t_{EE}) can be purposely increased to enhance the contrast between them.

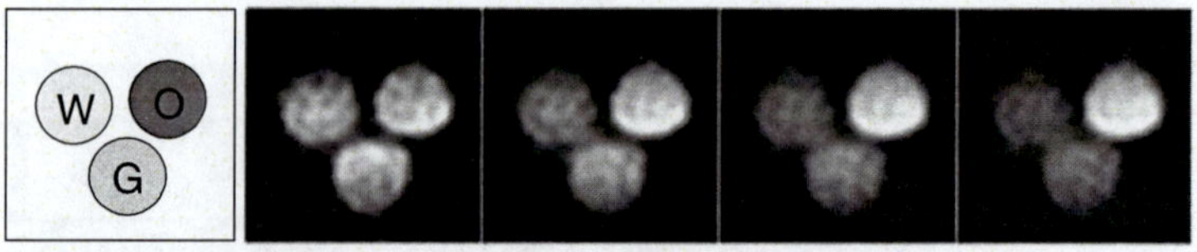

Fig. 5.15 Phantom composed of three tubes filled with water, oil, and gelatin. The imaging sequences allow to introduce contrast by diffusion by increasing the encoding echo time t_{EE}. From *left* to *right* the images were measured with $t_{EE} = 0.5$, 1, 2, and 5 ms. During the detection period 1,000 echoes were generated and co-added for sensitivity improvement. While the first image does not show any contrast between the tubes, for longer encoding times the contrast introduces signal attenuation for the water and gelatin sample, while almost no attenuation is observed for the oil tube

This approach is ideal, for example, to produce contrast by diffusion in liquid-like samples. In this case the signal intensity is strongly attenuated by diffusion during the encoding period, while for short enough detection echo times the CPMG decays might be similar. It must be noticed that increasing t_{ED} to enhance diffusion contrast during the CPMG decay is disadvantageous because it leads to SNR deterioration in the image due to a reduction in the total number of echoes that can be acquired during the detection period. Figure 5.15 shows images of three tubes filled with water, oil and gelatin. They were measured using encoding times of $t_{EE} = 0.5$, 1, 2, and 5 ms (from the left to the right) and 1,000 echoes during the acquisition train. It can be nicely observed that for short t_{EE} the method allows the measurement of density images, while for longer values important contrast can be produced keeping the sensitivity improvement achieved by adding the long echo train. The method has proven to be of great help discriminating fat and meat in biological tissue, results that are presented in Chap. 8.

5.4.3 3D Imaging

Combining the 2D phase-encoding method with slice selection, 3D spatial resolution is achieved [8]. Figure 5.16a shows an object with a 3D structure obtained by stacking a set of letters forming the word MOUSE cut from a sheet of 2 mm thick natural rubber. Having no spacer in between the letters the total structure is 10 mm high. For a better view, Fig. 5.16b shows the letters separated one from each other. After calibrating the frequency dependence with the position along the depth direction different 1 mm thick slices were selected, one inside each letter. The new multi-echo acquisition scheme was applied using 20 gradient steps and a FoV of 32 mm, giving a spatial resolution of about 1.6 mm. To have the same FoV at different depth both gradient amplitudes were calibrated. For this rubber sample, 96 echoes were added for sensitivity improvement. Using a recycle delay of 30 ms and two scans an experimental time of 45 s was needed to obtain the image of the first letter. To compensate the loss of sensitivity due to B_1 reduction as a function of depth, an increasing number of scans was used for the next letters needing a factor

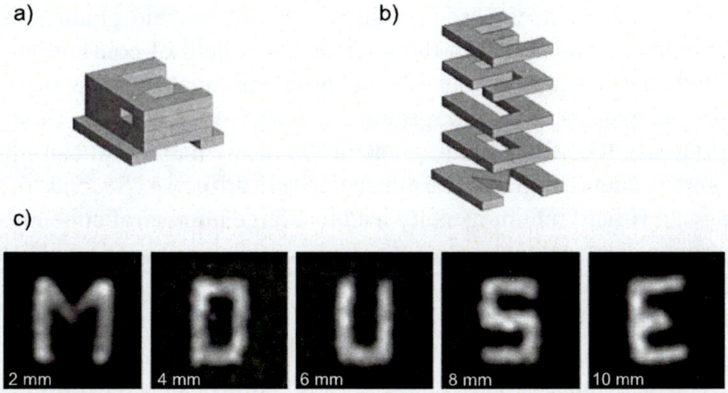

Fig. 5.16 (**a**) Object made by stacking the letters of the word MOUSE cut from a rubber sheet 2 mm thick. (**b**) Expanded view of the object obtained by drawing the letters separate from each other to see the object structure. (**c**) Images of each letter obtained by applying the multi-echo imaging method and selecting a 1 mm thick slice in the center of each letter. To compensate for the loss in sensitivity due to the B_1 decay, the number of scans used for each letter was increased proportional to depth. The total time to obtain each letter was 45, 45, 90, 120, and 180 s, respectively

of 4 for the last one. The 2D image of each letter can be observed in Fig. 5.16c. The flatness of the sensitive volume is good enough to observe no superposition of consecutive letters, and as no distortions can be observed in any of them we conclude that the uniformity of the pulsed gradients is acceptable in the complete range.

5.5 Displacement Encoding

Nuclear magnetic resonance has proved to be a powerful tool to characterize molecular motion non-invasively [1, 2]. It is suited to study displacement in a wide range of time and length scales. For opaque materials, for which optical methods are excluded, there are few experimental methods able to determine molecular displacements in any detail. NMR techniques offer a number of advantages compared to other experimental methods [9] and have been widely used in biology, medicine, and material science. NMR has enabled, for example, the measurement of vascular flow in stems and petioles at various stages of plant development, as well as vascular blood flow during the cardiac cycle. Also numerous applications in porous media such as natural sandstone have helped to elucidate models that describe the transport of fluids within the porous structure [10]. Combined with high-resolution imaging methods, it is a unique experimental method to check numerical solutions to the Navier–Stokes equation for Newtonian liquids or to validate constitutive equations used to describe the complex rheological behavior of non-Newtonian fluids [11, 12]. In contrast to other methods it is not affected by the vicinity of walls, providing, for example, the possibility to study effects such as the wall slip observed in the velocity shear of polymer melts and solutions [13].

A number of sequences based on the use of pulsed field gradients (PFG) have been implemented and tested in the homogeneous field of conventional superconducting and electromagnets, but several problems appear when they are implemented in the extremely inhomogeneous fields of open sensors. First, depending on the sequence, the strong static gradient G_0 of the main field can introduce significant signal attenuation due to molecular self-diffusion. Second, off-resonance excitation and rf field inhomogeneity lead to a large number of coherence pathways with improper displacement encoding. Third, the low magnetic field strength and the broadband signal inherent to these sensors yield low sensitivity and extend the experimental times to unviable limits. In this section a modification to the PFG-stimulated spin echo of Tanner [14], the so-called 13-interval sequence proposed by Cotts [15], is described. In particular, a phase cycling designed to eliminate spurious coherence pathways is given. This sequence reduces the signal loss due to diffusion in the presence of the strong background gradient and can be combined with the multi-echo detection CPMG-CP scheme to reduce the experimental time by several orders of magnitude, allowing the measurement of velocity distributions in a few minutes. Furthermore, this flow-encoding method can be combined with imaging techniques to spatially resolve velocity distributions and to obtain 2D velocity maps in situ.

5.5.1 PFG Methods in Inhomogeneous Fields

PFG methods measure coherent and diffusive displacement by means of spin echoes (SE) [16] and stimulated echoes (STE) [14] formed in the presence of pulsed field gradients. In systems where T_2 is much shorter than T_1, the PFG-STE sequence of Tanner has proven to be better suited than the conventional PFG-SE sequence. Figure 5.17a shows Tanner's sequence, where two gradient pulses of length δ separated by an evolution time Δ are applied to introduce phase shifts ϕ_i and ϕ_f proportional to the initial and final positions r_i and r_f [17]. As the magnetization is stored along the longitudinal axis during the evolution period Δ, PFG-STE allows longer evolution interval between the encoding and decoding gradient pulses (Fig. 5.17a). In this way the relaxation rate during the evolution period is given by T_1 and not by T_2 as in the SE sequence.

In the presence of a static magnetic field gradient G_0, the echo signal generated by the PFG-STE sequence is attenuated by diffusion as [14]

$$\begin{aligned}\ln(M(t)/M_0) = -\gamma^2 D\Big\{&\delta^2(\Delta+\tau-\delta/3)g_1^2\\&+\delta\big(2\tau\Delta+2\tau^2-2/3\delta^2-\delta(\delta_1+\delta_2)-(\delta_1^2+\delta_2^2)\big)g_1G_0\\&+\tau^2(\Delta+2/3\tau)G_0^2\Big\},\end{aligned}\tag{5.1}$$

where τ is the time between the first two 90° pulses and $\delta_{1,2}$ are the times between the rf and the gradient pulses. Depending on the G_0 value significant attenuation can

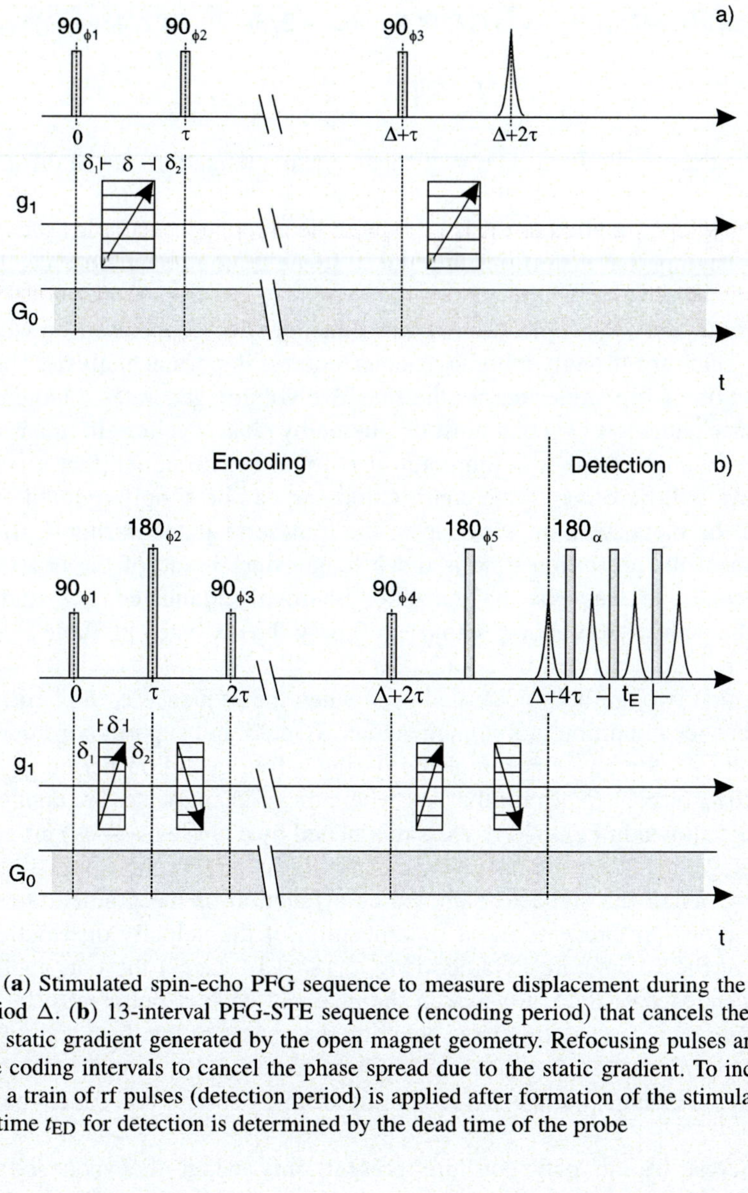

Fig. 5.17 (**a**) Stimulated spin-echo PFG sequence to measure displacement during the free evolution period Δ. (**b**) 13-interval PFG-STE sequence (encoding period) that cancels the effect of the strong static gradient generated by the open magnet geometry. Refocusing pulses are applied during the coding intervals to cancel the phase spread due to the static gradient. To increase the sensitivity a train of rf pulses (detection period) is applied after formation of the stimulated echo. The echo time t_{ED} for detection is determined by the dead time of the probe

be expected [15, 18]. This fact complicates, and in some cases prevents, the use of this method for displacement encoding in the presence of background gradients. To reduce the diffusive signal attenuation due to G_0, Cotts et al. developed the so-called 13-interval sequence, denoted as encoding period in Fig. 5.17b [15]. It includes 180° pulses applied during the coding intervals to cancel the phase spread introduced by G_0 during each of these periods, while the encoding phase is defined by means of bipolar gradient pulses. For this sequence the signal attenuation due to molecular diffusion is

$$\ln(M(t)/M_0) = -\gamma^2 D\Big\{\delta^2(4\Delta + 6\tau - 2/3\delta)g_1^2 + 2\tau\delta(\delta_1 - \delta_2)g_1 G_0 + 4/3\tau^3 G_0^2\Big\}. \tag{5.2}$$

In contrast to the PFG-STE sequence, signal attenuation due to diffusion under G_0 arises only during the two encoding periods, but not during Δ. The sequence has been widely applied to measure flow in heterogeneous samples such as porous media showing that distortions introduced by field inhomogeneities can be considerably reduced.

Unfortunately, the application of this sequence with an open sensor, where the B_0 and B_1 fields are strongly inhomogeneous requires important analysis. The complex distribution of flip angles across the sensitive volume generates a large number of coherence pathways (3^5) that must be filtered by phase cycling. In order to keep the number of phase steps to a minimum it is important to notice that several of this pathways will have negligible contribution and can be simply ignored. This is the case of the magnetization evolving in the transverse plane during Δ. In a typical experiment this evolution time is much longer than τ and of the order of several milliseconds, so transverse magnetization is strongly attenuated by T_2 and diffusion. From the pathways evolving with $q_3 = 0$ only the six listed in Table 5.3 fulfill the echo condition described by Eq. (2.12).

The first two are both generated even when the sequence is implemented under on-resonance condition and homogeneous B_1 field and superimpose during signal detection. As the phase accumulated during a free evolution period q_k is proportional to q_k (see Chap. 2), for the case where the gradient pulses are applied with the polarities shown in Fig. 5.17b, P_1 is modulated by a phase $(\phi_\mathrm{i} - \phi_\mathrm{f})$ proportional to the average displacement and P_2 by a phase $(-\phi_\mathrm{i} - \phi_\mathrm{f})$ proportional to the average position. When the signal is sampled as a function of the gradient strength both contributions are present, and a superposition of the velocity distribution and the image of the sample are obtained (Fig. 5.18a). To cancel the contribution of P_2, the dependence of the pathways on the phase of each rf pulse must be exploited. For example, it can be eliminated stepping the phases of the third and fourth pulses from 0 to $\pi/2$ and keeping the receiver phase constant [19] (Fig. 5.18b). In this case, P_1 does not change, but the signal coming from P_2 shift π and cancel out after two scans.

The rest of the pathways are present only when the sequence is implemented under non-ideal experimental conditions, so their contribution to the final

Table 5.3 Pathways interfering during the acquisition period of the 13-interval sequence

Pathway	Phase
$P_1 = M_{0,+1,-1,0,-1,+1}$	$+\phi_1 - 2\phi_2 + \phi_3 - \phi_4 + 2\phi_5$
$P_2 = M_{0,-1,+1,0,-1,+1}$	$-\phi_1 + 2\phi_2 - \phi_3 - \phi_4 + 2\phi_5$
$P_3 = M_{0,-1,-1,0,+1,+1}$	$-\phi_1 + \phi_3 + \phi_4$
$P_4 = M_{0,-1,0,0,0,+1}$	$-\phi_1 + \phi_2 + \phi_5$
$P_5 = M_{0,0,-1,0,0,+1}$	$-\phi_2 + \phi_3 + \phi_5$
$P_6 = M_{0,0,0,0,-1,+1}$	$-\phi_4 + 2\phi_5$

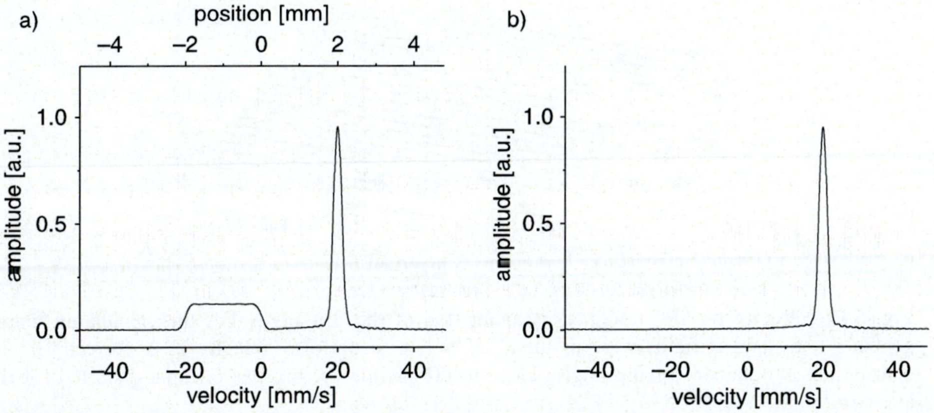

Fig. 5.18 (**a**) Simulated velocity profile for the 13-interval sequence (Fig. 5.17a). A single velocity of 20 mm/s inside a slice 2 mm thick and perpendicular to the flow direction was assumed. Not only a single velocity is obtained, but also the image of the slice is observed. (**b**) The unwanted image is completely removed by cycling the phases of the second rf pulse from 0 to $\pi/2$, keeping the received phase unchanged

signal depends on the particular sensor. To understand the performance of the pulse sequence, the signal response was calculated as a function of the effective flip angle (θ) assuming a single velocity v. This simulation cycles the phases needed to cancel the contribution of P_2, which at the same time eliminates P_3 (this path is not effectively encoded by the bipolar gradients because the 180° pulses applied in each coding period do not invert the sign of the accumulated phase). Figure 5.19 shows the velocity distribution obtained as a function of the flip angle. For θ values close to 90° the profile is undistorted and the single velocity is at the right spectral position. Nevertheless, when θ deviates from the ideal value (because either of B_1 or B_{off} variations), a spurious contribution at half the original velocity value is obtained. This peak is the magnetization encoded by only one of the two gradient pulses during each coding period, which is modulated with half of the correct frequency in the q-space.

The pathway responsible for this spurious signal is the P_5, which is encoded with a phase ($\phi_{\text{ini}}/2 - \phi_{\text{fin}}/2$). (As $q_2 = -1$ a positive initial phase is acquired during q_2 where the gradient pulse is negative, but a negative phase is acquired during the fifth period where $q_5 = +1$ and the gradient pulse is negative). In order to quantify the importance of this distortion for the conditions of a real experiment we included in the calculations the off-resonance excitation and the real B_1 distribution of the rf coil. Figure 5.20a shows the obtained velocity distribution. The peak at $v/2$ is clearly observed and its relative amplitude is almost 50% compared to the amplitude at the correct velocity. As observed in Fig. 5.19, the distortion is the largest for flip angles close to $\pi/4$. For this value a large component of the magnetization remains along the z-axis after the application of the first rf pulse and is not encoded by the first gradient pulse of the encoding period. This remnant component is eliminated by cycling the phase of the first rf pulse and the phase of the receiver from 0 to π

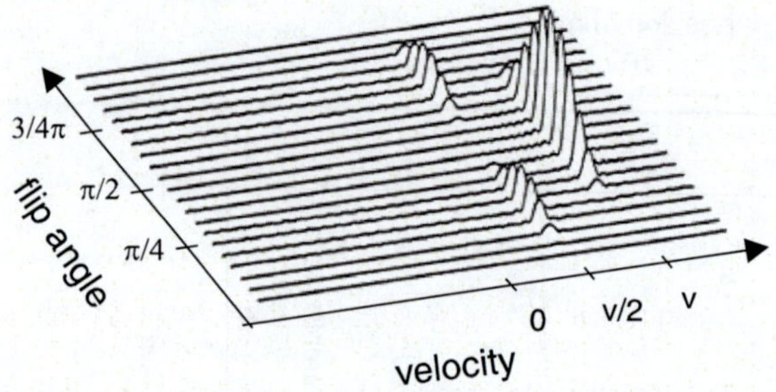

Fig. 5.19 Velocity profiles obtained as a function of the flip angle. For the calculation homogeneous static and rf fields were assumed as well as a uniform velocity v. A correct velocity distribution is obtained for flip angles close to 90°, while the profiles become distorted for flip angles close to 45 and 135°

(add–subtract phase cycling) [19]. This phase cycle also eliminates P_6 and the fresh magnetization created during Δ. Figure 5.20b shows the velocity profile obtained by including this phase cycle in the simulation. The distortion is clearly absent. To remove the remaining pathway P_4, which is encoded with a phase proportional to half of the average position $(-\phi_{\text{ini}}/2-\phi_{\text{fin}}/2)$, ϕ_2 must be cycled from $\pi/2$ to $-\pi/2$ keeping the receiver phase constant. However, it does not appear to introduce big distortions, so these steps are not included in Table 5.4.

Besides the mentioned distortions that appear as a consequence of the field inhomogeneity, the use of this technique is limited by the poor sensitivity of single-sided sensors. In Sect. 5.2.2 the use of the multi-echo acquisition scheme CMPG-CP for sensitivity improvement was described. Figure 5.17b shows that it can also be attached to the 13-interval sequence (encoding period). As already mentioned, the

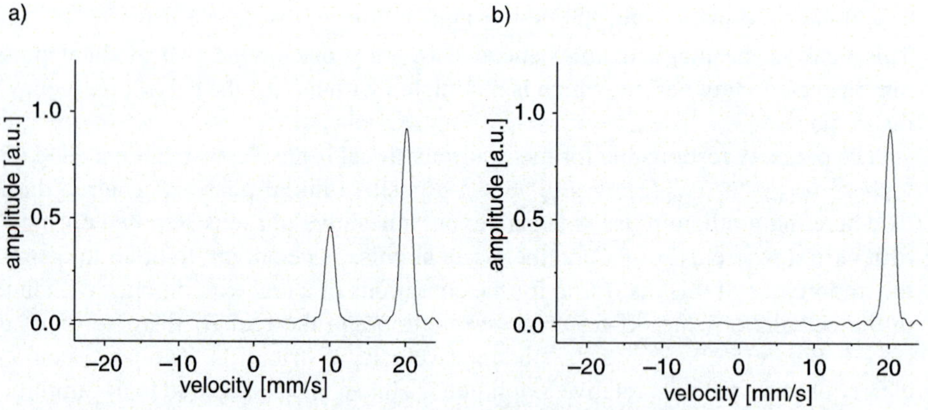

Fig. 5.20 (**a**) Velocity profiles calculated considering both the off-resonance due to the static gradient of the sensor and the B_1 field distribution provided by the rf coil. Although a single velocity of 20 mm/s was assumed in the simulation, a contribution at half of this value is also observed. (**b**) To effectively cancel this distortion, the phase of the first rf pulse and the receiver phase were cycled from 0 to π

Table 5.4 Phase cycle to filter unwanted pathways generated when the 13-interval sequence is applied in inhomogeneous fields

ϕ_1	ϕ_2	ϕ_3	ϕ_4	ϕ_5	ϕ_α	ϕ_{rec}
0	$\pi/2$	0	0	$\pi/2$	$\pi/2$	0
0	$\pi/2$	$\pi/2$	$\pi/2$	$\pi/2$	$\pi/2$	0
π	$\pi/2$	0	0	$\pi/2$	$\pi/2$	π
π	$\pi/2$	$\pi/2$	$\pi/2$	$\pi/2$	$\pi/2$	π
0	$\pi/2$	0	0	$\pi/2$	0	0
0	$\pi/2$	$\pi/2$	$\pi/2$	$\pi/2$	0	0
π	$\pi/2$	0	0	$\pi/2$	0	π
π	$\pi/2$	$\pi/2$	$\pi/2$	$\pi/2$	0	π

refocusing pulses only preserve the component of the echo signal parallel to the refocusing rf pulses, while the perpendicular component goes to zero after a transient period. For the reconstruction of the velocity distribution both magnetization components are necessary and the loss of one of them mirrors the spectrum. To sample both components two experiments are performed, switching the phase of the rf pulses of the refocusing train from 0 to $\pi/2$. In summary, the total experiment requires a minimum of eight scans per gradient step. The phase of the first rf pulse must be cycled from 0 to π to eliminate distortions due to the flip angle distribution, the phase of the third and fourth rf pulses from 0 to $\pi/2$ to eliminate the unwanted image from the velocity profile, and the sequence must be repeated changing the phase of the refocusing train to sample both components of the echo signal (see Table 5.4). It should be noticed that under strong background gradients and long enough Δ the unwanted pathways can be strongly attenuated because for them the 13-interval sequence is actually a stimulated echo sequence (only three of the five pulses act on the magnetization) where diffusion attenuation during Δ is observed.

5.5.2 Measurement of Velocity Distributions

The new flow-encoding method was implemented on the open tomograph described in Chap. 4. Its efficiency to remove the effect of the background gradient was tested first. The sequence was implemented without gradient pulses, and the echo intensity from water doped with Cu_2S was sampled as a function of Δ for different δ values. The acquired decays could be fitted by a mono-exponential function with a time constant independent of δ and equal to T_1 in all cases. This result proves that the stimulated echo stored as longitudinal magnetization during Δ is free of phase contributions introduced by the static gradient.

To illustrate the performance of the method, the water flow in a thin rectangular pipe was measured. The tube, 12 mm wide, 0.6 mm high, and 200 mm long, was placed at 10 mm from the sensor surface. When the ratio between the height a and the width b of the tube is $b/a \ll 1$, the velocity profile of laminar flow can be considered to be flat along the long side b while it changes quadratically with the height a [20]. A volume flow rate Q of 170 ml/h was driven by a precision pump Pharmacia

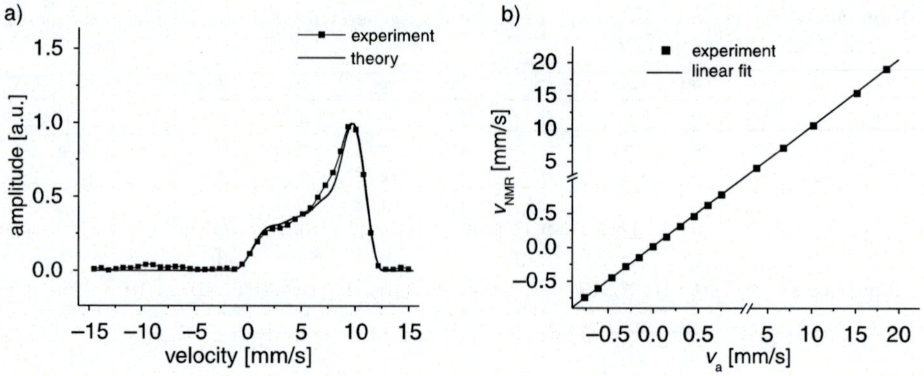

Fig. 5.21 (**a**) Velocity distribution of water flowing in a rectangular pipe obtained with the pulse sequence of Fig. 5.17b. *Dots* denote experimental data. The water used for the experiment was doped with Cu_2S to reduce the T_1 value to 0.9 s. The gradient pulse length δ was set to 0.5 ms and the evolution period Δ to 200 ms. To improve the signal-to-noise ratio 2,000 echoes were acquired with $t_{ED} = 0.11$ ms. The total experimental time to obtain the velocity distribution working at 10 mm from the sensor surface was 5 min. For comparison the theoretical distribution was calculated and is shown as a *continuous line*. (**b**) Correlation between the average velocity v_{NMR} measured using the single-step phase-encoding PFG-STE method with the value v_a, determined from the known flow rate and the geometry of the tube

P500 defining a maximum velocity $v_m = 3/2 \times Q/(ab)$ of about 10 mm/s. The dotted velocity profile in Fig. 5.21a was measured with the sequence of Fig. 5.17b. The q-space was sampled in 24 steps from negative to positive values and with the field of flow set to 30 mm/s, a velocity resolution of about 1.25 mm/s was defined. To excite all the spins across the tube, the length of rf pulses was set to 7 μs defining a slice thickness of about 1.4 mm. The shortest echo time $t_{ED} = 0.11$ ms, limited by the dead time of the probe, was employed. A train of 2,000 echoes was added to improve the sensitivity by a factor of 20 compared to single-echo detection [19]. Thanks to this dramatic sensitivity enhancement, only eight scans were required, determining a total time of about 5 min to measure the velocity distribution. The theoretical velocity profile for this pipe geometry was calculated and is shown in Fig. 5.21a as well. The good agreement between experimental and theoretical curves proves that the method is suitable to remotely measure velocity distributions, even in these extremely inhomogeneous fields.

Single-sided sensors are usually repositioned to scan the object at different spots. In some applications only the average velocity at each spot is required. When this is the case a single phase-encoding step can be used instead of the N steps needed to reconstruct the full propagator [21]. The sacrifice in velocity resolution leads to a direct reduction in experimental time by a factor of $N/2$. The procedure requires an initial experiment without gradient pulses to measure the reference phase of the echo signal, and a second experiment with the gradient pulses to introduce a phase shift in the echo proportional to the displacement. From this phase shift the average velocity inside the sensitive spot can be easily computed. The method was implemented to measure the average velocity $v_a = Q/(ab)$ in the rectangular pipe. Figure 5.21b shows the average velocity v_{NMR} measured with the single-step PFG-STE method as a function of the average velocity calculated via the known flow rate. The excellent correlation proves that the method is accurate in a wide velocity range. The time to measure the average velocity inside the selected volume, under the same conditions as in the previous experiment, was about 24 s.

5.6 Spatially Resolved Velocity Distributions

In this section the method developed to encode velocity is combined with a slice selection procedure to spatially resolve the velocity profile in the object along the depth direction. Flat slices at different depths can be selected by stepping the excitation frequency as was shown in the previous sections (Fig 5.16). To illustrate the performance of the method, the profile of water flowing in a rectangular pipe 30 mm wide, and 3 mm high was measured. A volume flow rate of 3,600 ml/h was driven by a precision pump, defining a maximum velocity in the pipe of about 17 mm/s. When the ratio b/a between the height and the width of the tube is much larger than 1, the velocity of laminar flow can be considered to be constant along the long side b while it changes as a quadratic function of the height a. The pipe was scanned along y by sweeping the position of the sensitive slice in steps of 0.3 mm from 7 to 10 mm, where the tube was placed. Figure 5.22a shows the spatially resolved velocity distribution where the parabolic profile is clearly visible [22]. Figure 5.22b shows the propagators measured at three different positions in the tube. They show how the averaged velocity in the slice goes to zero as the slice approaches the wall, while the width of the distributions increases, in agreement with the change in the local shear value.

As a second example the velocity distribution of water flowing in a circular pipe with 3 mm inner diameter was scanned with the same experimental parameters as before. Figure 5.22c shows the propagator spatially resolved along y [22]. In contrast to the rectangular geometry, for a circular pipe of radius R the velocity depends on both y and z as $v(x, y, z) = v_{\max} \times (1 - (y^2 + z^2)/R^2)$. This defines a velocity distribution in each slice with velocity ranging from zero to a maximum value that depends on the y-position. The parabolic behavior of the maximum velocity is clearly resolved in Fig. 5.22c, where it can be also observed that the velocity probability is non-zero from zero up to the maximum velocity in the slice. The complete propagator of the tube can be reconstructed by integrating the distributions along the depth direction. Figure 5.22d shows the characteristic hat function distribution expected for this geometry.

5.6.1 2D Velocity Maps

Once a slice is selected, the velocity distribution can be resolved along a lateral direction by including a pulsed field gradient g_z in the sequence (Fig. 5.23) [22]. Stepping the amplitudes of g_1 (in this case applied along x) and g_z independently in

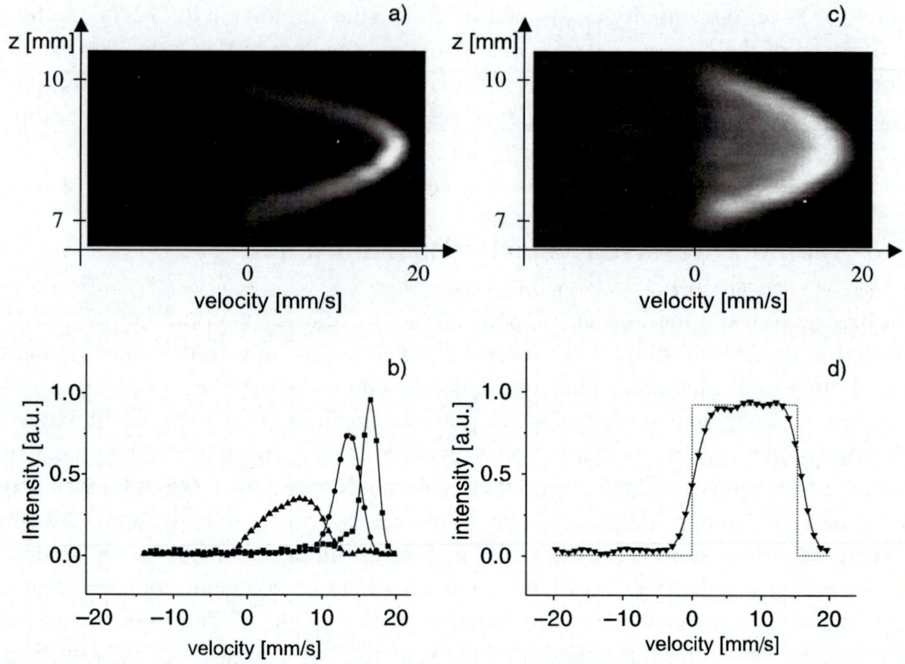

Fig. 5.22 (**a**) Velocity distribution of water undergoing laminar flow in a rectangular pipe 3 mm high and 30 mm wide spatially resolved along y. The velocity distribution of each slice was measured with the sequence of Fig. 4.1b using $\delta = 1$ ms, $\Delta = 50$ ms, FoF $= 40$ mm/s, 20 gradient steps, and eight scans per gradient value. To improve the signal-to-noise ratio 2,000 echoes were acquired with $t_{ED} = 0.11$ ms. The total experimental time to resolve the velocity distribution was about 30 min. The position $y = 0$ in the plot is at 7 mm from the surface of the sensor and corresponds to the position of the lower wall of the tube. (**b**) Propagators measured at the center (▲), between the center and the wall (•), and next to the wall (■). One can clearly see how the central velocity goes to zero when approaching the wall, while the width of the velocity distribution increases according to the increase in the velocity shear. (**c**) Velocity distribution of water flowing in a circular pipe, spatially resolved along y. The experimental parameters were the same as the ones used for the rectangular pipe. The flow rate was set to 190 ml/h defining a maximum velocity of 15 mm/s. The maximum velocity as a function of y shows the expected parabolic behavior. (**d**) Velocity distribution of the complete circular pipe showing the typical hat function corresponding to this geometry. The total velocity distribution was obtained by integration of the velocity profile (**c**) along y

a 2D experiment the velocity distribution in each pixel along z can be determined. By combining this method with the slice selection procedure the velocity distribution in each pixel of the 2D image can be obtained. This is a time-consuming 3D experiment. In general, a uniform velocity inside each pixel can be assumed, so that a single phase-encoding step can be used for g_1 instead of the N steps needed to reconstruct the full propagator [21]. The procedure requires two imaging experiments with different g_1 amplitudes, thus in each pixel of the 2D image a phase shift proportional to the displacement is introduced. From the phase shift the velocity in each pixel can be easily computed to reconstruct a 2D velocity map of the cross section perpendicular to the flow direction. As for the propagator measurement,

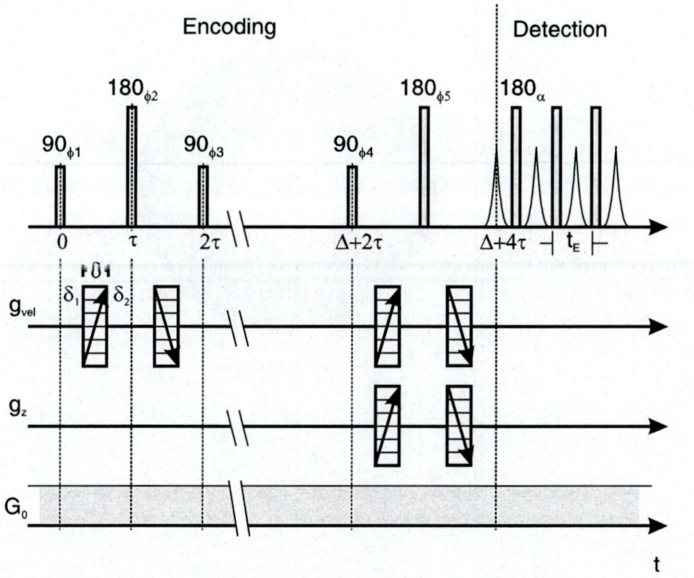

Fig. 5.23 Thirteen-interval stimulated spin-echo PFG sequence to measure displacement during the free evolution period Δ. The sequence includes a bipolar gradient pulse applied before the detection train to obtain spatial resolution along *z*

a minimum of eight scans is required for the phase cycle to remove the distortions and to sample the two components of the signal.

The method was implemented to map the velocity profile of a laminar water flow in a circular pipe with 6 mm inner diameter. The flow rate was set to 500 ml/h to define a maximum velocity of about 10 mm/s. The pipe was scanned along *y* moving the position of the sensitive slice in steps of 0.5 mm. A 1D profile along *z* of each slice was obtained by increasing the amplitude of the pulsed gradient in 16 steps, setting a FoV of 8 mm, the same spatial resolution along both spatial directions was defined. Figure 5.24a shows the 2D velocity map computed from the pixel phase shift measured with the single-step PFG-STE method. The plot clearly shows the annular pattern corresponding to the parabolic profile developed in this geometry. The agreement of the results with the theoretical profile can be judged in Fig. 5.24b, where the profiles along *y* and *z* are displayed together with the expected quadratic function.

The detailed information about fluid transport properties provided non-invasively, together with the simplicity of the measurement procedure, identifies single-sided NMR as a suitable technique to characterize flowing characteristics of complex fluids. In this chapter, an alternate PFG-STE sequence suitable to encode displacement in the presence of strong B_0 and B_1 field gradients has been presented. It has been successfully implemented on a single-sided NMR sensor to measure velocity distributions *ex situ*. As for imaging experiments, a key step to measure velocity *ex situ* is the incorporation of the multi-echo acquisition scheme following the displacement-encoding period.

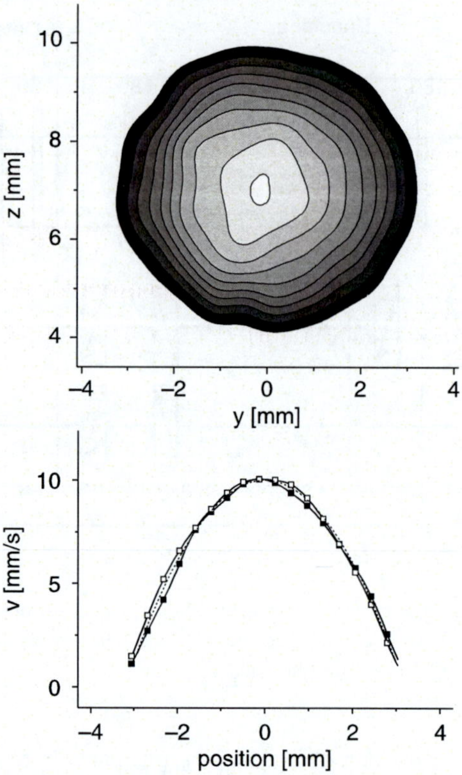

Fig. 5.24 (**a**) 2D velocity map of water flowing in a circular pipe, 6 mm in diameter. It was obtained with the pulse sequence of Fig. 5.23, using the single gradient step method to encode velocity. The total experimental time to obtain the velocity distribution with the tube placed between 4 and 10 mm from the sensor surface was 50 min. (**b**) Velocity profiles along *y* (□) and *z* (■) plotted together with the expected quadratic function (*dashed line*)

References

1. Callaghan PT (1991) Principles of nuclear magnatic resonance microscopy. Clarendon Press, Oxford
2. Blümich B (2000) NMR imaging of materials. Clarendon Press, Oxford
3. Perlo J, Casanova F, Blümich B (2005, Sep) Profiles with microscopic resolution by single-sided NMR. J Magn Reson 176(1):64–70
4. Prado PJ, Blümich B, Schmitz U (2000, June) One-dimensional imaging with a palm-size probe. J Magn Reson 144(2):200–206
5. Casanova F, Blümich B (2003, July) Two-dimensional imaging with a single-sided NMR probe. J Magn Reson 163(1):38–45
6. Meiboom S, Gill D (1958) Modified spin-echo method for measuring nuclear relaxation times. Rev Sci Instrum 29(8):688–691
7. Casanova F, Perlo J, Blümich B, Kremer K (2004, Jan) Multi-echo imaging in highly inhomogeneous magnetic fields. J Magn Reson 166(1):76–81
8. Perlo J, Casanova F, Blümich B (2004, Feb) 3D imaging with a single-sided sensor: an open tomograph. J Magn Reson 166(2):228–235

9. Xia Y, Callaghan P T, Jeffrey KR (1992, Sep) Imaging velocity profiles – flow through an abrupt contraction and expansion. Aiche J 38(9):1408–1420
10. Packer KJ, Tessier JJ (1996, Feb) The characterization of fluid transport in a porous solid by pulsed gradient stimulated echo NMR. Mol Phys 87(2):267–272
11. Arola DF, Powell RL, Barrall GA, McCarthy MJ (1999, Jan) Pointwise observations for rheological characterization using nuclear magnetic resonance imaging. J Rheol 43(1):9–30
12. Yeow YL, Taylor JW (2002, Mar) Obtaining the shear rate profile of steady laminar tube flow of newtonian and non-newtonian fluids from nuclear magnetic resonance imaging and laser doppler velocimetry data. J Rheol 46(2):351–365
13. Xia Y, Callaghan PT (1991, Aug) Study of shear thinning in high polymer-solution using dynamic NMR microscopy. Macromolecules 24(17):4777–4786
14. Tanner JE (1970) Use of stimulated echo in NMR-diffusion studies. J Chem Phys 52(5):2523–2526
15. Cotts RM, Hoch MJR, Sun T, Markert JT (1989, June) Pulsed field gradient stimulated echo methods for improved NMR diffusion measurements in heterogeneous systems. J Magn Reson 83(2):252–266.
16. Stejskal EO, Tanner JE (1965) Spin diffusion measurements – spin echoes in presence of a time-dependent field gradient. J Chem Phys 42(1):288–292
17. Kimmich R (1997) NMR: tomography, diffusometry, relaxometry. Springer, Berlin
18. Sun PZ, Seland JG, Cory D (2003, Apr) Background gradient suppression in pulsed gradient stimulated echo measurements. J Magn Reson 161(2):168–173
19. Casanova F, Perlo J, Blümich B (2004, Nov) Velocity distributions remotely measured with a single-sided NMR sensor. J Magn Reson 171(1):124–130
20. Gondret P, Rakotomalala N, Rabaud M, Salin D, Watzky P (1997, June) Viscous parallel flows in finite aspect ratio hele-shaw cell: analytical and numerical results. Phys Fluids 9(6):1841–1843
21. Xia Y, Callaghan PT (1992, Jan) One-shot velocity microscopy – NMR imaging of motion using a single phase-encoding step. Magn Reson Med 23(1):138–153
22. Perlo J, Casanova F, Blümich B (2005, Apr) Velocity imaging by ex situ NMR. J Magn Reson 173(2):254–258

Chapter 6
High-Resolution NMR in Inhomogeneous Fields

Vasiliki Demas, John M. Franck, Jeffrey A. Reimer, and Alexander Pines

6.1 Introduction

During recent years, portable and single-sided NMR devices and methodologies have drawn considerable attention. The ability to conduct NMR with portable and inexpensive systems, or *ex situ*, enhances a variety of fields and applications. For example, such NMR sensors enable scanning in the field and would provide access to immovable, arbitrary-sized objects. Presently, devices such as the NMR-MOUSE [1–6] and other single-sided or inside-out sensors [7–13] are used to reconstruct images or to determine the sample relaxation times, despite the strongly inhomogeneous fields of these devices. Such measurements generate information for applications ranging from materials and tissue evaluation [14–17] to flow measurements [18, 19], art preservation [20, 21], and well logging [22].[1] However, inhomogeneous field variations exceed chemical shift effects by orders of magnitude, so no chemical shift information can be extracted from the spectra. Therefore, the unavoidable spatial inhomogeneity of the static magnetic field has precluded the use of such devices for high-resolution spectroscopy.

An approach that circumvents the inherent field inhomogeneity present in single-sided systems makes use of the Earth's magnetic field [23]. Experiments in Earth's field NMR have been carried out and have measured relaxation, images [24–26], and brine diffusion through ice [23, 27–29] and other media [30, 12] and have detected ground-water [31–33]; recently, such experiments enabled high-resolution proton, lithium, and fluorine J-coupling spectroscopy [34, 64]. Unfortunately, the low intensity of the Earth's field results in a very poor nuclear spin polarization. It also hinders any kind of chemical shift-resolved spectroscopy. Furthermore, the Earth's field retains its homogeneity only in remote areas (far away from buildings, cars, etc.) which obviously results in additional practical complications.

V. Demas, J.M. Franck, J.A. Reimer, and A. Pines (✉)
University of California, Berkeley, CA 94720, USA

[1] The contributions to the field by oil companies such as, for example, Schlumberger-Doll and Chevron are immense.

F. Casanova et al. (eds.), *Single-Sided NMR*,
DOI 10.1007/978-3-642-16307-4_6, © Springer-Verlag Berlin Heidelberg 2011

Although the use of shimming hardware still remains a valid approach to the problem of field inhomogeneity in single-sided magnets, this chapter focuses on a number of schemes that rather rely on active spin manipulation after excitation. Such "*ex situ*" methodologies compensate for the dephasing of the spin ensemble that occurs between discrete acquisition points. They do this by periodically employing specialized radio frequency or gradient field waveforms [35–37]. The resulting signal retains chemical shift information, but the lines of the spectrum remain sharp, unaffected by broadening from inhomogeneous dephasing.

To date, only a handful of approaches have yielded reasonable means of recovering spectral information in the presence of strong field inhomogeneities [38–41]. Before the introduction of "*ex situ*" methodologies, it was considered impossible to refocus inhomogeneous dispersion without destroying the chemical shift information. Here we describe the use of simple composite pulse and adiabatic pulse versions of "hardware matching" pulses, as well as the highly flexible "shim pulses."

In addition to *ex situ* methods, a newer method, ultrafast NMR spectroscopy, can accommodate – at the expense of signal to noise – straightforward inhomogeneous compensation, also without the sophisticated hardware required by older methods.

Ex situ remains an active area of development. NMR spectroscopy and spectroscopic imaging information have been obtained in emulated *ex situ* conditions [35, 37, 42–46] and the *ex situ* matching methodology [35] obtained the first high-resolution spectrum in a one-sided low-field system [47]. Further development of *ex situ* will permit field deployment of this robust and inexpensive version of magnetic resonance.

6.2 Approaches Based on Spin Interactions

In single-sided systems, inhomogeneities can prove fatal to spectroscopic methods. Even in high-field superconducting magnets, however, field inhomogeneity imposes one of the strictest limitations for high-resolution NMR. Therefore, several research groups over the past decades have investigated approaches to high-resolution NMR in inhomogeneous fields.

Refocusing pulse sequences, such as Hahn echoes [48], have proven themselves a tremendous discovery for NMR. When applied to inhomogeneous environments they refocus inhomogeneous broadenings, making possible the measurement of relaxation, diffusion, and homonuclear J-couplings. In fact, spin echo modulation as a function of interpulse spacing first indicated the existence of J-couplings, previously obscured by field inhomogeneities [49]. However, these sequences simultaneously remove chemical shift information and heteronuclear J-couplings.

An early approach removed inhomogeneous broadening by means of multiple quantum coherence transfer echoes [38, 50, 51]. This technique takes advantage of the total spin coherence, which evolves only under the Zeeman term. The creation of an echo usually requires the application of two excitation pulses. The first generates a coherence which defocuses due to inhomogeneities and the second inverts the accumulated effects of the inhomogeneous interaction and refocuses the signal. During the interval separating the two pulses, coherence can be transferred from one coherence level to another (from the maximum quantum coherence level, which is coherence N for a system of N coupled spins, to any other quantum coherence level n). The two coherences involved evolve under inhomogeneities at different rates, with that during the maximum quantum coherence allowing evolution only under the inhomogeneous Zeeman term. If the spins evolve for a period in the maximum quantum coherence level and then at any other level, the term due to static field inhomogeneities and susceptibility broadening in the n quantum spectrum will vanish [38, 51]. The requirement of a total spin coherence limits the applicability of this technique, as does the fact that it does not excite the lines of uncoupled systems.

A second multiple quantum-based approach capitalizes on intermolecular zero-quantum coherences (iZQCs) in solution [52]. iZQCs give high-resolution spectroscopic information in inhomogeneous fields so long as the field varies negligibly over the dipolar correlation length scale (about $10\,\mu m$) [39]. The HOMOGENIZED detection method [39, 53] developed in the laboratory of W. S. Warren[2] worked in magnets with fields up to 18 T with small inhomogeneities (on the order of 1 ppm). HOMOGENIZED gave high-resolution proton spectra of liquid mixtures [39, 41]. Unfortunately a few drawbacks prevent the routine application of this technique to *ex situ* systems. Primarily, the necessary spin coupling must be achieved via distant dipole–dipole interactions. The signal comes from iZQCs within some correlation distance [54–56] that depends on the pulsed field gradients applied during the pulse sequence. Larger inhomogeneities require a smaller correlation distance, which yields a smaller signal. In addition, typical signals are 5% of the conventional single-quantum signals. Already low signal levels in *ex situ* systems (typical field strengths less than 1–2 T and field gradients of 0.1–1 T/m) would make iZQCs impossible to observe.

In yet another approach, the nuclear Overhauser effect (NOE) between solvent and solute yields high resolution. One-dimensional experiments utilize a hole burning approach, limiting the amount of signal to the portion of spins with "correct" frequencies. Therefore, for a larger correction, the amount of signal will decrease. Two-dimensional experiments use hetero-NOESY or HOESY sequences [40]. NOE-based approaches deal with small inhomogeneities (like those due to susceptibility variations on a fine scale) that can be "filtered out" by means of NOE. The main drawback for NOE techniques is that the strength of the solvent–solute NOE can be weak and depends on the longitudinal relaxation of solvent and solute.

6.3 *Ex Situ* NMR: Spatially Dependent "*z*-Rotations"

Spin interaction techniques have facilitated spectral resolution improvements despite the presence of small static field inhomogeneities. However, all of these

[2] HOMOGENIZED is similar to the CRAZED sequence also developed by Warren et al. [57].

methods would fail in the weak, highly inhomogeneous fields present in systems of interest for practical situations. Therefore, we proceed to describe a series of approaches, developed in recent years, that allow high-resolution NMR in such *ex situ* systems. The *ex situ* procedure generally requires a field map of the inhomogeneities (by MRI techniques or Hall probes). Pulse sequences then employ the information from the field map to refocus the effects of the inhomogeneities and yield a corrected FID and narrowed spectral lines.

The different *ex situ* methodologies utilize rf, static field gradient pulses, or combinations of both to impart position-dependent phase corrections (position-dependent z-rotations) to the sample. These corrections are independent of chemical shift and scalar couplings; therefore, this information is preserved. Currently, such methodologies fall into two main categories: hardware matching techniques, which occur in both composite and adiabatic pulse variants, and shim pulses, including those based on both the adiabatic double passage [37] and chirp pulses [36].

Ex situ methodologies significantly decrease the precision required of magnet design, though hardware design and pulse sequence design do still remain interdependent. In general, the magnet design (since not all open magnets will be tractable) will dictate the ideal pulse scheme, which should incorporate rf field matching, rf efficiency, short length, and broadband behavior, together with adjustability. In practice, the field in *ex situ* environments varies dramatically with a nonlinear profile. In the event that the available rf power allows the pulses to act over a large inhomogeneity bandwidth, composite pulses can perform very well in the presence of large gradients. The following factors determine the corrective strength of *ex situ* methodologies:

- In the case of hardware matching, the rf to static field matching quality.
- Rate of applied correction – this depends on the gradient strength, the rf amplitude, and the pulse design. Everything else equal, the stronger the rf power, the faster the pulse-based correction.
- The offset dependence of the pulse.
- The rates of diffusive and stochastic (i.e., T_2) dephasing that compete with the rephasing induced by the refocusing pulses.

More inhomogeneous magnets require larger corrections and typically more rf power, while in the case of hardware matching more care in matching both fields needs to be given. Large rf power is not, however, desirable or frequently available in portable sensors, and matching can become as tedious as conventional shimming. Pulse sequence design can help overcome this limitation; adiabatic pulses operate over extremely large bandwidths without the need for a very high rf power, thereby maximizing the efficiency of the available rf power. Pulse sequences based on adiabatic pulses are, however, detrimentally long (pulses on the order of milliseconds) and therefore sensitive to diffusion and relaxation. Pulse sequences work over a specific offset bandwidth; therefore, for large gradients only small portions of the sample yield desired information. A valid approach limits the signal acquisition – at the expense of signal to noise – to these regions. Specialized pulses (correlation-selective pulses) or hardware (separate coils for excitation and detection) might also perform such a volume selection. However, the mutual optimization of the magnet and rf coil geometry remains both a compatible and a far more desirable option.

6.3.1 Ex Situ Matching: Compensating Static Field Inhomogeneities via Spatially Matched rf

The first *ex situ* correction reported was that of *ex situ* hardware matching. Meriles et al. [35] constructed an rf coil generating a field with the same spatial gradient as the static field. They then employed composite rf pulses to induce a rotation about the z-axis, contrary to the rotation induced by the static field during evolution. The spatial variation of the rf rotation canceled the spatial variation of the rotation induced by the static field (similar to the effects observed in [65]). For the first time a sample in the presence of a static field gradient generated a signal yielding a high-resolution NMR spectrum. Figure 6.1a shows the "standard" one-pulse NMR spectrum in a homogeneous B_0 field where five peaks are resolved. In the presence of a B_0 gradient of 0.12 mT/cm, the inhomogeneously broadened spectrum is rendered featureless and extends over a range of 20 ppm (Fig. 6.1b). The spectrum obtained by the *ex situ* sequence under the same conditions resolves all five proton NMR peaks (Fig. 6.1c).

6.3.1.1 Composite "z-Rotation" Pulses and Nutation Echoes

Initial *ex situ* experiments employed pulses, designed according to a simple graphical model, to generate spatially dependent z-rotations. In this scheme, the magnetization initially spreads over the yz-plane, driven by a β_x pulse of duration τ_β.[3] A $\pi/2_y$ pulse then tips the magnetization into the transverse plane before it freely evolves for a period τ. This initial composite z-rotation pulse both excites and z-rotates the spins. At the end of this period, a "nutation echo" forms, a single point is acquired, another constant-rotation composite $\pi/2$ pulse tips the magnetization

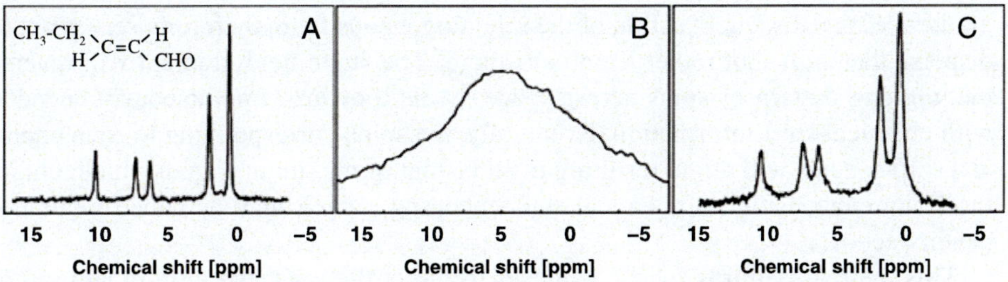

Fig. 6.1 From *left* to *right*: **(a)** A standard NMR spectrum corresponds to the "chemical fingerprint" of the sample. **(b)** In the presence of inhomogeneities, the spectrum is broadened and the chemical shift information obscured. **(c)** The lost information is recovered via "*ex situ*" matching (from [35])

[3] β is a space dependent pulse, $\gamma B_1(x) \cdot \tau_\beta = \phi_{10} + \Delta\phi_1(x)$.

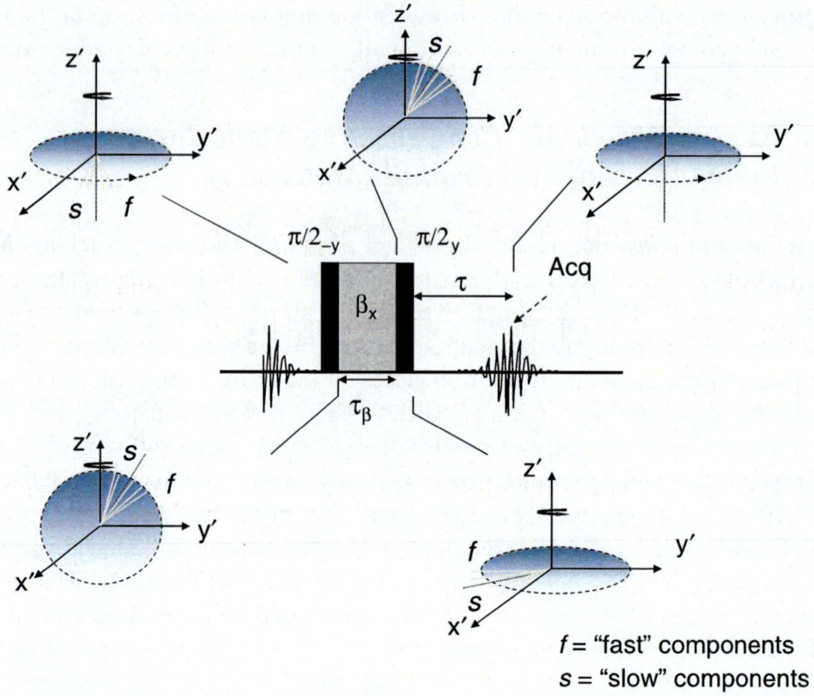

Fig. 6.2 **(a)** A β pulse gives a position–dependent $\Delta\omega_1$ spread. A hard $\frac{\pi}{2}$ pulse then tips the magnetization on the transverse plane and a nutation echo is formed at time τ proportional to the length of the β pulse and κ, the ratio between $\Delta\omega_1$ and $\Delta\omega_0$. **(b)** A series of such z-rotation pulses can be applied to acquire stroboscopically an FID that is modulated by chemical shift

back onto the yz-plane, and the cycle repeats. In this fashion, the entire effectively homogeneous FID can be acquired through stroboscopic acquisition (Fig. 6.2).

The z-rotation pulse, therefore, deposits the spins out of phase. Along the direction of the static field gradient, they are rotated increasingly further away from the x-axis; that is, they are wound into a spiral along the direction of the static field gradient. The subsequent action of the inhomogeneous field, therefore, serves not to disperse the spins, but rather to refocus them. The static field gradient will reverse the winding pattern of spins across space, while they are simultaneously encoded with chemical shift information. Eventually, the spins corresponding to each chemical shift species will align, resulting in an instant of maximum signal which, unlike the Hahn echo, includes chemical shift encoding, which may be called a "homogeneity echo."

This method requires robust $\frac{\pi}{2}$ pulses to direct the magnetization in and out of the xy-plane in an ideal manner; the β pulse, however, exploits the rf inhomogeneity to encode a spatially dependent phase shift onto the spins. The high-field experiments manipulated composite pulses to minimize the detrimental effects of static and rf inhomogeneities in the $\frac{\pi}{2}$ pulse performance, as described in [42]. These "self-compensated z-rotation" *ex situ* pulses were based on the initial work by Levitt and Freeman [58], who introduced combinations of rf pulses to induce inversion of the z-magnetization over a wide range of rf and/or offset imperfections. As shown in [58] composite π pulses perform exactly like a π pulse applied on a selected axis in the xy-plane. The rf inhomogeneity acts to rotate the initial axis by the angle Δ. Because the axis of inversion depends on the rf, spatially variable phase shifts of the resultant signal can be attained if an appropriate rf coil delivers pulses crafted according to these concepts.

In the case of the low-field experiments as described in [47, 59], however, self-compensated z-rotations were not used. To achieve a significant change in the direction of this axis, the misset, Δ, must be somewhat large. The rf profile in these systems and the small volumes used make the relative rf variation negligible (i.e., the ratio of ΔB_1 to $B_{1,0}$ is small). The misset during the $\frac{\pi}{2}$ and π rotations produces only negligible effects.

Variations of the high-field matching experiments have been performed [42, 44] by a collaboration between the groups at the University of California, Berkeley, and the RWTH-Aachen University. This collaboration produced the first high-resolution NMR spectrum by a one-sided system [47]. In these experiments, the sequence $\beta - \frac{\pi}{2}$ – evolution was used. Changes to the duration of the β pulse vary the z-rotation angle, thus shifting the echo formation time. The chemical shift information can be obtained in an indirect dimension, consisting of consecutive scans with nutation echoes at different points along the FID. As was done in these experiments, this indirect detection setup allows a π pulse train to repeat the nutation echo (Fig. 6.3) [47]. The subsequent averaging of these nutation echoes provides an important gain in signal particularly crucial at such low fields.

The magnet employed for these studies generated a low field (0.2 T) at the surface – with a static gradient of 0.04 T/m. A main U-shaped permanent (NdFeB) magnet provided the primary static field. A magnet of opposing polarity shimmed the static magnetic field profile at the sample position. An rf coil designed to match the field dependence in 3D space generated all rf pulses (Fig. 6.4).

A series of field maps identified the position of the rf coil that corresponded to the best matching between the rf and static fields. The field maps were also used to

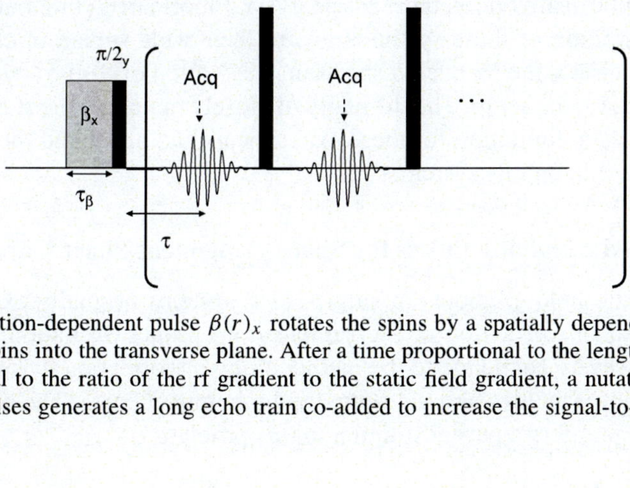

Fig. 6.3 A position-dependent pulse $\beta(r)_x$ rotates the spins by a spatially dependent angle. A $\frac{\pi}{2}$ pulse tips the spins into the transverse plane. After a time proportional to the length of the β pulse and proportional to the ratio of the rf gradient to the static field gradient, a nutation echo forms. A train of π pulses generates a long echo train co-added to increase the signal-to-noise ratio by a large factor

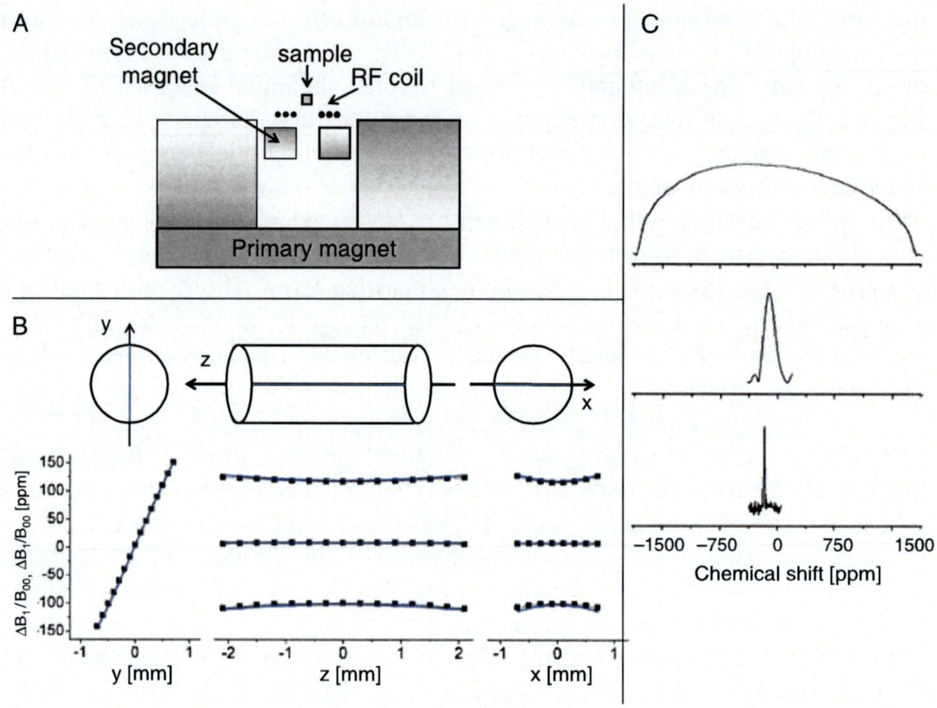

Fig. 6.4 **(a)** Magnet array schematic with permanent magnet element shimming. The inner or "shim" magnet opposes the field created by the outer magnet reducing the magnitude of the static gradient and defining a magnetic field with adjustable spatial dependency. **(b)** The surface coil and static field have a matched space dependence along three dimensions (for a range of a few millimeters). The two field maps around the sweet spot are shown **(c)** From *top*: spectra of C_6F_6 before and after optimizing the position of the inner magnet. The final spectrum is after using the refocusing sequence of the *upper inset*

adjust shimming permanent magnet units to produce a static field that best matched the field profile produced by the rf coil. A 1 mm diameter, 3 mm long capillary tube held a mixture of fluorinated compounds (hexafluorobenzene, perfluorohexane, and perfluorinated polyether) for analysis. Fluorinated compounds provide an easy demonstration of these methods due to their wide spread in chemical shifts. Figure 6.4 contrasts the recovered spectrum (resolution of 8 ppm) with the standard spectrum in the same setup. The difficulty of closely matching the rf and static fields imposed the main limitation for these experiments and prevented the acquisition of a proton spectrum on this system.

6.3.1.2 Scaled Adiabatic Pulses for Space-Dependent Phase Corrections

The large static field gradients in single-sided systems normally exclude all but a small, on-resonance fraction of the sample from proper excitation and encoding. This poses a severe challenge to the methodology of correction by spatial matching of the rf inhomogeneities and the static field inhomogeneities. The scaled adiabatic double passage offers a partial solution to this problem.

Adiabatic pulses behave different from most MR pulses, in that they act on all spins within their bandwidth in a fashion that is almost exactly uniform. A general adiabatic pulse is specially designed with time-dependent amplitude and frequency modulation that changes the direction of the effective magnetic field (the combined effect of offset in B_0 and $B_1(t)$, i.e., B_{eff}) adiabatically.

The scaled adiabatic double passage pulse sequence consists of two adiabatic full passage pulses with the same frequency ($\omega(t)$) and amplitude ($B_1(t)$) modulation, but with relative amplitude of the second full passage scaled by a factor λ. Scaled adiabatic double passages impose a phase correction that linearly depends on rf over a large offset range.

A relatively simple physical picture demonstrates how adiabatic passages function. The effective field during a pulse takes the simple form

$$B_{\mathrm{eff}}(t) = B_1(t)\hat{x} + \frac{\Delta\omega(t)}{\gamma}\hat{z} \tag{6.1}$$

where $\Delta\omega(t) = \omega(t) - \omega_0$ and $\hat{x}$, $\hat{z}$ are unit vectors in the rotating frame. B_{eff} begins aligned with the static field, then continuously deflects from the z-axis as the ratio between $B_1(t)$ and $\Delta\omega(t)$ changes until it comes to its final orientation, opposed to the direction of the static field. During a continuous excitation that varies sufficiently slowly (the rate of precession about B_{eff} far exceeds the rate of reorientation of B_{eff}), the spins that fall within the bandwidth of the pulse will always reorient with B_{eff} as it deflects.

The adiabatic *ex situ* matching pulse (scaled adiabatic double passage) employs adiabatic pulses in a unique fashion. Unlike the most common applications of adiabatic full passages, which are magnetization inversion, the adiabatic *ex situ* matching pulse acts on an initial state where the spins have already been excited into the transverse plane. In this context, the magnetization will precess about the effective field, locked into the plane perpendicular to it (i.e., the effective field takes the place of the static field as the direction of the quantization axis)[4] (Fig. 6.5). That is, in addition to inverting during a full passage, spins will also pick up a phase from precessing about the effective field. Since the first full passage effects inversion, a second, identical pulse would cancel the effects of the first. However, since the second pulse of the scaled adiabatic double passage has a different amplitude it applies a slightly different phase to the spins. Therefore, the second pulse fortunately fails to cancel some rf-dependent portion of the phase. This miscancelation leaves a net phase proportional to the local rf field. More importantly, this pulse sequence will apply this desired phase robustly and uniformly across a wide bandwidth of offsets (Fig. 6.6).

We can derive and analyze the specific effects of adiabatic pulses by means of a propagator analysis. During an adiabatic full passage, a rotation of angular fre-

[4] In order for the adiabatic pulses to perform the desired modulation of the effective field, the adiabatic condition should be met, i.e., $|\gamma B_{\mathrm{eff}}(t)| \gg |\Omega|$, where Ω is the angular frequency at which the direction of B_{eff} deflects.

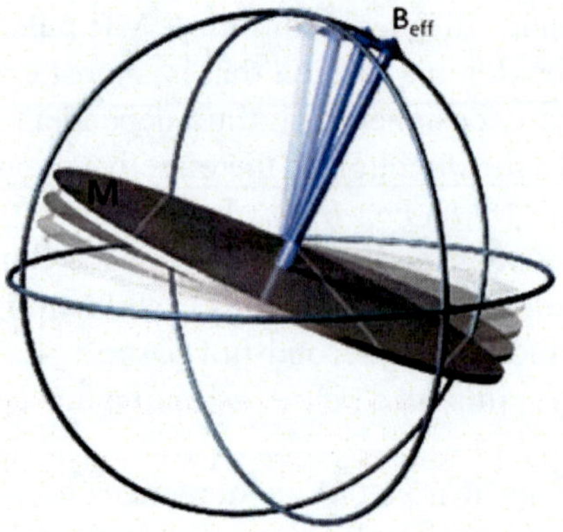

Fig. 6.5 A "snapshot" of the effective field shown by the *arrow* and attendant precessing nuclear magnetization, represented by a *disk*. The adiabatic pulses slowly rotate the effective field. The magnetization components in the transverse plane (represented by a *disk* here) remain in the plane perpendicular to the effective field and rotate around it acquiring a phase that can be easily calculated

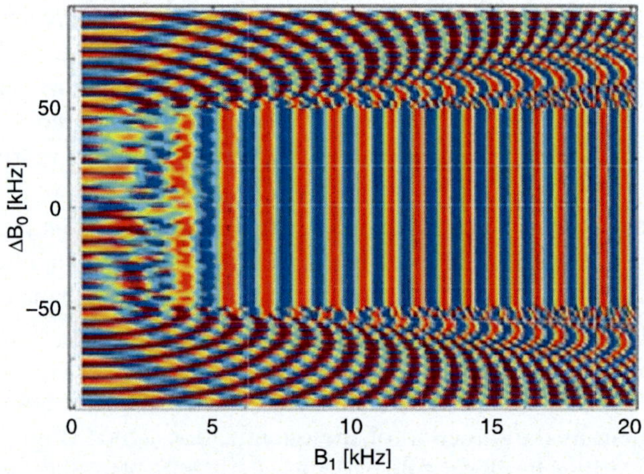

Fig. 6.6 Modulation of one of the transverse magnetization components by the adiabatic double passage. The two pulses have hyperbolic secant/hyperbolic tangent frequency and amplitude modulation and the relative maximum rf amplitudes are scaled by 0.5. The modulation is almost linear on rf amplitude and independent over a 100 kHz offset range (from [42])

quency $\omega_{\text{adia}}(t) = -\gamma B_{\text{eff}}(t)$ initially rotates spins about the z-axis. The axis of rotation then deflects at a rate Ω through the transverse plane and back onto the z-axis. We can more compactly represent this rotation by the Hamiltonian

$$\mathbf{H}(t) = e^{i\mathbf{I}_y \int_0^t \Omega(\tau)\tau} \omega_{\text{adia}}(t)\mathbf{I}_z e^{-i\mathbf{I}_y \int_0^t \Omega(\tau)\tau}. \tag{6.2}$$

A simple transformation will elucidate the actions of this Hamiltonian. We start by observing a general Hamiltonian for a spin state in one frame of reference $|\psi\rangle$, as well as a unitary operator $O(t)$ which transforms this state into a new frame of reference, in which the state is designated by $|\phi\rangle$. The Hamiltonian $\tilde{H}$ for the state $|\phi\rangle$ in the new frame of reference can be calculated as follows:

$$|\phi\rangle = \mathbf{O}(t)\,|\psi\rangle\,, \tag{6.3}$$

$$\frac{\partial\,|\phi\rangle}{\partial t} = \frac{\partial \mathbf{O}}{\partial t}\,|\psi\rangle + \mathbf{O}\frac{\partial\,|\psi\rangle}{\partial t}, \tag{6.4}$$

$$= \left(\frac{\partial \mathbf{O}}{\partial t} - i\mathbf{O}\mathbf{H}\right)\mathbf{O}^{+}\,|\phi\rangle\,, \tag{6.5}$$

$$= -i\tilde{\mathbf{H}}\,|\phi\rangle\,. \tag{6.6}$$

From Eq. (6.6), we can easily calculate the evolution under the Hamiltonian from (6.2) in the frame which "tracks" $\mathbf{B}_{\text{eff}}$, $O(t) = e^{-i\mathbf{I}_\mathbf{y}\int_0^t \Omega(\tau)d\tau}$.[5] A subsequent transformation back into the standard frame of reference yields the propagator for an adiabatic pulse:

$$e^{-i\mathbf{I}_\mathbf{y}\int_0^t \Omega(\tau)d\tau}\tilde{T}e^{i\int_0^t \tilde{\mathbf{H}}(\tau)d\tau} = e^{-i\mathbf{I}_\mathbf{y}\int_0^t \Omega(\tau)d\tau}\tilde{T}e^{-i\int_0^t \omega_{\text{adia}}\mathbf{I}_\mathbf{z}+\mathbf{I}_\mathbf{y}\Omega(\tau)d\tau}, \tag{6.7}$$

where $\tilde{T}$ is the Dyson operator which transforms all propagators that do not commute with themselves at all times into a correctly time-ordered product of infinitesimal time-slice propagators. In the case of adiabatic deflection (reorientation), this simplifies to

$$\lim_{\frac{\Omega}{\omega_{\text{adia}}}\to 0}\left(e^{-i\mathbf{I}_\mathbf{y}\int_0^t \Omega(\tau)d\tau}\tilde{T}e^{-i\int_0^t \mathbf{H}(\tau)d\tau}\right) = e^{-i\mathbf{I}_\mathbf{y}\int_0^t \Omega(\tau)d\tau}e^{-i\int_0^t(\omega_{\text{adia}}\mathbf{I}_\mathbf{z})d\tau}, \tag{6.8}$$

$$\equiv e^{-i\mathbf{I}_\mathbf{y}\int_0^t \Omega(\tau)d\tau}e^{-i\phi(t)\mathbf{I}_\mathbf{z}} \tag{6.9}$$

The effective Hamiltonian in this frame is

$$\tilde{\mathbf{H}}(t) = \omega_{\text{adia}}(t)\mathbf{I}_z - \Omega(t)\mathbf{I}_y, \tag{6.10}$$

which does not commute with it self at all points in time. However, in the limit that the reorientation frequency Ω is much smaller than the precession frequency ω_{adia}, the effect of the second term is negligible, and the propagator becomes

$$|\phi(t)\rangle = e^{-i\mathbf{I}_z\int_0^t \omega_{\text{adia}}(\tau)d\tau}\,|\phi(0)\rangle\,. \tag{6.11}$$

By incorporating a transformation back to the original rotating frame, the net propagator in this adiabatic limit becomes simply

[5] Adiabatic pulses "reorient" the quantization axis, as opposed to "hard" rf excitation pulses which transfer populations.

$$\mathbf{O}^{+}(t)\left|\varphi(t)\right\rangle = \mathbf{U}(t)\left|\psi(0)\right\rangle = \mathbf{O}^{+}(t)e^{-i\mathbf{I}_z\int_0^t \omega_{\mathrm{adia}}(\tau)d\tau}\mathbf{O}(0)\left|\psi(0)\right\rangle, \tag{6.12}$$

$$\mathbf{U}(t) = e^{-i\mathbf{I}_y\int_0^t \Omega(\tau)d\tau}e^{-i\mathbf{I}_z\int_0^t \omega_{\mathrm{adia}}(\tau)d\tau}, \tag{6.13}$$

$$\equiv e^{-i\mathbf{I}_y\int_0^t \Omega(\tau)d\tau}. \tag{6.14}$$

In the case of a full adiabatic passage, the hamiltonian sweeps through a full half-circle:

$$\int_0^{t_{\mathrm{pulse}}} \Omega(\tau)d\tau = \pi. \tag{6.15}$$

Therefore an "adiabatic double passage" generates a net propagator

$$\left(e^{-i\pi\mathbf{I}_y}e^{-i\phi_2(t)\mathbf{I}_z}\right)\left(e^{-i\pi\mathbf{I}_y}e^{-i\phi_1(t)\mathbf{I}_z}\right) = e^{i(\phi_1(t)-\phi_2(t))\mathbf{I}_z}, \tag{6.16}$$

where we can write the net rotation about $\mathbf{B}_{\mathrm{eff}}$ as

$$\phi(t) = \int_0^t \omega_{\mathrm{adia}}(\tau)d\tau. \tag{6.17}$$

The qualitative characteristics of adiabatic pulses remain relatively constant. The rf frequency starts above and ends below the center of the excitation bandwidth. Adiabatic pulses operate in a consistent manner across their bandwidth. The maximum difference between the carrier and center frequencies typically reaches 10–50 kHz. The frequency sweep determines the bandwidth of the pulse, and the amplitude of the rf (combined with the sweep rate) determines the maximum bandwidth over which the pulse can remain adiabatic.

All adiabatic pulses exhibit this general behavior and many adjustable characteristics will determine the properties of specific pulses. The sweep width of $\Delta\omega$ typically ranges from 10–100 kHz and determines the bandwidth of the pulse. The rf power of the pulse, meanwhile, determines the rate at which the pulse can change frequency while remaining adiabatic. Furthermore, the adiabatic pulse literature demonstrates several pairs of amplitude and frequency modulation functions [60].

Up to now, the hyperbolic secant/tangent pair has been employed in *ex situ* NMR. That is, $B_1(t)$ and $\Delta\omega(t)$ take the forms

$$B_1(\mathbf{r},t) = B_1^{\mathrm{max}}(\mathbf{r})\mathrm{sech}(\beta t), \tag{6.18}$$

$$\Delta\omega(\mathbf{r},t) = \omega_0(\mathbf{r}) - \omega_{\mathrm{rf}}(t) = \omega_0(\mathbf{r}) - \mu\beta\tanh(\beta t), \tag{6.19}$$

where $2\mu\beta$ gives the width of the frequency sweep. A pulse whose bandwidth exceeds the inhomogeneities across a sample will act uniformly over that sample, even in the presence of an ω_0 gradient.

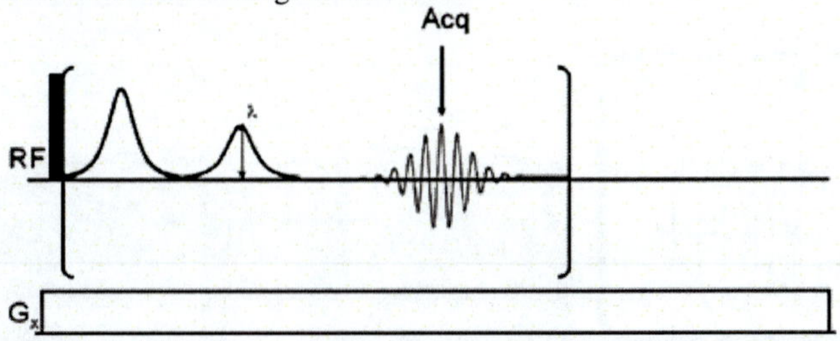

Fig. 6.7 A series of scaled adiabatic double passages generate phase correction and, in the presence of a static field gradient, subsequent nutation echoes for individual dwell points. Therefore, a stroboscopically sampled FID contains homogeneous chemical shift evolution information

In the first experiments employing these types of pulses, the double passage was applied periodically, and the relative scaling λ was set to center the echo in the acquisition window (Fig. 6.7). A stroboscopic repetition of this scaled adiabatic double passage every dwell period generates an FID which contains chemical shift evolution information, despite large inhomogeneities. Proof of principle experiments were conducted in a high-field (180 MHz proton Larmor frequency) setup, with a 20 kHz/cm gradient and an rf amplitude of 6 kHz on an acetone/benzene mixture (Fig. 6.8). A conical solenoid excited the sample and detected the signal. This solenoid produced a constant gradient in the rf field. This rf gradient matched the static field gradient, i.e., $B_0(x, y, z) \propto B_1(x, y, z)$. The scaled adiabatic double passage in this experiment consisted of two hyperbolic secant/tangent adiabatic pulses, each 20 ms long and sweeping over 28 kHz. Despite the inhomogeneities in this setup, resolved spectra were obtained.

As an alternative to stroboscopic acquisition, the scaling ratio λ between the two pulses of a single adiabatic double passage can vary between consecutive scans. For each scaling value, B_1 inhomogeneities compensate for a different amount of inhomogeneous evolution under B_0. Therefore, each differently scaled adiabatic double passage generates a nutation echo at a different point in time. The change of the phase with respect to B_1, and therefore the time that elapses before the nutation echo, generally increases in proportion with λ. This fact clearly and rapidly indicates the exact sequence of pulses needed for the entire experiment. Incrementation of λ thus introduces a new time-domain dimension which, after Fourier transformation and skewing, indicates the rf strength. The Fourier-transform of the direct and λ-scaling dimensions yields the correlation of B_0 with B_1 (Fig. 6.9a). In the case of good matching, this correlation map illustrates narrow, linear traces like those shown. A shearing (skewing) transformation [61] or, equivalently a first-order phase correction applied before the final Fourier transform with respect to t_1, removes the tilting in (ω_1, ω_2). Upon addition, the sheared data produce high-resolution chemical shift spectra (Fig. 6.9).

An optimized version of the above pulse sequence can generate chemical shift correlated images. A time delay inserted between the pulses gives time to generate an echo and perform phase encoding. Such a pulse sequence consists of an excitation pulse followed by a delay τ and two adiabatic π pulses, separated by the double of the initial time interval 2τ (Fig. 6.10). This experiment generated

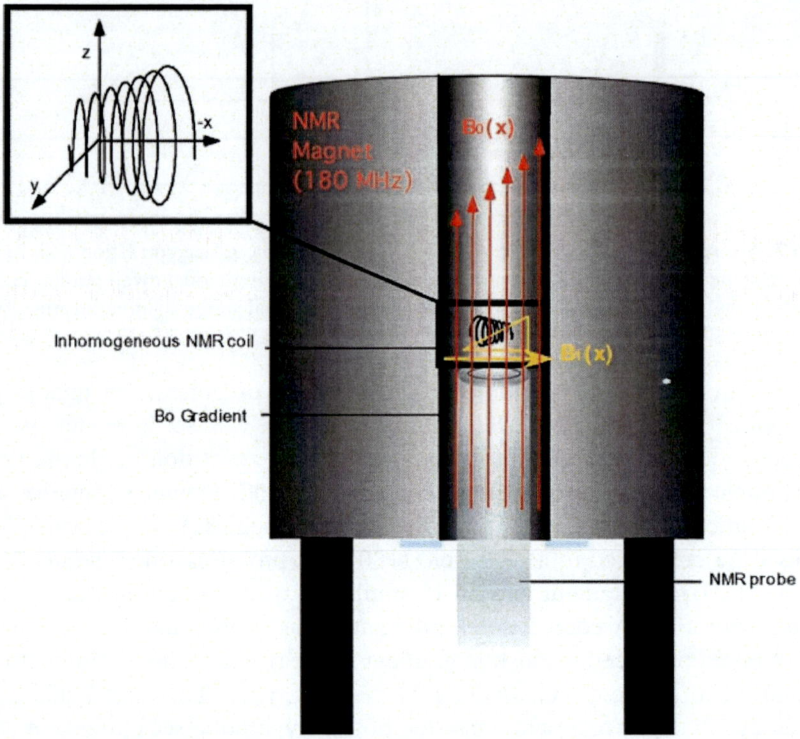

Fig. 6.8 The experimental arrangement consists of a conventional 4.2 T magnet (180 MHz proton Larmor frequency) with the application of a constant, linear static magnetic field gradient ($G_x = 3$ G/cm) that emulates inhomogeneities inherent to *ex situ* situations. The coil used for excitation and signal detection was a conical solenoid

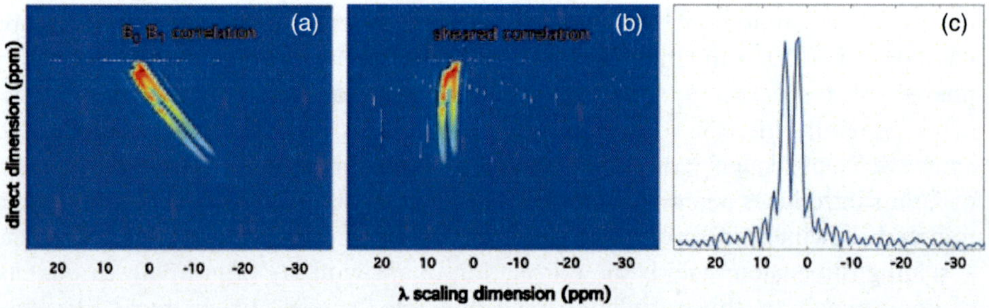

Fig. 6.9 **(a)** The direct and scaling dimensions are selected to yield the correlation of the B_1 field and the static magnetic field. They show linear correlation indicated by the *two lines* for both chemical species in the plot on the *left*. **(b)** The data are then sheared, as in the plot on the *right*, so that the spectrum can be recovered upon projection. **(c)** The recovered one dimensional proton spectrum (from [45])

images for every select chemical species. Figure 6.11 shows three of such images: the sum over all chemical shifts, as well as the chemical shift-selected water and oil images.

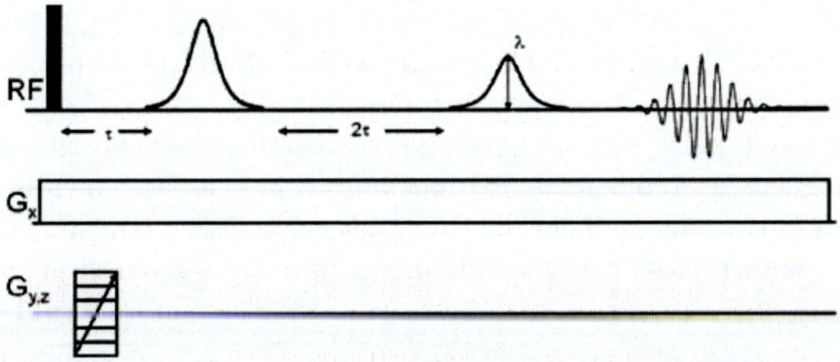

Fig. 6.10 Pulse sequence for the static and rf gradient fields: two scaled adiabatic π pulses follow an excitation pulse. The static field gradient along x ($G_x = 3$ G/cm) is applied during all experiments in order to emulate the natural inhomogeneity of an *ex situ* environment. Phase encoding is used to image the other two dimensions and the overall experiment is repeated for different scaling ratios, λ, of the two adiabatic pulses

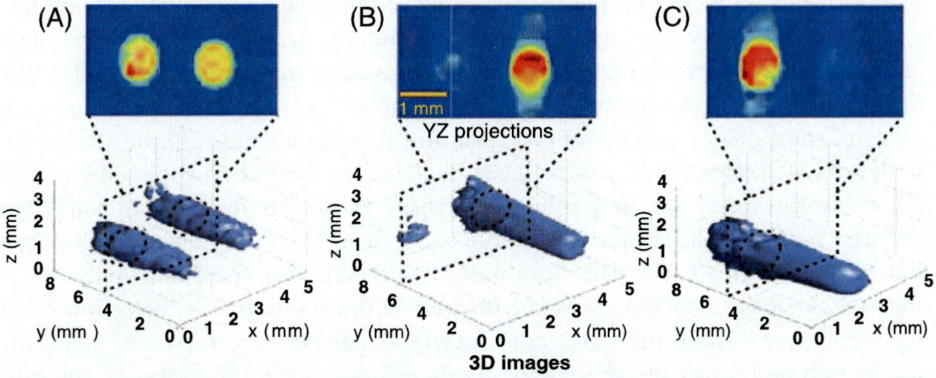

Fig. 6.11 *yz* Projections and three-dimensional images of the sample (as a subset of the 4D data set) correlated with chemical shift. **(a)** Either both tubes (trace at 6.5 ppm), or **(b)** the water (trace at chemical shift of 4.7 ppm for water), and **(c)** the oil (trace at chemical shift 8.5 ppm for oil) tubes are selectively displayed depending upon chemical shifts (from [45])

6.3.2 Shim Pulses: Corrections Based on Gradient Modulations During an Adiabatic Double Passage

The hardware matching method described in the previous section capitalizes on precise rf coil design and construction to recover high-resolution images with or without chemical shift resolution. Alternatively, *ex situ* "shim pulses" can also correct for inhomogeneous spatial fields [37]. In this scheme, static field gradients from imaging gradient coils apply a spatially dependent phase during the course of an adiabatic double passage. A time-dependent modulation of the static field gradients controls the specific spatial variation of this phase. For a given inhomogeneity profile, a computer program calculates the appropriate gradient modulation that cancels the inherent static field inhomogeneities, yielding either an effectively homogeneous field or an effectively constant spatial field gradient for imaging in one or more dimensions. Thus, shim pulses can correct for complicated spatially nonlinear field profiles, even where complicated hardware, such as shim coils or precisely matched rf coils, remain either unavailable or inconvenient [37].

This adiabatic pulse method shares several similarities with and demonstrates some differences from the adiabatic matching technique, as shown in [62]. Equation (6.20) easily accommodates the shim pulses gradient modulation as a spatially dependent frequency modulation, yielding the phase of a given shim pulse as

$$\phi(\mathbf{r}, t) = \gamma \int_0^t \sqrt{B_1(\mathbf{r}, t')^2 + \left(\mathbf{r} \cdot \mathbf{G}_{\text{shim}} + B_0(\mathbf{r}) - \frac{\omega_{\text{rf}}(t')}{\gamma}\right)^2}\, dt'. \tag{6.20}$$

Just as in hardware matching pulse sequences, the phase correction from sequential shim pulses will add constructively. Because of this attribute, as well as the offset independence of the pulses, the field strength of the imaging gradient does not necessarily need to exceed the strength of the inhomogeneity. The only limitation of the strength of the corrective phase arises from the signal decay imposed by relaxation during the pulses. Like hardware-matched adiabatic pulses and adiabatic pulses in general, shim pulses operate relatively uniformly over a bandwidth of offsets. For both types of corrective pulses, as the strength of the correction applied by a pulse increases, the bandwidth of the pulse decreases. However, the bandwidth for shim pulses, typically 6 kHz, shrinks significantly from that of comparable matching pulses, typically 40 kHz. The shim pulse bandwidth also decreases more rapidly than the adiabatic matching pulse bandwidth as the strength of a correction scales up. Such issues with shim pulses must inevitably balance their increased flexibility, as the imposed offset of the imaging gradients strains the robustness of the original adiabatic passage. Therefore, the robustness of the unaltered adiabatic pulses primarily determines the robustness and strength of the shim pulse corrections.

In the case of *ex situ* "shim pulses" for spectroscopy, the combination of "*ex situ* matching" and "shim pulses" relaxes the hardware requirements on both techniques. Matching does not have to be "perfect," and the strength of the shim pulse gradients does not have to be as high, so the combination of these techniques to generate pulse sequences such as those described in [62] was natural.

Practically, the general "*ex situ* shim pulse" scheme works as follows. First, the static magnetic field is mapped via NMR. Second, a coil geometry that matches the static field to a first degree is built (it could be inhomogeneous if a single-sided coil is desired or a homogeneous coil can be used). For a given set of gradient coils, rf geometry, and static field profile, the shim pulses are optimized to either produce a homogeneous or linearly varying field for imaging purposes. An example "shim pulse" is shown in Fig. 6.12.

Shim pulses have already demonstrated significant offset correction under artificially imposed *ex situ* conditions, and similar experiments are underway in *ex situ* systems with naturally occurring gradients. The experiments by Topgaard et al. [37] applied shim pulses periodically, stroboscopically acquiring high-resolution spectra (Fig. 6.12) in a high-field magnet unshimmed by a few parts per million. Shim pulses

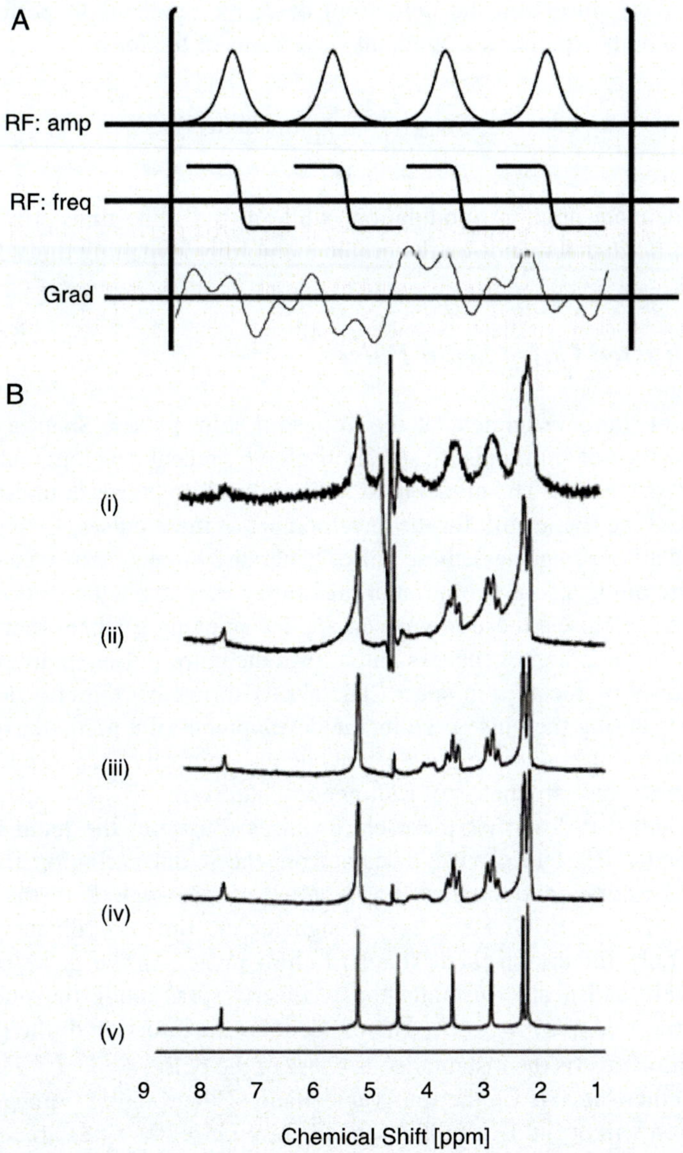

Fig. 6.12 (**a**) Shim pulse sequence for recording 1D NMR spectra. The adiabatic rf pulses and the numerically optimized gradient modulation (with linear and quadratic correction) are shown. A total of 1, 024 time-domain points were recorded stroboscopically during windows in a pulse train of shim pulses with length of 2 ms and rf sweep width 40 kHz. Effective chemical shift evolution takes place during a dwell time of 150 μs for each detected point. (**b**) The ^{1}H NMR spectra for vitamin B1 in D_2O water. (*i*) Inhomogeneous magnetic field without applied gradients, (*ii*) inhomogeneous magnetic field with linear correction, (*iii*) inhomogeneous magnetic field with linear and quadratic correction, (*iv*) homogeneous magnetic field without applied gradients, and (*v*) standard spectrum obtained with a simple $\frac{\pi}{2}$ acquire experiment (from [37])

also effectively linearized the field from defective gradients to perform accurate imaging. In both experiments, gradient waveforms of the form

$$G_{\text{shim}}(t) = \sum_i a_i \sin(\omega_i t) \tag{6.21}$$

allowed adequate gradient modulation, while at the same time providing convenience and limited demands on the gradient amplifier switching times [37].

6.3.3 Adjusted Chirp Shim Pulses

Shortly after the development of the Topgaard shim pulses, Shapira et al. developed a new type of shim pulses [36], motivated by recent developments in ultrafast spectroscopy [66, 67]. The model used to describe chirp pulses in ultrafast 2D spectroscopy laid the framework for the development of these pulses. Like the Topgaard pulses (and all *ex situ* pulses), these pulses generate a homogeneity echo, which refocuses the inhomogeneous evolution of previously excited magnetization. Unlike the Topgaard shim pulses, these rely on an applied imaging gradient that overwhelms the effects of the inhomogeneities and allows the chirp pulses to address the spins predominantly by location in space. The ability to neglect the effects of the inhomogeneities during the pulse permits the development of a particularly simple and physical model of how these pulses generate an arbitrarily shaped inhomogeneity-compensating phase by means of rudimentary hardware.

The model that describes these chirp pulses illustrates the main ideas simply and eloquently. The chirp pulse sweeps from the initial excitation frequency F_{ini} to the final excitation frequency $F_{\text{ini}} + Rs(t_p)$ at the rate R in the presence of a gradient of strength G. Here, $s(t)$ designates the time coordinate of the pulse; $s(t) = t$ only for the original, standard chirp pulse. As the pulse progresses, it consecutively and nearly instantaneously inverts spins along the plane given by $F_{\text{ini}} + Rs(t) + \Omega_{\text{inh}}(z) = \gamma z G$, where z denotes the direction of the gradient. The magnetization inverts the instantaneous phase of the pulse, $\phi_{\text{rf}} = \int_0^{t_p} F_{\text{ini}} + Rs(t)dt$, and determines the axis (in the transverse plane of the rotating frame) about which the inversion will occur. During the second chirp pulse, the spins are again inverted at the appropriate instant in time. Therefore, the net result of both pulses can be described by the propagator[6]:

$$\begin{aligned} &(e^{-i\phi_{\text{rf}}^{(2)} I_z} e^{-i\pi I_x} e^{+i\phi_{\text{rf}}^{(2)} I_z} e^{-i\gamma(t[z]^{(2)})zGI_z}) * \\ &(e^{-i\gamma(t_p^{(1)}-t[z]^{(1)})zGI_z} e^{-i\phi_{\text{rf}}^{(1)} I_z} e^{-i\pi I_x} e^{+i\phi_{\text{rf}}^{(1)} I_z}) = \\ &= e^{-i\left(2\phi_{\text{rf}}^{(2)} - (t[z]^{(2)} + t_p^{(1)} - t[z]^{(1)})zG - 2\phi_{\text{rf}}^{(1)}\right) I_z}. \end{aligned} \tag{6.22}$$

---

6 The superscripts (1) and (2) denote the first and second pulses, respectively.

Therefore, the chirp pulses invert the spins along a plane in space corresponding to the resonant frequency at a given point in time. As the pulses scan through the different frequencies, that plane moves through space. A differing period elapses between the arrival of the two planes, corresponding to the first and second inversions, at any given point in space. This period becomes an effective "backward evolution" period. The length of this period, augmented by twice the difference in phase of the first and second pulses (at the point in time of the first and second inversions), gives the phase of the shim pulse at a given point in space. Therefore, denoting the time at which inversion at z occurs by $t(z)$,[7] the shape of the corrective phase can be very easily adjusted by tweaking the length of the "backward evolution" period for the different points in space. Mathematically, this simply amounts to choosing the function $s(t)$. For a given inhomogeneity, a choice of the appropriate $s(t)$ will generate the appropriate corrective pulse.

We have seen, so far, various methodologies that generate spatially correlated phase corrections which compensate for inhomogeneous dephasing. Ultrafast spectroscopy, however, offers an entirely different means of inhomogeneous field correction. Frydman developed the technique of ultrafast spectroscopy [63], which employs either a series of frequency-shifted rf pulses or a pair of chirp pulses, both in the presence of a strong gradient. These pulses separate (encode) the sample in slices (along the direction of the gradient) or voxels (in the case of orthogonal axis 3D gradients). Each slice can be encoded with an amplitude or phase modulation corresponding to a different indirect evolution t_1 [36], effectively eliminating the need for a true indirect time dimension and dramatically speeding up experiment

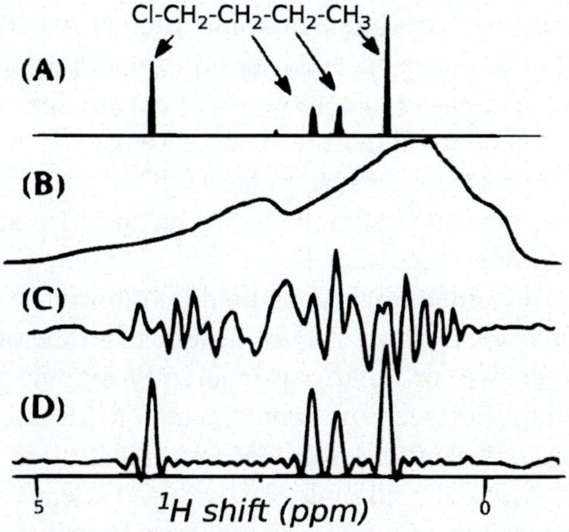

Fig. 6.13 (**a**) ^{1}H FT NMR spectrum of $C_4H_9Cl/CDCl_3$ recorded in a homogeneous 11.75 T field. (**b**) Identical experiment, but in the presence of an artificial 1.5 kHz field inhomogeneities. (**c**) Spatially encoded ^{1}H NMR spectrum recorded under identical inhomogeneous conditions as (**b**). (**d**) Same experiment but with phase corrections to the spatially encoding rf excitation pulses. The figure is taken from [36]

[7] $t[z]$ is determined from $F_{\text{ini}} + Rs(t[z]) + \Omega_{\text{inh}}(z) = \gamma z G$.

acquisition speed. It turns out, furthermore, that an ultrafast experiment can easily accommodate an automatic deconvolution of the inhomogeneities along its indirect dimension.

Between the advent of adiabatic matching and shim pulses, Shapira and Frydman [36] showed how to compensate for inhomogeneities along the indirect dimension. Starting from a standard ultrafast experiment, the lengthening or shortening of the evolution time of a voxel can compensate for the variance of the field at that point, thereby imparting the correct phase or amplitude modulation to a voxel and effectively compensating for the inhomogeneities. Thus, at the cost of the signal lost from dividing the sample into separate voxels, the experiment itself can effectively deconvolve the inhomogeneities (Fig. 6.13).

6.4 Summary

NMR probes non-invasively and in a powerful way for a variety of data, including the structure and dynamics of biochemical compounds, chemical reaction progress, and a variety of industrial quality control indicators. NMR's spatially enhanced offspring, magnetic resonance imaging (MRI), typically generates crucial graphical representations of fluid flow, rates of diffusion, and the interiors of materials and living organisms. Therefore, the general technique of magnetic resonance (MR) plays an indispensable role in fields as varied as pharmaceuticals, geotechnical research, genetics, molecular biology, materials science, and diagnostic radiology.

Previously, MR demanded a highly homogeneous magnetic field. However, hardware and methodology advances over the last decade have transformed unconventional MRI and relaxation measurements into routine experiments with compact one-sided sensors [1–6]. More recently, despite inhomogeneous magnetic fields, very practical new techniques have allowed for *ex situ* spectroscopy [35, 47] and even recovered the chemical shift information associated with MR images [45]. These preliminary results have persuaded researchers to high hopes, and slowly the proof of principle experiments continue to move toward practical *ex situ* systems with naturally occurring gradients.

Ex situ MR, or MR in inhomogeneous fields, continues to grow rapidly because a variety of fields and applications would benefit from the ability to conduct MR with portable, inexpensive, or otherwise inherently inhomogeneous systems. The complete realization of this technique would extend MR to field medical diagnosis, on-site identification of hazardous substances (typically via signatures from production of nuclear, chemical, and biological weapon agents, narcotics, explosives, toxins, and poisons), characterization of geological conditions (such as in an oil field), cheap industrial control, space exploration, and more.

References

1. Eidmann G, Savelsberg R, Blümler P, Blümich B (1996) J Magn Reson A 122:104–109
2. Blümich B, Blümler P, Eidmann G, Guthausen A, Haken R, Schmitz U, Saito K, Zimmer G. (1998) Magn Reson Imaging 16:479–484

3. Prado PJ, Blümich B, Schmitz U (1998) J. Magn. Reson. A 144:200–206
4. Guthausen G, Guthausen A, Balibanu F, Eymael R, Hailu K, Schmitz U, Blümich B (2000) Macromol Mater Eng 276/277:25–37
5. Balibanu F, Hailu K, Eymael R, Demco DE, Blümich B (2000) J Magn Reson A 145:246–258
6. Casanova F, Blümich B (2003) J Magn Reson 163:38–45
7. Kenyon WE, Howard JJ, Sezinger A, Straley C, Matteson A, Horkowitz K, Ehrlich R (1989) Transactions of the SPWLA 30th annual logging symposium, Denver, CO
8. Murphy DP (1995) World Oil 216(4):65–70
9. Freedman R, Morriss CE (1995) Presented at the 70th SPE annual technical conference and exhibition, Dallas, TX
10. McDonald PJ, Newling B (1998) Rep Prog Phys 61:1441
11. Mallet MJD, Halse MR, Strange JH (1998) J Magn Reson 132:172–175
12. Fukushima E, Jackson J (1999) NMR News Lett 490:40–42
13. Callaghan PT, Dykstra R, Eccles CD, Haskell T, Seymour JD (1999) Cold Reg Sci Technol 29:181–202
14. Guthausen A, Zimmer G, Blümler P, Blümich B (1997) J Magn Reson 130:1–7
15. Haken R, Blümich B (2000) J Magn Reson 144:195–199
16. Kuehn H, Klein M, Wiesmath A, Demco DE, Blümich B, Kelm J, Gold PW (2001) Magn Reson Imaging 19:497–499
17. Anferova S, Anferov V, Rata DG, Blümich B, Arnold J, Clauser C, Blümler P, Raich H (2004) Magn Reson Eng B 23 B:26–32
18. Casanova F, Perlo J, Blümich B (2004) J Magn Reson 171:124–130
19. Perlo FC, Blümich B (2005) J Magn Reson 173:254–258
20. Blümich B, Anferova S, Sharma S, Segre AL, Federici C (2003) Magn Reson Imaging 161:204–209
21. Proietti N, Capitani D, Pedemonte E, Blümich B, Segre AL (2004) J Magn Reson 170:113–120
22. Kleinberg RL (2001) Concepts Magn Reson, 13(6):396–403
23. Callaghan PT, Le Gros M. (1982) Am. J. Phys 50(8):709–713
24. Weichman PB, Lavely EM, Ritzwoller MH (1999) Phys Rev Let 82(20):4102–4105
25. Weichman PB, Lavely EM, Ritzwoller MH (2000) Phys Rev E. 62(1):1290–1312
26. Mohoric A, Plainisic G, Kos M, Duh A, Stepisnik J (2004) Instrum Sci Technol 32:655–667
27. Callaghan PT, Eccles CD, Seymour JD (1997) Rev Sci Instrum 68(11):4263–4270
28. Callaghan PT, Eccles CD, Haskell TG, Langhorne PJ, Seymour JD, (1998) J Magn Reson 133:148–154
29. Mercier OR, Hunter MW, Callaghan PT (2005) Cold Reg Sci Technol 42(2):96–105
30. Stepisňik J, Erzen V, Kox M (1990) Magn Reson Med 15:386–391
31. Shushakov OA (1996) Geophysics 61(4):998–1006
32. Semenov AG, Burshtein AI, Yu Pusep A, Schirov MD (1988) USSR patent 1079063
33. Semenov AG, Schirov MD, Legchenko AV, Burshtein AI, Yu Pusep A (1989) Great Britain patent 2198540B
34. Appelt S, Kuehn H, Haesing FW, Bluemich B (2006) Nat Phys 2:105–109
35. Meriles CA, Sakellariou D, Heise H, Moule AJ, Pines A (2001) Science 293:82–85
36. Shapira B, Frydman L (2004) J Am Chem Soc 126:7184–7185
37. Topgaard D, Martin RW, Sakellariou D, Meriles CA, Pines A (2004) Proc Natl Acad Sci USA 101:17576
38. Weitekamp D, Garbow JR, Murdoch JB, Pines A (1981) J Am Chem Soc 103:3578–3580
39. Vathyam S, Lee S, Warren WS (1996) Science 272:92–96
40. Balbach JJ, Conradi MS, Cistola DP, Tang C, Garbow JR, Hutton WC (1997) Chem Phys Lett 277:367–374
41. Lin YY, Ahn S, Murali N, Brey W, Bowers CR, Warren WS (2000) Phys Rev Lett 85(17):3732–3735
42. Sakellariou D, Meriles CA, Moule AJ, Pines A (2002) Chem Phys Lett 363:25–33

43. Meriles CA, Sakellariou D, Pines A (2003) J Magn Reson 164:177–181
44. Antonijevic S, Wimperis S (2003) Chem Phys Lett 381:634–641
45. Demas V, Sakellariou D, Meriles C, Han S, Reimer J, Pines A (2004) Proc Natl Acad Sci USA 101(24):8845–8847
46. Demas V, Meriles CA, Sakellariou D, Han S, Reimer J, Pines A (2006) Toward ex-situ phase-encoded spectroscopic imaging. Concepts Magn Reson Part B Magn Reson Eng B 29B(3):137–144
47. Perlo J, Demas V, Casanova F, Meriles C, Reimer J, Pines A, Blümich B (2005) Science 308:1279
48. Hahn EL (1950) Phys Rev 80(4):580–594
49. Hahn EL, Maxwell DE (1952) Phys Rev 88:1070–1084
50. Weitekamp D (1982) PhD thesis, University of California-Berkeley
51. Garbow JR, Weitekamp DP, Pines A (1983) J Chem Phys 79(11):5301–5310
52. He Q, Richter W, Vathyam S, Warren WS (1993) J Chem Phys 98(9):6779–6800
53. Warren WS, Ahn S, Mescher M, Garwood M, Ugurbil K, Richter W, Rizi RR, Hopkins J, Leigh JS (1998) Science 281:247–250
54. Bowtell R, Robyr P (1996) Phys Rev Lett 76(26):4971–4974
55. Lee S, Richter W, Warren WS (1996) J Chem Phys 105(3):874–900
56. Bouchard L-S, Rizi RR, Warren WS (2002) Magn Reson Med 48:973–979
57. Warren WS, Richter W, Andreotti AH, Farmer BT II (1993) Science 262:2005–2009
58. Levitt MH, Freeman R (1981) J Magn Reson 43(3):65–80
59. Grunin L, Blümich B (2004) Chem Phys Lett 397:306–308
60. Garwood M, DelaBarre L (2001) J Magn Reson 153:155–177
61. Grandinetti PJ, Baltisberger JH, Llor A, Lee YK, Werner U, Eastman MA, Pines A (1993) J Magn Reson A 103:72–81
62. Franck JM, Demas V, Martin RW, Bouchard L-S, Pines A (2009) Shimmed matching pulses: simultaneous control of RF and static gradients for lnhomogeneity correction. J Chem Phys 131(23):234506
63. Frydman L, Scherf T (2002) Lupulescu A Proc Natl Acad Sci USA 99:15858
64. McDermott R, Trabesinger AH, Muck M, Hahn EL, Pines A, Clarke J (2002) Science 295:2247
65. Ardelean I, Kimmich R, Klemm A (2000) J Magn Reson 146:43–48
66. Shrot Y, Shapira B, Frydman L (2004) J Magn Reson 171:163–170
67. Tal A, Shapira B, Frydman L (2005) J Magn Reson 176:107–114

Chapter 7
High-Resolution Spectroscopy in Highly Homogeneous Stray Fields

Ernesto P. Danieli

Single-sided sensors from permanent magnets are portable, inexpensive, and provide unlimited access to study samples of arbitrary size. The price paid for working in the stray field of an open magnet is more inhomogeneity than inside the magnetic field. Although resistive coils [1] are widely used in high-resolution superconducting magnets to shim the field, this solution must be discarded due to the excessive currents required to correct the inhomogeneities of stray fields. For example, with conventional surface coils, a current of 1,000 A is needed to generate a magnetic field with a gradient of 1 T/m, and this is a small gradient for a single-sided sensor (see Chap. 4). Alternative magnet geometries have been tried out mainly to reduce the strong gradient along the depth direction [2]. However, off-resonance problems remain and make it difficult to implement conventional NMR experiments, thus making it nearly impossible to conduct high-resolution spectroscopy.

By accepting inhomogeneous magnetic fields as unavoidable, samples have been characterized in terms of NMR parameters such as signal amplitude and relaxation times. These parameters can be obtained thanks to the large efforts invested in adapting and developing pulse sequences which work in inhomogeneous fields. An example of this approach is presented in Chap. 6 where it was demonstrated how chemical shift information can be retrieved in an inhomogeneous magnetic field. Through the precise matching between the spatial dependence of B_0 and B_1, it is possible to generate chemical shift-modulated nutation echoes, which are formed by a nutation of the magnetization in the B_1 field and a precession in the B_0 field [3]. An important step that led to the experimental demonstration of this technique in a single-sided magnet was the realization that small magnets properly placed in the main magnet can be used to tune the spatial dependence of the magnetic field B_0 to match the one generated by the rf coil. This concept triggered the use of permanent magnets to shim the intrinsic inhomogeneities of a U-shaped magnet. Instead of adjusting the current in a coil to obtain the desired harmonic correction of the field, this approach does not require a power supply as the positions of the shim magnets

E.P. Danieli (✉)
Institut für Technische Chemie und Makromolekulare Chemie, RWTH Aachen University, D-52074 Aachen, Germany

F. Casanova et al. (eds.), *Single-Sided NMR*,
DOI 10.1007/978-3-642-16307-4_7, © Springer-Verlag Berlin Heidelberg 2011

are adjusted mechanically. In this way, it was possible to homogenize the magnet field of a single-sided sensor to the sub-ppm regime, and the first proton spectra could be obtained in the stray field of a magnet [4].

In this chapter, the concept of this shim unit is described in detail. In particular, the spatial movements of the shim magnets required to generate the different harmonic corrections are explained. In addition, further steps toward improving the field homogeneity and the size of the working volume are discussed. A major problem that arises with sensors built from permanent magnets is the strong temperature dependence of the magnetic field strength. This issue is considered at the end of the chapter in the context of typical NMR experiments and methods for its alleviation are described.

7.1 Sensor Design

Several magnet geometries can be used to build a single-sided sensor (see Chap. 4). The spatial dependence of the magnetic field can be modified by incorporating into the main magnet assembly a set of small movable permanent magnets which form a so-called shim unit. In order to apply this concept in the design of a particular sensor, it is important to take the following considerations into account:

- The magnet is composed of two parts or units, i.e., the main unit generating the main magnetic field and the shim unit used to correct the inhomogeneity of the main field up to a certain order.
- The shim unit is specially designed to correct the spatial dependence of the main field, so the shim unit will work only in combination with the particular main unit.
- The numerical optimization of the shim unit requires detailed knowledge of the main field.
- The position of the magnet blocks forming the shim unit needs to be adjustable because of unavoidable inaccuracies in the polarization, size, and position of individual magnet pieces. Small displacements of the magnet blocks around their optimum positions allow one to experimentally reach the performance calculated numerically.

This section gives a detailed description of the steps that have to be followed to end up with a single-sided spectroscopy magnet [4]. First, the features of the main unit are discussed, and second, the geometry of the shim unit required to obtain a uniform magnetic field is presented.

7.1.1 Main Unit

For the discussion presented in this chapter, a main magnet with dimensions $30 \times 10 \times 27$ cm^3 and built in a U-shaped geometry has been chosen. It consists of two permanent magnet blocks with anti-parallel polarization placed on an iron yoke separated by a gap between them [5, 6]. The direction along the gap is called x, the direction along the depth y, and z is the direction across the gap. All distances are measured from a reference system whose origin is placed at the geometric center of the magnet with the xz-plane placed at the magnet surface. The region of interest is a slice parallel to the magnet surface located at a depth y_0, with the size of the slice and the depth y_0 small compared to the magnet gap. In good approximation, the magnetic field in this region can be considered to be oriented along the z-axis with its dominant gradient component along the depth direction. Due to the symmetry of the magnet along z, the spatial dependence of the field along this direction has only even terms in its Taylor expansion. As a function of position the field has a minimum at the center of the magnet ($z = 0$) and increases as the permanent magnet blocks are approached. This is mainly due to the inverse dependence of the field on the distance to the magnetic source. This suggests that the field is shaped like an arms-up parabola centered at $z = 0$. Also due to symmetry arguments the field along x is an even function of x. But in this case the field decreases when moving away of the center, reflecting the finite size of the magnet along x. Therefore, the field behaves like an arms-down parabola along x. This spatial dependence gives rise to the well-known horse saddle shape of the magnetic field in the xz-plane. The field shape described above can formally be expressed by expanding the magnitude of the magnetic field $B_0(\mathbf{r})$ around $\mathbf{r_0} = (0, y_0, 0)$ as

$$B_0(\mathbf{r_0}) = B_{00} + G_y(y_0)(y - y_0) + \alpha_z(y_0)z^2 + \alpha_x(y_0)x^2 + \cdots , \qquad (7.1)$$

where $B_{00} = B_0(\mathbf{r_0})$, $G_y(y_0) = \frac{\partial B_0(\mathbf{r})}{\partial y}|_{r_0}$, and $\alpha_k(y_0) = \frac{\partial^2 B_0(\mathbf{r})}{\partial k^2}|_{r_0}$ with $k = x, z$.

Some specifications of the setup used in [4] are the following: at a depth $y_0 \simeq 5$ mm, the average field is around $B_{00} \simeq 0.25\,\mathrm{T}$ (10 MHz ^{1}H Larmor frequency), and the inhomogeneities are characterized by $G_y \simeq -1,000$ ppm/mm, $\alpha_z \simeq 300$ ppm/mm^2, and $\alpha_x \simeq -30$ ppm/mm^2. Those data reveal that the main source of inhomogeneity is the gradient along the depth direction. This is why a slice in the xz-plane was chosen as the region of interest for shimming. In the presence of this field inhomogeneity, the size of the volume where a field variation of 1 ppm is obtained (without shims) is about $180 \times 60\ \mu\mathrm{m}^2$ along the lateral directions and 1 μm thick along the depth direction. These dimensions provide some insight into the magnitude of the inhomogeneities to be corrected with single-sided magnets.

7.1.2 Shim Unit

The field generated by the shim unit must reproduce the spatial dependence of the main field, in principle having the smallest possible average field strength. One possibility to meet this requirement is to build the shim unit with the geometry of the main magnet (in this case, a U-shaped geometry but with a different aspect ratio and a smaller size (Fig. 7.1a). By setting the polarization of the shim unit opposite to

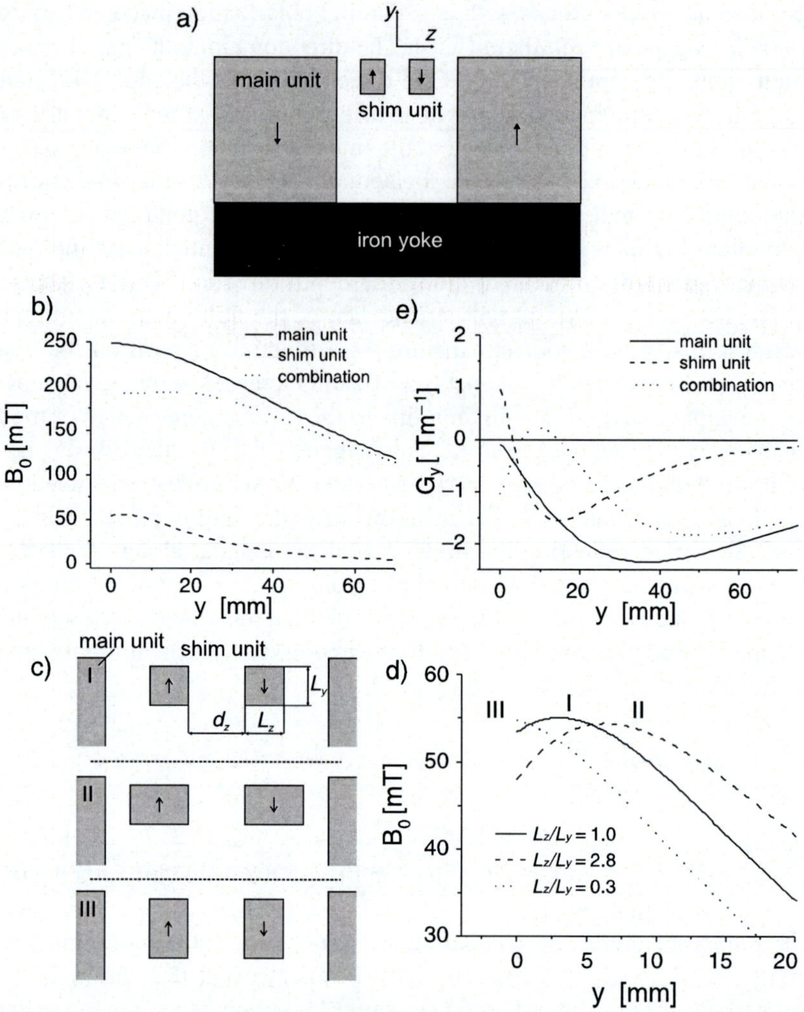

Fig. 7.1 (**a**) Sensor array: the main unit is a conventional U-shaped magnet [5, 6], and the shim unit consists of a pair of magnet blocks also in a U-shape configuration but without the iron yoke. The *arrows* show the direction of polarization of the magnets. (**b**) Dependence of the magnetic field strength (magnitude) for the main and the shim units, shown by *solid* and *dashed lines*, respectively. The shim unit dimension is $300 \times 15 \times 15\ \text{mm}^3$ and a gap $d_z = 40\,\text{mm}$. The *dotted line* shows the combination of both fields. (**c**) Different configurations I, II, and III of the shim unit which originates the maximum of the magnetic field at different values of $y_{\max}$ as can be seen in (**d**). (**e**) Magnetic field gradient along the depth direction corresponding to the fields of (**b**).

that of the main field, the inhomogeneities of the main field can be corrected while the total field strength is maintained at an acceptable magnitude.

7.1.2.1 Matching Field Inhomogeneities Along the Depth

Equation (7.1) contains three coefficients, G_y, α_z, and α_x, that characterize the field inhomogeneities associated with the main unit. A similar equation holds for the shim unit. Let us start by focusing on how to cancel the gradient G_y along the depth direction. Figure 7.1b shows the field strength as a function of depth for the main magnet ("big" U-shaped) and the shim unit ("small" U-shaped). For each unit the field has a maximum along the depth direction that depends on the height L_y of the magnets and the distance between their centers measured along the gap direction $Lz + d_z$ (see Fig. 7.1c). These variables determine the position y_{max} for the maximum field of each unit. An increase in the block's height shifts the maximum closer to the surface, whereas an increase in the magnet's width (alternatively, the gap d_z) effects the opposite, as it can be seen in Fig. 7.1d. The use of an iron yoke has mostly two effects, i.e., an increment in the field strength and a displacement of the maximum to smaller depths. G_y is positive for $y < y_{\mathrm{max}}$ and becomes negative for $y > y_{\mathrm{max}}$ (Fig. 7.1e). The main unit has the field maximum at the surface of the magnet, which explains why G_y^{main} does not assume positive values in the depth range shown in Fig. 7.1e.

However, G_y^{shim} takes positive values for $y < 3\,\mathrm{mm}$ and becomes negative for $y > 3$ mm. At $y_0 \simeq 5\,\mathrm{mm}$ the gradients of both fields match. When both units are brought together and their polarizations are in opposite directions, the gradient of the total field becomes zero at this particular depth. Thus,

$$G_y(y_0) = G_y^{\mathrm{main}}(y_0) + G_y^{\mathrm{shim}}(y_0) = 0. \qquad (7.2)$$

Following this idea, it is possible to design a sensor with no linear variation of the magnetic field along the depth direction at a particular position y_0, which is called "sweet spot" [2].

7.1.2.2 Matching Field Inhomogeneities Across the Gap

The second largest source of inhomogeneity is related to the field variation across the gap direction which is characterized by α_z. The field generated by the shim unit has the same spatial dependence as the one generated by the main magnet and can also be characterized by a coefficient α_z^{shim} which can be varied to match α_z^{main}. Let us consider that the shim unit designed above to cancel the main gradient at a depth y_0 generates at this particular depth $|\alpha_z^{\mathrm{shim}}| > \alpha_z^{\mathrm{main}}$. Then, if the size of the shim unit is shrunk, at the same time reducing its gap (Fig. 7.2a), G_y^{shim} remains unchanged and a smaller $|\alpha_z^{\mathrm{shim}}|$ is obtained (Fig. 7.2b).

Hence, there must be an intermediate shim unit that fulfil $|\alpha_z^{\mathrm{shim}}| = \alpha_z^{\mathrm{main}}$. Thus, as the polarization of the main unit and that of the shim unit are opposite to one another, the quadratic coefficients along z cancel each other. If the initial condition is $|\alpha_z^{\mathrm{shim}}| < \alpha_z^{\mathrm{main}}$, the inverse procedure should be followed. Figure 7.2 also shows that it is possible to generate a shim field which satisfies

$$\alpha_z(y_0) = \alpha_z^{\mathrm{main}}(y_0) + \alpha_z^{\mathrm{shim}}(y_0) = 0. \qquad (7.3)$$

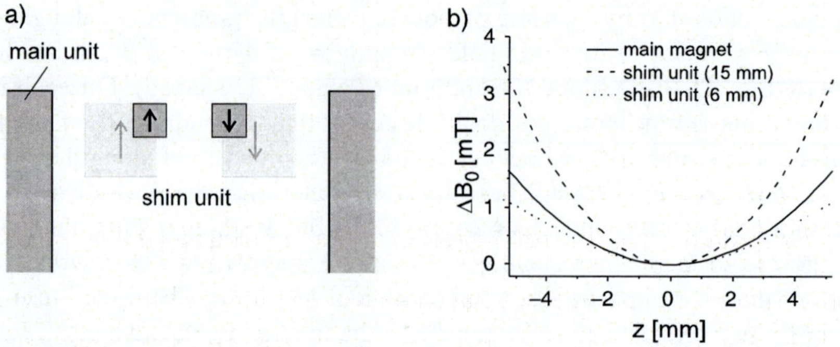

Fig. 7.2 **(a)** Adjusting the size of the shim unit the value of α_z can be varied by fulfilling simultaneously at $y_0 = 5$ mm the condition $G_y^{\text{main}} = G_y^{\text{shim}}$. The big shim unit (*light gray*) has a size of $15 \times 15\,\text{mm}^2$ and a gap $d_z = 40$ mm, whereas the small one is $6 \times 6\,\text{mm}^2$ in size and has a gap of 30 mm. **(b)** Dependence of the magnetic field strength (magnitude) as a function of the lateral direction z for the two shim unit sizes and the main magnet. The offset field value at $z = 0$ was subtracted in order to facilitate the comparison

7.1.2.3 Matching Field Inhomogeneities Along the Gap

The last coefficient to be canceled in Eq. (7.1) is α_x. Although this coefficient is at least one order of magnitude smaller than α_z, it must be canceled while keeping the values G_y^{shim} and α_z^{shim} constant. Fortunately, it is possible to vary α_x^{shim} in a relatively large range almost without modifying G_y^{shim} and α_z^{shim} by changing the length L_x of the shim unit along x (see Fig. 7.3a). Figure 7.3b shows that the relation $|\alpha_x^{\text{shim}}| > |\alpha_x^{\text{main}}|$ is inverted for a long L_x (300 mm) $|\alpha_x^{\text{shim}}| < |\alpha_x^{\text{main}}|$ but for a sufficiently short L_x (58 mm). Consequently, for an intermediate L_x value, the condition $|\alpha_x^{\text{shim}}| = |\alpha_x^{\text{main}}|$ can be achieved. By again considering that the polarization of the main unit and that of the shim unit is opposite to each other, we can write

$$\alpha_x(y_0) = \alpha_x^{\text{main}}(y_0) + \alpha_x^{\text{shim}}(y_0) = 0. \qquad (7.4)$$

7.2 Shimming Magnetic Fields with Movable Permanent Magnets

Up to now it was shown how to design a shim unit to compensate for the field inhomogeneity of an ideal U-shaped magnet within a certain region around $\mathbf{r_0} = (0, y_0, 0)$. Imperfections in the magnet blocks used to build either the main magnet or the shim unit are expected to lead to variations in the values of G_y, α_z, and α_x. Therefore, in a real magnet, control variables are required to accordingly adjust the shim terms y, x^2, and z^2. Even though it is true that such imperfections unavoidably affect the mentioned field coefficients they surely also introduce inhomogeneities that are not considered in the expansion of Eq. (7.1). For instance, the

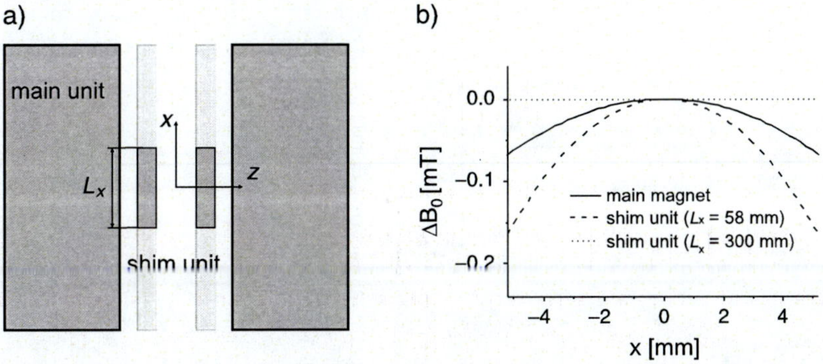

Fig. 7.3 (**a**) *Top view* of the sensor array of Fig. 7.1a. (**b**) Dependence of the magnetic field strength (magnitude) as a function of x for the two different sizes L_x of the shim unit and the main magnet. The offset field value (at $x = 0$) was subtracted in each case to facilitate the comparison. For $L_x = 73$ mm, the field curvatures of the main and shim units are matched

symmetry arguments used to assume an even dependence of the field along the z- and x-axis are no longer valid. Upon considering, for example, that one of the magnet blocks of the main unit has higher polarization than the other one, a constant gradient along z is expected. In general, any imperfection in the magnet pieces introduces an asymmetry in the field that can be characterized (up to second order) by linear components x and z, and by cross-terms xy, xz, and yz. The following describes the control variables required in the magnet to generate each of these shimming terms.

7.2.1 Generation of Linear Terms Along y

We have already shown that the combination of two nested units can be used in a numerical simulation to generate a so-called sweet spot. The crucial step toward matching the gradient of both units at one depth in a real magnet requires a variable to control the field dependence of at least one of the units. Such control can be acquired by considering the movement of the whole shim unit along the depth direction (Fig. 7.4a). When the shim unit is displaced a distance Δy in the positive y-direction, the dashed line in Fig. 7.1e shifts to the right. Notice that the reference system is fixed to the main magnet. Thus, at y_0 the gradient generated by the shim unit is smaller in magnitude than the one of the main magnet; in this case, the gradient of the total field becomes negative. If, however, the shim unit is displaced a distance Δy in the negative y-direction, the dashed line in Fig. 7.1e shifts to the left. In this case, at the position y_0 the gradient of the shim unit dominates over the one of the main unit, so that the gradient of the total field becomes positive. In this way the y shim component can be varied and its amplitude is proportional to Δy, which becomes the control variable for this term, as can be seen in Fig. 7.4b.

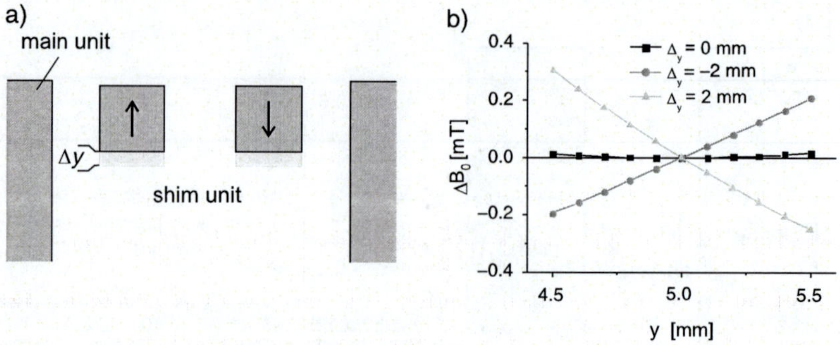

Fig. 7.4 (**a**) Schematic representation of the shim unit movement in order to obtain a linear correction along the depth. (**b**) Total magnetic field behavior as a function of the depth for three different positions of the shim unit indicated by the Δy parameter (see text)

7.2.2 Generation of Linear Terms Along x and z

The shim unit was designed to cancel the main field inhomogeneities described by Eq. (7.1), thereby fulfilling the requirements specified by Eqs. (7.2), (7.3) and (7.4). The same unit can be used to compensate for linear field variations along x and z. Let us consider the z-direction for the present analysis (a similar argument holds for the x-direction). A displacement Δz of the whole shim unit leads to a total field given by

$$\begin{aligned} B_0(0, y_0, z) &= B_0^{\mathrm{main}}(0, y_0, z) + B_0^{\mathrm{shim}}(0, y_0, z - \Delta z) \\ &= B_{00}^{\mathrm{main}} + \alpha_z^{\mathrm{main}} z^2 + B_{00}^{\mathrm{shim}} + \alpha_z^{\mathrm{shim}}(z - \Delta z)^2 \\ &= (B_{00}^{\mathrm{main}} + B_{00}^{\mathrm{shim}} + \alpha_z^{\mathrm{shim}} \Delta z^2) + \\ &\quad (\alpha_z^{\mathrm{main}} + \alpha_z^{\mathrm{shim}}) z^2 - (2\alpha_z^{\mathrm{shim}} \Delta z) z. \end{aligned} \tag{7.5}$$

The first term in brackets represents the magnitude of the total field independent of z (spatially homogeneous term). The second term in brackets vanishes if Eq. (7.3) is fulfilled, and the last one is the desired shim field with a linear variation along the z-axis. The shim field coefficient depends on the magnitude and direction of the displacement Δz enabling positive as well as negative z-corrections. Similarly, a linear term along the x-axis is achieved by displacing the whole unit along the x-axis.

7.2.3 Generation of Quadratic Terms x^2 and z^2

Once the optimum dimensions of the shim magnets are found in the simulations, the control of the z^2 component in a real magnet is obtained by varying the gap of the shim unit. Let us remember that the curvature of the magnetic field along the z-axis corresponding to the main unit is positive ($\alpha_z^{\mathrm{main}} > 0$). Increasing the gap size of the

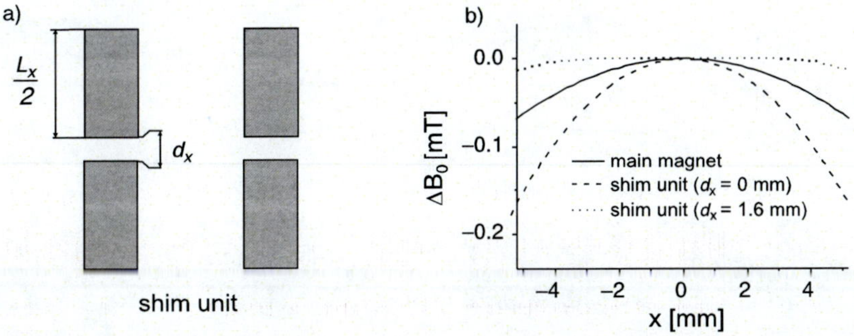

Fig. 7.5 (**a**) The shim unit is split into two identical pairs to introduce the gap d_x along the x-axis. (**b**) Dependence of the magnetic field strength (magnitude) as a function of x for two different values of d_x with $L_x = 58$ mm. The offset field value (at $x = 0$) was subtracted in each case to facilitate the comparison. For $L_x = 58$ mm and $d_x = 1$ mm, the field curvatures of the main and shim units are matched

shim unit leads to a lower value of the curvature $|\alpha_z^{\text{shim}}|$ of the field it produces. Since both main and shim unit fields are subtracting each other, the result is a positive z^2 shim field correction since ($|\alpha_z^{\text{shim}}| < \alpha_z^{\text{main}}$). Reducing the gap, however, leads to a negative z^2 shim field ($|\alpha_z^{\text{shim}}| > \alpha_z^{\text{main}}$). Note here that a change in the gap value leads to a small but undesired variation of G_y^{shim}, which needs to be readjusted by varying the position of the shim unit along the y-axis. This repositioning slightly modifies the z^2 term requiring an iterative procedure to cancel the field variations along both the y- and z-axes.

The natural control variable for the shim term x^2 is the length of the magnet blocks. However, it cannot be taken as a real variable, because it would require a large collection of blocks with different lengths to be interchanged at the time of shimming the magnet. Even in this case, this becomes a time-consuming and thus impractical procedure, since for each block pair, one would need to precisely adjust its position in order to cancel the inhomogeneities along the other axes. Another way of controlling the x^2 coefficient is to split the blocks, thereby generating a second gap d_x along the x-axis (Fig. 7.5). For small increments d_x, the effect on the field is equivalent to increasing the length of the blocks. Since d_x cannot assume negative values, it is important to design the shim unit to cover a sufficiently large range for d_x. The dependence of the shim term x^2 on d_x is opposite to that of z^2 on d_z. Thus, an increment of d_x leads to a total negative x^2 shim field and a reduction of the gap to a positive x^2 shim field.

7.2.3.1 Generating Cross-Terms xy, xz, and yz

In contrast to the generation of linear terms, cross-terms are obtained by asymmetrically moving the magnet block pairs. For example, whereas the generation of an x term requires a movement of all four magnets along the x-axis, an xz term is achieved by, respectively, displacing along the positive and negative x-axis the shim unit blocks with polarization up and the blocks with polarization down (see Fig. 7.6). The dependence of the field after an asymmetric displacement of the shim unit can

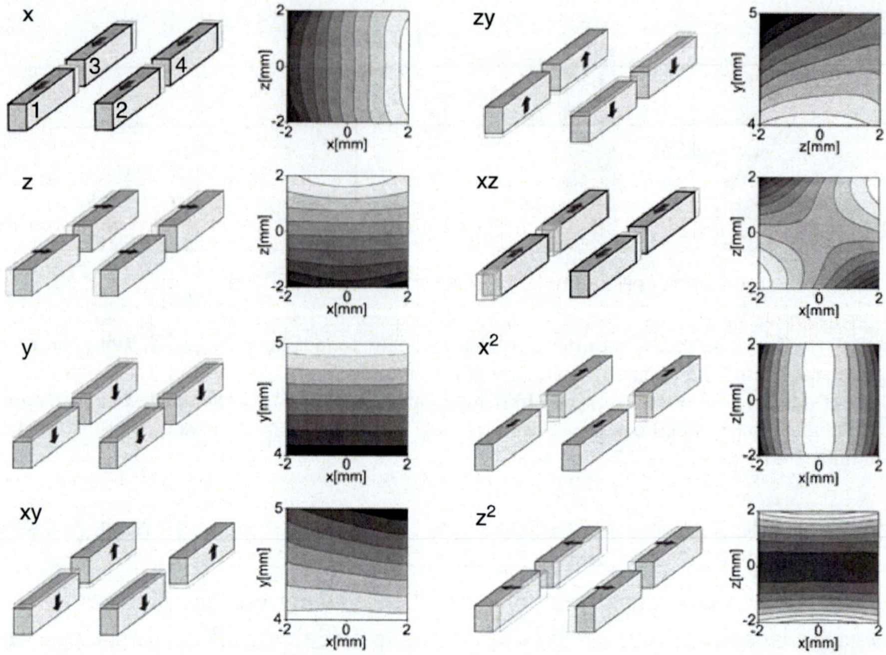

Fig. 7.6 The effect of moving the shim magnets in the directions indicated by the *arrows* is shown by means of 2D maps of the total magnetic field strength. The field strength (magnitude) increments are indicated by color changes from *black* (weak) to *white* (strong). The *magnet numbers* are shown in the plot for the linear x term

be understood by considering the effect of each pair on the total field. The magnet pair displaced along the positive x-axis (left pair) contributes with a G_x^{-z} that is negative[1] $\forall z$ and its magnitude decreases when moving from $z < 0$ (left) to $z > 0$ (right). The pair moved along the negative x-axis (right pair) generates a G_x^{+z} that is positive $\forall z$, and its magnitude decreases when moving from $z > 0$ (right) to $z < 0$ (left). If the amplitude of the displacement of each pair is the same, in terms of symmetry, both gradients have identical strength at $z = 0$, thereby canceling each other and resulting in a $G_x = 0$ at this position. For $z > 0$, $|G_x^{+z}| > |G_x^{-z}|$; thus a $G_x > 0$ is generated by the shim magnets. On the other hand, for $z < 0$, $|G_x^{+z}| < |G_x^{-z}|$, the opposite behavior is obtained. This dependence of the magnetic field shows that in a first approximation, the magnetic field is proportional to xz. The proportionality constant or, in other words, our control variable is the amplitude of the displacement along the x-axis. The other two cross-terms can be generated using the displacements described in Table 7.1 in which all the aforementioned movements are summarized as explained above. The shim unit displacements together with 2D maps of the total field obtained after the corresponding movement are shown in Fig. 7.6. The next section delineates the setup built to implement this shim concept as well as the results of experiments in ^{1}H high-resolution spectroscopy.

[1] According to Eq. 7.5, the linear variation of the field along x is given by $-\,(2\alpha_x^{\text{shim}}\delta x)x$, with $\alpha_x^{\text{shim}} > 0$.

Table 7.1 Shim components and required magnet displacements. The magnet numbers correspond to the ones in Fig. 7.6

	Magnet			
Shim component	1	2	3	4
x	$-\Delta x$	$-\Delta x$	$-\Delta x$	$-\Delta x$
y	$-\Delta y$	$-\Delta y$	$-\Delta y$	$-\Delta y$
z	Δz	Δz	Δz	Δz
x^2	Δx	Δx	$-\Delta x$	$-\Delta x$
z^2	Δz	$-\Delta z$	Δz	$-\Delta z$
xy	$-\Delta y$	$-\Delta y$	Δy	Δy
zy	Δy	$-\Delta y$	Δy	$-\Delta y$
xz	Δx	$-\Delta x$	Δx	$-\Delta x$

7.3 Experimental Results

By using the described approach, a main magnet built with a classical U-shaped geometry with dimensions of $280 \times 120 \times 280$ mm^3 and a gap $d_z^{\text{main}} = 100$ mm (Fig. 7.7) was shimmed [4]. It was made from a NdFeB alloy with a remnant flux density of 1.33 T, coercive field strength of 796 kA/m, and a temperature coefficient $\kappa = -1,200$ ppm/°C. Within the working volume, situated at a depth $y_0 \simeq 5$ mm, the average field measured was $B_{00} \simeq 0.25$ T (10 MHz ^{1}H Larmor frequency), and the field inhomogeneities were characterized by $G_y \simeq -1,000$ ppm/mm, $\alpha_z \simeq 300$ ppm/mm^2, and $\alpha_x \simeq -30$ ppm/mm^2. This magnet was previously used for imaging [6] and single-sided spectroscopy with the field-matching techniques [7]. The shim unit was designed according to the procedure described above. Although Sects. 7.1 and 7.2 showed how four magnet blocks (two identical pairs) can be designed and used to cancel the spatial variations of the main field, Ref. [4] indicated that it was more convenient to use two sets of four magnets (Fig. 7.7) denoted, respectively, as the lower and upper shim unit. The lower shim unit (the bulky one) was located at the bottom of the gap, whereby its main function is to cancel the strong gradient along the depth direction. The positions of the magnet blocks were fixed and, in practice, this set could be considered to be part of a new main unit with reduced gradient along the depth direction. The upper shim unit was also placed in the gap but directly underneath the sensor surface. These four blocks were movable along the three Cartesian directions in order to generate the desired shim components (Table 7.1). Thanks to the presence of the lower shim unit, the upper shim unit was built smaller to correct weaker inhomogeneities. This was advantageous, because the magnet displacements required for shimming are larger, leading to more precision in shimming the field. The dimensions and positions of all shim magnets were computer simulated by considering the scanned magnetic field of the main magnet as the target field to be shimmed. A four-turn surface rf coil with an outer diameter of 7 mm placed at 3 mm from the magnet surface was used for excitation and detection. The natural lateral selection of the rf coil was combined with a 90° soft pulse for excitation (slice) to achieve volume selection.

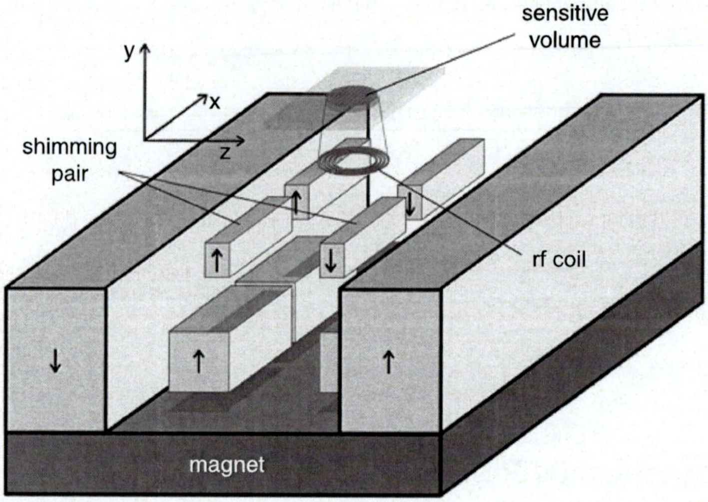

Fig. 7.7 Magnet array used to generate a volume of highly homogeneous magnetic field external to the magnet. The *arrows* show the direction of polarization of the magnets. The four pairs of shim magnets placed in the gap compensate for the inhomogeneity of the magnetic field of the main magnet

The sensitive volume generated by this magnet configuration and having lateral dimensions of about $5 \times 5\ \text{mm}^2$ and a thickness of 0.5 mm was located 5 mm from the surface of the magnet, whereby the measured ^{1}H Larmor frequency was 8.33 MHz (see Fig. 7.7). The sensor was also equipped with three single-sided shim coils placed inside the main gap (not shown in Fig. 7.7) that can produce pulsed gradients along the Cartesian directions. The coils producing the gradients along the x- and z-axes were similar to the ones described in [6] but with different aspect ratios. The coil generating the gradient along the depth was similar to the z-axis gradient coil but it was wound in an anti-parallel configuration. These coils were utilized in the final stage of the shimming procedure. Once the shim units were mounted and placed in their optimum positions (calculated), the final shim configuration was found iteratively. First, the total magnetic field was scanned by measuring the resonance frequency of a tiny water sample ($\sim$1 mm^3), displaced all over the sensitive volume. Then, the new positions of the shimming magnet pairs required to correct the remnant inhomogeneities were calculated numerically and readjusted in the real sensor. This procedure was repeated several times until deviations from the average field of less than 10 ppm were obtained. To further improve the resolution, the current through the single-sided shim coils was optimized by maximizing the signal peak in the frequency domain of a large sample. This procedure offered a fine correction of the linear terms. Figure 7.8a shows the spectrum of a water sample much larger than the sensitive volume (an arbitrarily large sample) placed on top of the sensor, whereby the line width is 2.2 Hz, corresponding to a spectral resolution of about 0.25 ppm. For comparison, the state-of-the-art spectrum for single-sided

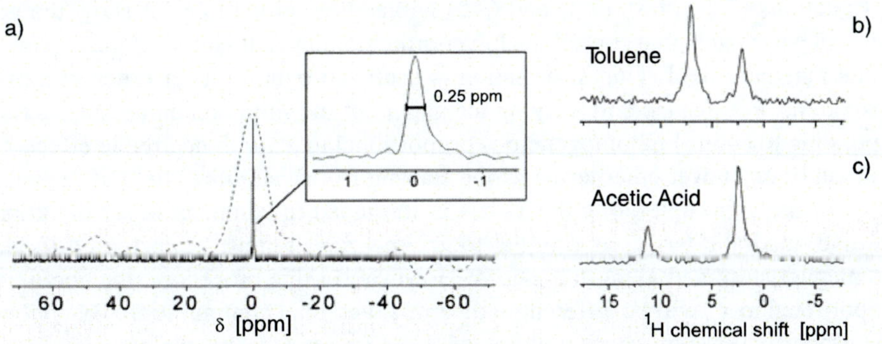

Fig. 7.8 **(a)** *Solid line*: magnitude spectrum of a water sample much larger than the sensitive volume, placed on top of the rf coil. The spectrum is the Fourier transform of the Hahn-echo signal from 64 scans acquired with a repetition time of 5 s to improve the SNR. The full width at medium height of the line (*inset*) is 0.25 ppm at a proton resonance frequency of 8.33 MHz. *Dashed line*: best spectrum obtained by *ex situ* spectroscopy by means of the nutation-echo method. The line width is 8 ppm [7]. Both data sets were obtained in the same measuring time, using sensors of the same size and working at comparable depths. Therefore, spectral resolution and sensitivity can be compared quantitatively. **(b)**, **(c)** ^{1}H NMR spectra of different liquid samples obtained within a measuring time of 1 min. The chemical shift differences and the relative peak intensities are in good agreement with the results obtained using conventional high-resolution NMR spectrometers

NMR, measured recently by nutation echoes in the presence of spatially matched static and rf fields [7], is shown as well, whereby the spectral resolution has been improved by a factor of about 30 with a concomitant fivefold extension of the excited volume along both lateral directions. Together with a sensitivity-optimized surface rf coil, the increased size of the sensitive volume resulted in an appreciably higher signal-to-noise ratio (SNR). The sub-part per million resolution achieved in this work allowed us to resolve different molecular structures such as toluene and acetic acid (Fig. 7.8b, c). The laboratory experiments were performed without the need of a temperature chamber. The insulation given by a 1 cm thick polystyrene foam layer in addition to the large thermal inertia of the magnet, allowed for averaging over 1 min without significant frequency drift. The toluene ^{1}H spectrum exhibited two lines at 7.0 and 2.1 ppm with relative intensities of 5:3, corresponding to the aromatic and the methyl protons, respectively. In the case of acetic acid, the two lines corresponded to methyl and carboxylic protons, which appear at 2.3 and 11.3 ppm, with an intensity ratio of 3:1 [4].

7.4 Shimming the Magnet to Higher Order

Applying a movable shim unit has allowed high resolution to be obtained even by using imperfect magnet pieces [4]. The first question that arises now is how can the configuration presented in the previous section be optimized so as to improve resolution, magnetic field intensity, and sensitive volume size?

Equations (7.2), (7.3), (7.4) and (7.5) and additionally Table 7.1 show the geometrical parameters that control the lower order homogeneity terms of the expansion of the magnetic field. This information is useful to obtain higher order shimming corrections. For instance, by varying the gap d_x of the shim unit along the x-axis, it is possible to control inhomogeneities proportional to x^2 with negligible effects (up to an order of 2) over other terms. Upon starting from a shimmed field, if the length L_x^{shim} of the shim unit along the x-axis is shortened to compensate for the arising positive quadratic term proportional to x^2, the gap d_x must be increased to keep the magnet shimmed along the x-axis. However, in this procedure, the coefficient proportional to x^4 will be different in the initial and final configuration. By applying this property, different configurations of the shim unit can be explored in order to obtain higher order shimming (configuration that minimizes the fourth order). In this sense, the optimum shim configuration is not searched randomly in the variable space defined by the geometrical parameters of the shim unit, such as L_x, L_y, L_z, d_x, d_z, z_0, but rather it is obtained in a controlled way by varying coupled variables as, for example, L_x^{shim} and d_x.

7.4.1 Improving Resolution and Working Volume Size

The signal acquired during an NMR experiment is proportional, among other parameters, to the amount of spins precessing at the Larmor frequency. Any deviation from this frequency value, e.g. resulting from inhomogeneities of the magnetic field, will cause a line broadening that determines the sensor resolution and, indirectly, the working volume. To quantify the performance of a given sensor, the spatial dependence of its magnetic field must be analyzed. The behavior of the magnetic field for the shimmed sensor described in Fig. 7.7 can be seen in the simulations shown in Fig. 7.9. Figure 7.9a shows the field variation along the depth for $x_0 = z_0 = 0$, where variations are measured with respect to the field value at $y_0 = 5$ mm. Although the sensor is shimmed, it can be seen that in a region of 4 mm centered at y_0 the field variations are within 100 ppm. A zoom of the figure shows that variations lower than 1 ppm are expected in a layer thinner than 500 μm. On the other hand, Fig. 7.9b, c shows the spatial dependence of field variations for the z- and x-axes, respectively. In each case three different depths $y - y_0 = 1, 0, -1$ mm are shown. Let us focus on the curve corresponding to $y - y_0 = 0$ (black squares). Both plots (Fig. 7.9b, c) show that within a region of 4 mm the inhomogeneities are one order of magnitude lower than the ones along the depth (Fig. 7.9a). Moreover, the variations along the z-axis (Fig. 7.9b) are more pronounced than the ones along the x-axis (Fig. 7.9c), corresponding to the fact that the gap direction is more critical in terms of homogeneity than the x-axis. It can also be seen from this depth ($y - y_0 = 0$) that the lateral range where the field variation stays within 1 ppm (zoom of Fig. 7.9b, c) is much larger than that corresponding to other depth values such as $y - y_0 = -1$ or $y - y_0 = 1$ mm, where the field variation easily exceeds 1 ppm in the lateral range of 2 mm. This plot shows proof for the sensitivity of the field homogeneity with respect to the depth variable. Whereas the lateral field variations are bounded for

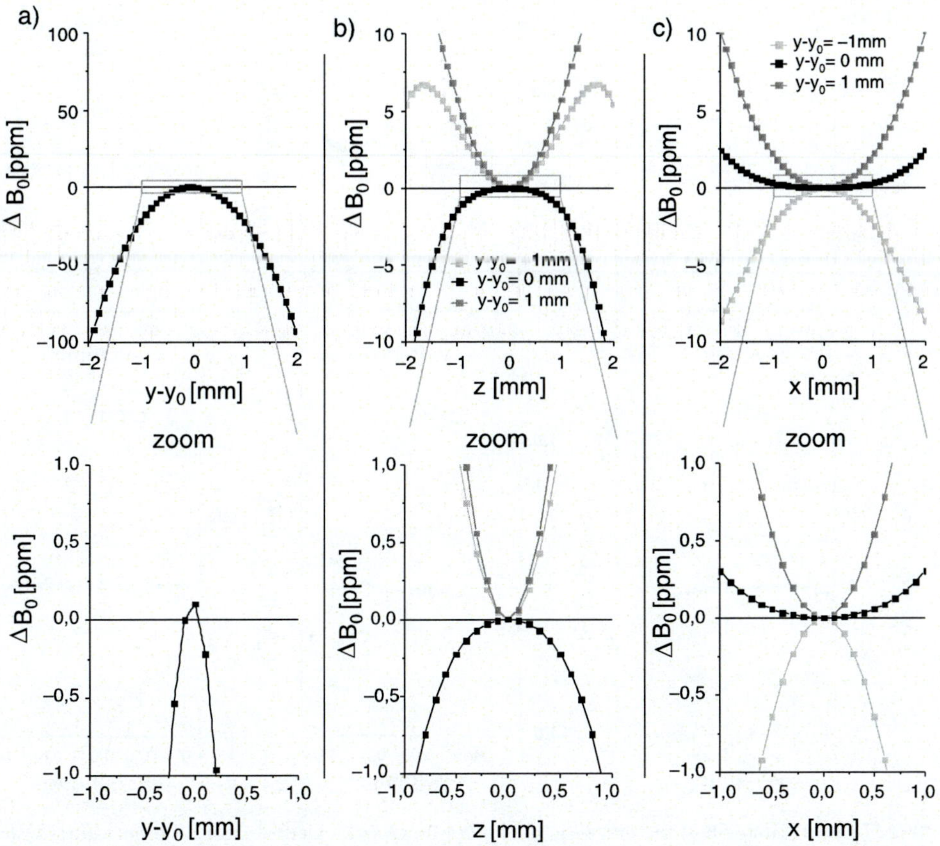

Fig. 7.9 Variations of the magnetic field strength as a function of the depth (**a**), and the lateral axes z (**b**) and x (**c**) calculated for the magnet shown in Fig. 7.7. The variations are computed with respect to the center of the sensitive volume located at $\mathbf{r_0} = (0, 5, 0)$ mm. The variations along the lateral axes are calculated for three different values of the depth coordinate. A zoom of the region within the *square box* is shown at the bottom of each figure

the working depth $y = y_0$, they immediately become out of control when moving a millimeter away, thereby restricting the size of the sensitive volume.

To optimize the sensor performance, it is then required to cancel the field variations rather in a depth range than at a single point y_0. For this purpose, it is convenient to monitor the behavior of the derivatives of the magnetic field instead of its magnitude. In a first approach, the variation of the field along the y-axis can be quantified by the magnitude of the field gradient G_y along the depth for $x_0 = z_0 = 0$. For the lateral directions the lower field variations are given by the coefficients α_x and α_z of the second-order terms proportional to x^2 and z^2, respectively, which is reflected in the plots of Fig. 7.9b, c. There, the parabolic dependence of the field along the x- and z-axes is clearly identified for $y - y_0 = -1$ and $y - y_0 = 1$ mm showing an important dependence on y, where in Fig. 7.9c $\alpha_x(y - y_0 = -1) < 0$, whereas $\alpha_x(y - y_0 = 1) > 0$. A detailed behavior of G_y, α_x and, α_z as a function

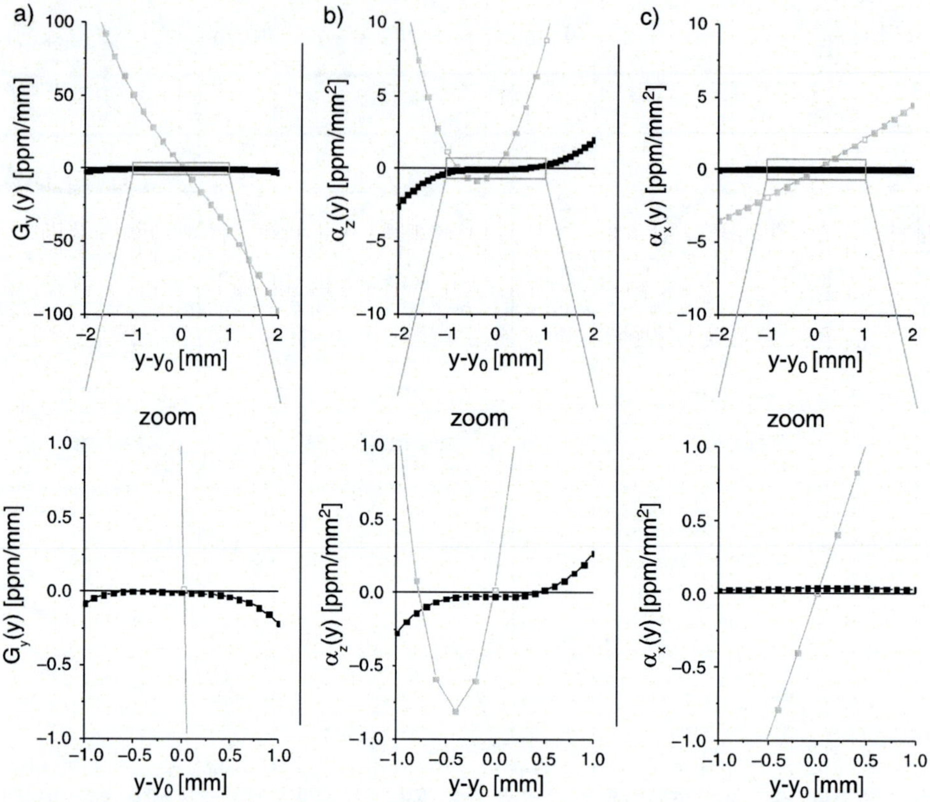

Fig. 7.10 (**a**) Gradient of the magnetic field along y, $G_y(y)$, calculated for different values of the depth. *Gray squares* correspond to the sensor of Fig. 7.7, whereas *black squares* correspond to a new sensor design. (**b**) Curvature of the magnetic field along z, $\alpha_z(y)$, calculated for different depth values. The *gray squares* indicate the behavior of the sensor of Fig. 7.7. The *open gray squares* correspond to the curves of Fig. 7.9b. *Black squares* display the behavior of an optimized sensor. (**c**) Idem as (**b**) but along the x-axis. For all plots, $y_0 = 5$ mm for the sensor of Fig. 7.7, while $y_0 = 4$ mm for the new sensor design

of y is shown in Fig. 7.10 with the gray line-dot curve. The particular values of these gradients corresponding to the curves of Fig. 7.9 are shown as empty gray squares. It can be seen that all of them are simultaneously zero at $y - y_0 = 0$, in accordance with Eqs. (7.2), (7.3), (7.4) and with the results shown in Fig. 7.8, where the spectral information obtained from protonated samples comes from a thin slice along the depth. However, for values of y that are different from y_0, these parameters (G_y, α_x, and α_z) deviate rapidly from zero.

A new sensor was optimized following, in a first step, the procedures described in Sect. 7.1. In this way, the dimensions of the main unit are first established depending on the final application of the sensor, and the dimensions and position of the shim unit, composed of an upper and a lower part, are calculated in order to satisfy Eqs. (7.2), (7.3) and (7.4). At this point, to increase the homogeneous region, the optimization process involves calculating the gradient G_y and the curvature of the field α_x and α_z for different depth values, in an attempt to reduce the slope at y_0. The ideal situation would be the one in which the slope and offset of each curve are zero. The values of these coefficients (G_y, α_x, and α_z) are monitored by exploring the space of variables obtained from the different sizes and position of the pieces that build the shim units. The black curves in Fig. 7.10 show the behavior of G_y, α_x, and α_z as a function of the depth for a particular configuration of these parameters (L_x, L_y, L_z, d_x, d_z, z_0). To compare the new simulations with the results obtained for the sensor of Fig. 7.7, the curves are plotted with respect to the center of the sensitive volume y_0 associated with each sensor. The difference between both sensors can be seen in more detail in the zooms of Fig. 7.10a–c). The plateau in the G_y curve shows that at least in a 2 mm range along the depth, the field variation is lower than 0.2 ppm. Moreover, since the values of α_z for this depth range are also very small (field variations that are lower than 0.2 ppm), this suggests that the volume of the sensitive region is not a thin slice but is rather a cylinder centered at y_0 with its axis along x. This change in the geometry of the sensitive volume would lead to an important increase in the sensitivity of the sensor.

In order to obtain the results shown by the black curves in Fig. 7.10, moving the lower shim unit is also necessary. In addition to the control variables mentioned in the last section, there are new variables such as the gaps along the x- and z-axes, d_x^{L} and d_z^{L}, respectively, and the position of the lower shim unit along the y-axis with respect to the upper one, $y_0^{\mathrm{U}} - y_0^{\mathrm{L}}$, whereby L and U denote the lower and upper shim units, respectively.

7.5 Temperature Compensation

Although the high homogeneity obtained by implementing movable permanent magnets allowed the resolution of proton spectra of molecules in solution, a particular temperature stability during the experiment is necessary to acquire the results shown in Fig. 7.8. The reason for this is the strong temperature dependence of the remnant magnetization of permanent magnets. Depending on the magnetic material, a temperature variation of just 1°C leads to a drift in the resonance frequency of several hundreds of parts per million, thus eliminating the possibility of averaging to improve sensitivity or eliminating any spectroscopy experiment that requires more than one attempt. To overcome this problem, a new concept based on the combination of materials with different magnetic properties can be considered. A straightforward solution would be to combine two magnetic materials with opposite temperature coefficients. In this way, whereas the field generated by one material increases with the temperature, the other decreases thereby keeping the total field constant. Even though the great majority of magnetic materials have negative temperature coefficients, the same cancelation effect can be attained if the two materials are polarized in opposite directions. Notice that the two materials do not need to be combined to build every single block in the magnet. Actually, the two materials are only required to generate fields that oppose each other in the region of interest, and no restriction is imposed on their spatial localization. By naming the unit A_1 and unit A_2 the sets of pieces with different magnetic materials, the total field can then be expressed as

$$B(\mathbf{r}, T) = B^{A_1}(\mathbf{r}, T) + B^{A_2}(\mathbf{r}, T), \tag{7.6}$$

where B^{A_1} and B^{A_2} each represents the main component of the magnetic fields generated by the units A_1 and A_2, respectively. In a first approximation, the magnetic field can be assumed to depend linearly on temperature,

$$B^{A_1}(\mathbf{r}, T) = B^{A_1}(\mathbf{r}, T_0)(1 - \kappa^{A_1}\Delta T), \tag{7.7}$$

where a similar expression holds for B^{A_2}. Here, $\Delta T = T - T_0$ and κ^{A_1} (κ^{A_2}) is the respective temperature coefficient.

Placing Eq. (7.7) into Eq. (7.6), the dependence of the total field with the temperature T is given by

$$\begin{aligned} B(\mathbf{r}, T) = {} & [B^{A_1}(\mathbf{r}, T_0) + B^{A_2}(\mathbf{r}, T_0)] \\ & -[B^{A_1}(\mathbf{r}, T_0)\kappa^{A_1} + B^{A_2}(\mathbf{r}, T_0)\kappa^{A_2}]\Delta T. \end{aligned} \tag{7.8}$$

In the case in which units A_1 and A_2 generate perfect homogeneous fields, the second term is responsible for the drift of the spectral lines but does not introduce any line broadening. By nullifying this term, a temperature-independent magnetic field can be generated. This condition can be fulfilled by setting the strength of the magnetic field of each unit according to the proportion of the temperature coefficients of the magnetic materials used.

The described approach can be also applied to the case of single-sided sensors, where the units A_1 and A_2 can be directly associated with the "main" and "shim" units, respectively. However, in this case, the spatial dependence of the magnetic field generated by both units must be taken into account. As previously stated in Sect. 7.1, the inhomogeneity along the depth is at least three orders of magnitude larger than in the other directions; thus, the space-dependent factor on the right-hand side of Eq. (7.7) can be expressed as

$$B^{\text{main}}(\mathbf{r}, T_0) = B^{\text{main}}(\mathbf{r_0}, T_0) + G_y^{\text{main}}(\mathbf{r_0}, T_0)\Delta y, \tag{7.9}$$

where the superscript A_1 was replaced by "main." A similar relation holds for the "shim" unit. By placing Eqs. (7.7) and (7.9) into Eq. (7.6), the dependence of the total field with the temperature T is given by

$$\begin{aligned} B(\mathbf{r}, T) = {} & [B^{\text{main}}(\mathbf{r_0}, T_0) + B^{\text{shim}}(\mathbf{r_0}, T_0)] \\ & -[B^{\text{main}}(\mathbf{r_0}, T_0)\kappa^{\text{main}} + B^{\text{shim}}(\mathbf{r_0}, T_0)\kappa^{\text{shim}}]\Delta T \\ & -[G_y^{\text{main}}(\mathbf{r_0}, T_0)\kappa^{\text{main}} + G_y^{\text{shim}}(\mathbf{r_0}, T_0)\kappa^{\text{shim}}]\Delta T\,\Delta y, \end{aligned} \tag{7.10}$$

where the condition $G_y^{main}(\mathbf{r_0}, T_0) + G_y^{shim}(\mathbf{r_0}, T_0) = 0$ (Eq. 7.2) was used, since the combination of both units generates a homogeneous field. As mentioned above, when the temperature varies, the second term of Eq. (7.10) modifies the value of the magnetic field in the central region of the sensitive volume. However, the third term of Eq. (7.10) produces a line broadening due to the effective gradient $[G_y^{main}(\mathbf{r_0}, T_0)\kappa^{main} + G_y^{shim}(\mathbf{r_0}, T_0)\kappa^{shim}]\Delta T$ along a slice of width Δy of the sensitive volume. In order to contemplate the importance of each of these contributions, consider, for example, the main unit of the sensor of Fig. 7.7 whose magnetic field is $B^{main}(\mathbf{r_0}, T_0) = 0.25$ T and the gradient along the depth is $G_y^{main}(\mathbf{r_0}, T_0) = 0.4$ T/m [6]. For a slice along the depth of $y = 0.5$ mm, we have that $G_y^{main}(\mathbf{r_0}, T_0)\Delta y = 0.0002$ T, which means that up to a multiplicative factor $(\kappa^{main}\Delta T)$, the magnitude of the third term is *three orders of magnitude* smaller than that of the second term. Then, to measure spectra under temperature variations, it is first necessary to eliminate the huge drift of the spectral line; only then will the line broadening become observable. To eliminate the line drift, the following condition should be fulfilled:

$$B^{main}(\mathbf{r_0}, T_0)\kappa^{main} + B^{shim}(\mathbf{r_0}, T_0)\kappa^{shim} = 0. \tag{7.11}$$

As the polarization of the fields of the two respective units points in opposite directions, to satisfy Eq. (7.11) it suffices to vary the dimensions of the shim unit so as to modify the average field strength, thereby keeping the matching conditions (Eqs. (7.2), (7.3) and (7.4). At this point, one may argue that an arbitrary increase in the dimensions of the shim unit is not possible, because overlapping with the main unit can occur (notice that the shim unit is placed inside the main gap). This limitation would impose a boundary for the maximum value of $B^{shim}(\mathbf{r_0}, T_0)$ consequently restricting the use of Eq. (7.11). It is possible, however, to overcome this by realizing that since the field polarizations of the main and shim units oppose one another, the regions where the overlapping occur are equivalent to holes in the main unit. To clarify this point, let us momentarily assume that both main and shim units are made from the same material that guarantees equal remnant polarization for both units. In this case, the fields produced by the overlapping materials exactly cancel each other, thereby reproducing a situation in which no magnetic material is placed in the volume of intersection (see Fig. 7.11a). This concept can also be applied in a situation where the remnant polarizations of the shim and main units are different. Consider that the remnant polarization of the shim unit is 10% higher than the one of the main unit. Values approximate a real situation if the shim and main units are made of NdFeB and SmCo, respectively. Under these conditions, the final result from the overlap of materials is not a hole in the main unit but rather an equivalent magnet with the size of the overlapped region and a remnant polarization strength 10% of the shim unit and pointing in the same direction of the strongest polarization (see Fig. 7.11b). This set of three different magnets (shim, fictitious, and main magnets) can be replaced by a main unit with a hole plus a shim unit with smaller dimensions (right plot in Fig. 7.11b). In this way, it is possible to "transfer"

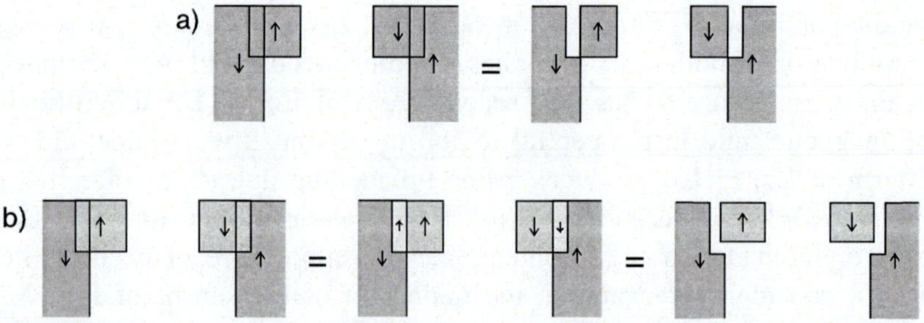

Fig. 7.11 (**a**) Scheme showing the superposition of main and shim units (*left*) and the equivalent configuration in which the main unit has a "hole" and the shim unit is left without the portion of material superposed with the main unit (*right*). The same *gray scale* for both units indicates equal remnant polarizations. (**b**) Similar case as in (**a**) but for main and shim units whose remnant polarizations are different. The plot furthest *left* shows the superposition case. The superposition region can be thought as a fictitious magnet with lower polarization (*middle plot*) which finally results in a hole in the main unit and a shim unit with modified volume (*right plot*). The strength of the remnant polarization is indicated by different *gray scales* changing from *dark gray* (strong) to *white* (no polarization)

material from the shim unit to the main unit until Eq. (7.11) and the homogeneity conditions given by Eqs. (7.2), (7.3), and (7.4) are simultaneously satisfied.

It can be seen from Eq. (7.11) that the total field of the sensor could be increased if the ratio of the thermal coefficients $\kappa^{\text{shim}}/\kappa^{\text{main}}$ increases. For the case in which the main unit is built of SmCo and the shim unit of NdFeB, the ratio $\kappa^{\text{shim}}/\kappa^{\text{main}} = 4$. This means that the final sensor produces a magnetic field one-fourth smaller than that generated by the main unit.

To eliminate the third term in Eq. (7.10), it is not possible to apply the same argument used with the second term (Eq. 7.11), because the restriction $G_y^{\text{main}}(\mathbf{r_0}, T_0) + G_y^{\text{main}}(\mathbf{r_0}, T_0) = 0$ has already been imposed to the main and shim units in order to satisfy homogeneity (see Eq. 7.2). Then, in this case, the solution would be to design a sensor whose units satisfy $G_y^{\text{main}}(\mathbf{r_0}, T_0) = G_y^{\text{shim}}(\mathbf{r_0}, T_0) = 0$.

By following the strategy presented in Sects. 7.4 and 7.5, a new sensor design was optimized for improving the performance of the previously built magnet [4] in terms of homogeneity, magnetic field intensity, and homogeneous region, fulfilling the temperature compensation condition. This sensor consists of a U-shaped main unit made of SmCo with similar dimensions as the one shown in Fig. 7.7, but with the particularity that it has two holes in each block of the U-shaped magnet. These holes can be interpreted as the overlap of a certain portion of the main magnet and a smaller U-shaped magnet whose opposite polarization results in a zero magnetic field in the superposition volumes as was explained previously. In addition to the main unit, the sensor has two movable shim units which are made of NdFeB in order to satisfy Eq. (7.11). The magnitude of the field generated at the center of the sensitive volume, which is placed at $y = 4$ mm from the surface, is 9.6 MHz. This represents an increase of 30% with respect to the field that the sensor of Fig. 7.7 would produce if its main unit were built from SmCo.

A good indicator of the homogeneity of the magnetic field generated by the sensor in a whole volume is provided by the line width measured at half height of the Fourier transform of the free-induction decay. Simulations of the spin evolution in the magnetic field using the conventional Bloch equation within a cylindrical volume of radio $R = 1.5\,\text{mm}$ in the yz-plane and length $L_x = 4\,\text{mm}$ centered at $\mathbf{r_0} = (0, 4, 0)\,\text{mm}$ were done. The line width obtained was lower than 0.15 ppm. It can be said from this line that more than 70% of the signal comes from a bandwidth smaller than 1.6 Hz. The temperature variation of the magnetic field was also calculated and found to be 1.7 ppm/°C which improves the sensor performance by three orders of magnitude as was reported in the previous section.

7.6 Conclusions

This chapter has extensively described the strategy to shim the field of single-sided magnets to high homogeneity by means of small, movable permanent magnet blocks. Particular attention has been paid to explain the combination of different movements of the magnetic pieces forming the shim unit required to generate and control the lower shim order terms (x, y, z, x^2, z^2, xy, yz, and xz) of the magnetic field. The experimental implementation of this procedure leads to a spectral resolution of 0.25 ppm. This resolution allows the use of *ex situ* NMR to determine the molecular composition of liquids by analyzing the ^{1}H NMR spectra.

To increase the magnet performance, it was convenient to match the inhomogeneities of the main and shim units (Eqs. (7.2), (7.3) and (7.4)) over a volume instead of a particular depth position. This higher order shimming was obtained by monitoring the derivatives of the magnetic field as a function of the depth. These functions were then analyzed for different values of the geometric parameters of the shim unit of the sensor such as length, width, height, and length of gaps. Thus, optimal configurations for the magnetic pieces were obtained which improves the sensitive volume size and generates a higher magnetic field (both relevant issues concerning sensitivity). To exert control over these higher order shim terms, it is necessary to move the pieces of the lower shim unit.

Finally, a novel approach to overcome the major problem introduced by the temperature dependence of the remnant polarization of magnetic materials has been discussed. This approach consists of combining two magnetic materials with different thermal coefficients arranged properly in order to build a magnet generating a temperature-compensated magnetic field with the same performance as the one described to obtain sub-part per million resolution.

References

1. Golay MJE (1958) Field homogenizing coils for nuclear spin resonance instrumentation. Rev Sci Instrum 29(4):313–315
2. Fukushima E, Jackson JA (2002) Unilateral magnet having a remote uniform field region for nuclear magnetic resonance. US Patent 6489872

3. Meriles CA, Sakellariou D, Heise H, Moule AJ, Pines A (2001, July) Approach to high-resolution *ex situ* NMR spectroscopy. Science 293(5527):82–85
4. Perlo J, Casanova F, Blümich B (2007, Feb) *Ex situ* NMR in highly homogeneous fields: H-1 spectroscopy. Science 315(5815):1110–1112
5. Popella H, Henneberger G (2001) Design and optimization of the magnetic circuit of a mobile nuclear magnetic resonance device for magnetic resonance imaging. COMPEL 20:269–278
6. Perlo J, Casanova F, Blümich B (2004, Feb) 3D imaging with a single-sided sensor: an open tomograph. J Magn Reson 166(2):228–235
7. Perlo J, Demas V, Casanova F, Meriles CA, Reimer J, Pines A, Blümich B (2005, May) High-resolution NMR spectroscopy with a portable single-sided sensor. Science 308(5726):1279–1279

Chapter 8
Applications in Biology and Medicine

Bernhard Blümich

Diagnostic imaging has revolutionized medicine. Computed tomography (CT), magnetic resonance imaging (MRI), and ultrasound are standard diagnostic tools in medicine [1]. Ultrasound sensors are sufficiently compact to be operated in any medical doctors office, and CT scanners have become available that can be moved to operating rooms and intensive care patients. Given the success of medical MRI, the use of mobile NMR for medical diagnostics is an exciting topic [2, 3]. Mobile instruments are being introduced for MRI [4], in particular for neurosurgery [5], and dedicated limb scanners have been on the market for some time [4–6]. Small imagers with closed magnets have been developed [7–9] for different kinds of biomedical studies [10, 11]. Early on, portable NMR devices with open magnets have been proposed in the context of the development of well-logging sensors for use as a medical diagnostic tool [2], and a miniaturized well-logging sensor has made it all the way into the clinical testing stage as an endovascular endoscope [12–14]. Medical applications of larger portable, single-sided NMR devices [15, 16] started to be investigated with the advent of the NMR-MOUSE [17–19]. The reported studies focus on surface-near tissue like the Achilles tendon [16, 20–22] and skin [16, 23, 24] measuring tissue anisotropy and depth profiles. Two-dimensional imaging has been explored on the finger and on biological phantoms [16, 25–27]. *Ex vivo* studies of biological tissue have been conducted on mummies and bones [28], and a silicone breast implant has been analyzed in view of testing the use of single-sided NMR as a quality control tool of the sterile implant in a sealed wrapping as well as in view of identifying a shell rupture by changes of the filler through contact with body fluid [29]. In the following, studies of skin, tendon, and bone with single-sided NMR are reviewed as well as some imaging studies.

B. Blümich (✉)
Institut für Technische Chemie und Makromolekulare Chemie, RWTH Aachen University, D-52074 Aachen, Germany

F. Casanova et al. (eds.), *Single-Sided NMR*,
DOI 10.1007/978-3-642-16307-4_8,

8.1 Skin

Images detailing the stratigraphy of skin are hard to obtain by conventional MRI machines, and special coils [30–33] or dedicated scanners are needed [34–39]. Yet skin is the largest organ of humans, covering about 2 m^2 in area. It is the interface of the body to the environment, which protects the body and plays a defining role in the perceived identity of individuals. Single-sided NMR is particularly well suited for skin studies, as only a few millimeters of penetration depths are of interest and the skin of nearly each part of the body can be accessed with a small sensor [16, 23, 24]. Topics of interest in skin studies concern pathological changes like cancer, healing of skin, formation of scars, and cosmetic skin treatment. The first studies were conducted with the original U-shaped NMR-MOUSE [17] providing inferior depth resolution by way of its curved sensitive volume [23]. Yet cutis and subcutis could already be identified due to their different relaxation behavior. Most subsequent studies employed the profile NMR-MOUSE [19]. Its high depth resolution of better than 5 μm is particularly appreciated when different layers of the cutis need to be resolved.

As the mama carcinoma is the most frequent type of carcinoma in women living in the western world, breast cancer screening tests are well established. Nevertheless, some diagnostic problems remain, such as the identification of carcinoma without calcerous deposits in the dense gland tissue and the identification or exclusion of malignoma in scar tissue following an operation or radiation therapy. In these cases, NMR tomography of the breast should be conducted together with the application of contrast agents. For some patients the localization of the area of interest is well known, so that an imaging analysis is not required. In an early pilot study with the U-shaped NMR-MOUSE, the dignity of breast cancer was assessed (Fig. 8.1a) [40–42]. Following the intravenous application of a contrast agent (Gd-DTPA, Magnetvist®) the amplitude of the partially saturated NMR signal was monitored at a fixed recovery time as a function of time after delivery of the contrast agent. The signal buildup times provide the information about the tissue dignity [43]. T_1 was significantly shortened by the contrast agent, but the signal-to-noise ratio was low. For tissue discrimination, the echo amplitude was monitored at a fixed recovery time t_R of 200–250 ms following saturation. The value of t_R was selected in such a way that a low signal was observed for healthy tissue without contrast agent. With contrast agent the signal increased due to faster T_1 relaxation. As malign tissue grows fast, the blood supply is good, and the signal buildup is fast following delivery of the contrast agent. In this way malign tissue and skin could be differentiated by the NMR-MOUSE (Fig. 8.1b). Even scar tissue and skin could already be differentiated. Scar formation is a medically relevant issue in the context of burn injuries and skin transplants. With the profile NMR-MOUSE, the healing process of the skin can be monitored through the bandage. In a clinical study it has been found that the signal amplitude (Fig. 8.2a) and the relaxation time $T_{2\mathrm{eff}}$ (Fig. 8.2b) follow similar modulations through the depth of the skin. Both are able to clearly differentiate healthy skin and skin scarred from taking a split-skin graft,

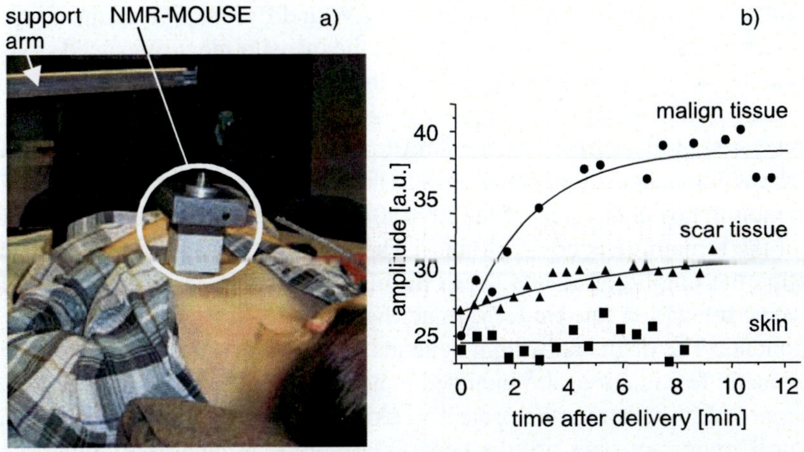

Fig. 8.1 Clinical study of contrast-agent-induced T_1 change of an inflammatory mama carcinoma. (**a**) Setup of the NMR-MOUSE on a healthy volunteer. (**b**) Signal recovery curves acquired after delivery of contrast agent to a patient with breast cancer. The signal buildup for different tissues is observed at a depth of 0–1 mm (*bottom*) with the saturation recovery method at constant recovery times t_R. Malign tissue ($t_R = 250\,\text{ms}$), scar tissue ($t_R = 200\,\text{ms}$), and skin ($t_R = 200\,\text{ms}$) can be discriminated. The buildup times extracted from the experimental data are 2.5 and 5.7 min for the malign and scar tissues, respectively, and essentially infinity for the skin

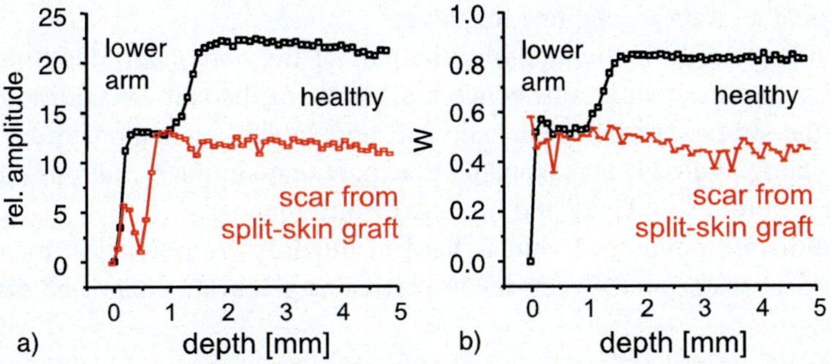

Fig. 8.2 Comparison of CPMG amplitude profiles (**a**) through the skin of both lower arms of a patient where skin has been split off from one arm for replacement at another part of the body and corresponding $T_{2\text{eff}}$ profiles (**b**b)

although the CPMG amplitude is more sensitive to the structure of the outer layer of the scar than $T_{2\text{eff}}$.

As skin plays a defining role in the perceived identity of individuals, it is of great interest to apply the NMR-MOUSE in studies of skin treated with cosmetics. NMR has been shown to be a suitable method for resolving different skin layers and following moisture uptake and drying of skin by magnetic resonance imaging with clinical imagers [30–33] and with the half-open GARfield magnet [34–39]. The cost of clinical MRI for such studies is high, and the GARfield methodology can only access the skin on the side of the hand. The profile NMR-MOUSE appears to be the most versatile and least expensive NMR device for mapping skin and studying the effects of cosmetics on skin in vivo at different parts of the body. For the measurement of skin depth profiles in vivo, the measurement time needs to be short. This can be achieved by limiting the number of points across the depth range and by increasing the signal-to-noise ratio through addition of CPMG echoes. It turned out that for skin measurements, a contrast parameter w derived from a CPMG train with about 300 echoes is advantageous to be calculated as the quotient of two sums. The denominator is the sum of the first 50 echo amplitudes and the numerator the sum of the remaining echo amplitudes. The resultant number w can be understood to define the amplitude of the depth profile in terms of spin-density weighted by transverse relaxation rate time $T_{2\mathrm{eff}}$ and the self diffusion coefficient of the mobile components. The diffusion weight depends on the value of the echo time.

Cosmetic formulations are intended to act on the outer layers of the skin and are not supposed to penetrate the protective skin barrier. Starting from the outside, the two most important ones are the superficial epithelial layer, the epidermis, and a deep connective tissue layer, the dermis. The strata of the epidermis are the outer stratum *corneum*, where the cells are keratinized in a squamous structure, and the stratum *basale*, where the cells are generated. The dermis is subdivided into two layers. The thinner more superficial one that lies adjacent to the epidermis is the stratum *papillare*, and the deeper one known as the stratum *reticulare*. Underneath the dermis lies the hypodermis which consists of adipose tissue. It is not part of the skin although some of the epidermal appendages including hairs and sweat glands often appear as if they penetrate this layer.

Depending on the individual and the part of the body, skin depth profiles vary (Fig. 8.3a, b). These variations concern all strata of the skin. In addition, the diameters of the strata vary between male and female skin as demonstrated in Fig. 8.3 for two young adults. It is reasonable to expect that similar variations can be found between different skin types and for skin of different ages.

The moisture content of skin is hard to quantify by relaxation measurements, as protons of water and glycerides in particular glycerine cannot be distinguished

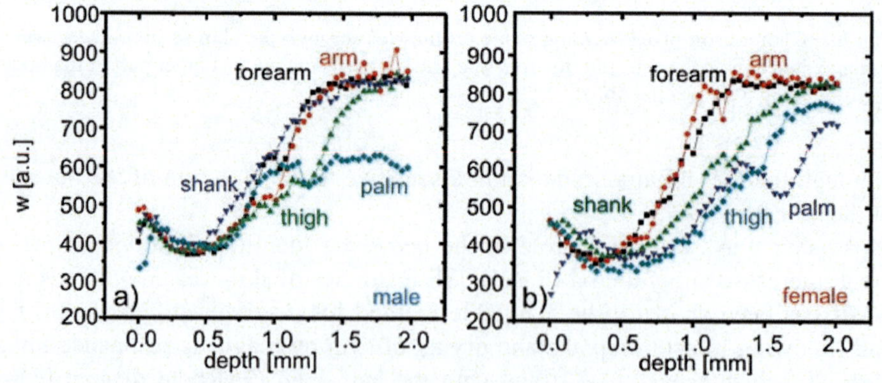

Fig. 8.3 Layers of the skin and depth profiles through the skin of a male (**a**) and a female (**b**) volunteer showing considerable differences at similar body sections

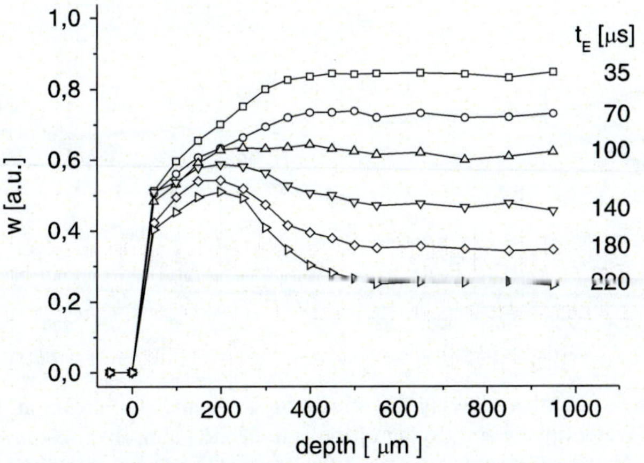

Fig. 8.4 Skin profiles measured in the palm of the hand as a function of the echo time. The nominal spatial resolution was 100 μm according to an acquisition window of 20 μs. The position of the sensor was moved in steps of 50 μm in the first 500 μm and in steps of 100 μm up to 1,000 μm. The acquisition time per point was 20 s using 64 scans and a repetition time of 300 ms

due to their strong intermolecular interactions. Yet the mobile components can be characterized by their translational diffusion. The shape of the depth profile changes considerably when increasing the diffusion weight by prolonging the echo time t_E in CPMG measurements (Fig. 8.4). It turns out that for t_E short enough to neglect attenuation by diffusion, the amplitude of the profile changes only in the region of the epidermis. This is in agreement with an expected change from dead cells in the stratum *corneum* to young and soft cells in the stratum *basale* and the highly moisturized subcutis. According to the profiles depicted in Fig. 8.4, the region corresponding to the epidermis is the least affected by diffusion, while the deeper lying layers are more affected.

Given the sensitivity of the parameter w to diffusion and relaxation, it is a suitable quantity to study the effects of moisture and skin-care products by means of depth profiles. As skin-care products do not penetrate the skin, depth ranges of less than 1 mm are sufficient, yet high resolution is required to resolve the layer structure of the outer skin. At such low depths, smaller rf coils can be employed that generate a stronger B_1 field to excite slices 100 μm thick with sufficiently short rf pulses. Spatial resolution across the slice is obtained by Fourier transformation of the echo. Figure 8.5 depicts a series of profiles obtained by exciting 100 μm thick slices and shifting the sensor in increments of 100 μm to map the effects of hydration with water and skin cream at different times [24]. The skin of the palm dries from being soaked in pure water in a time longer than 30 min (Fig. 8.5a) while the moisturizing effect of skin cream vanishes in much shorter time after wiping off the cream (Fig. 8.5b). The changes in the skin depth profiles of the palm caused by exposure to water and skin cream are up to 250 μm in both cases, which is the region associated with the epidermis.

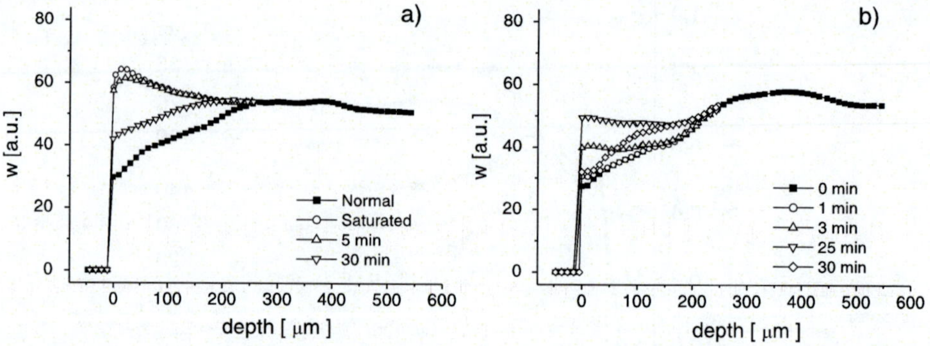

Fig. 8.5 Skin-depth profiles showing the effects of water and skin cream at different times after application. (**a**) Water uptake and drying of the skin of the palm. (**b**) Moisturizing of the palm by skin cream and drying following removal of the cream. The acquisition time for one profile was 1 min

These examples demonstrate that the skin structure is different for male and female subjects and at different parts of the body. Correspondingly, the moisturizing depths differ. Given the free access provided by the open geometry of the NMR-MOUSE, such investigations can be conveniently performed in vivo to study the effects of different skin-care products on the skin at different parts of the body.

8.2 Tendon

Tendon is a high-performance biological material of exceptional strength with a high degree of macroscopic molecular order, which has been studied in detail by different NMR methods [22, 44–52]. It consists mainly of collagen triple helices which are arranged in bundles in a hierarchical structure with crimps and twists to provide optimum mechanical stability (Fig. 8.6a) [53]. Due to the high order in tendon and cartilage, the magic angle effect is observed as the tensorial spin interactions for fibers are reduced when oriented at the magic angle [54–56]. With conventional tomography in humans, tissue anisotropy [44] can be followed in vivo only by spectroscopic imaging and by investigations of the anisotropy of translational diffusion with pulsed gradient fields [57–59], while the relaxation anisotropy can only be studied successfully on small samples in vitro or on animal models [22, 45–47, 52].

Single-sided NMR sensors with the field parallel to the surface such as the original NMR-MOUSE and the profile NMR-MOUSE are well suited to study the angle dependencies of NMR parameters from large objects in terms of relaxation rates and other quantities like double-quantum NMR and, more generally, NMR of spin modes [60, 61], as the direction of the polarization field B_0 can be changed within the sensitive slice by rotating the sensor (Fig. 8.6b). Studies of Achilles tendon in vivo with single-sided sensors are facilitated by the fact that the tendon is close to the surface (Fig. 8.6c).

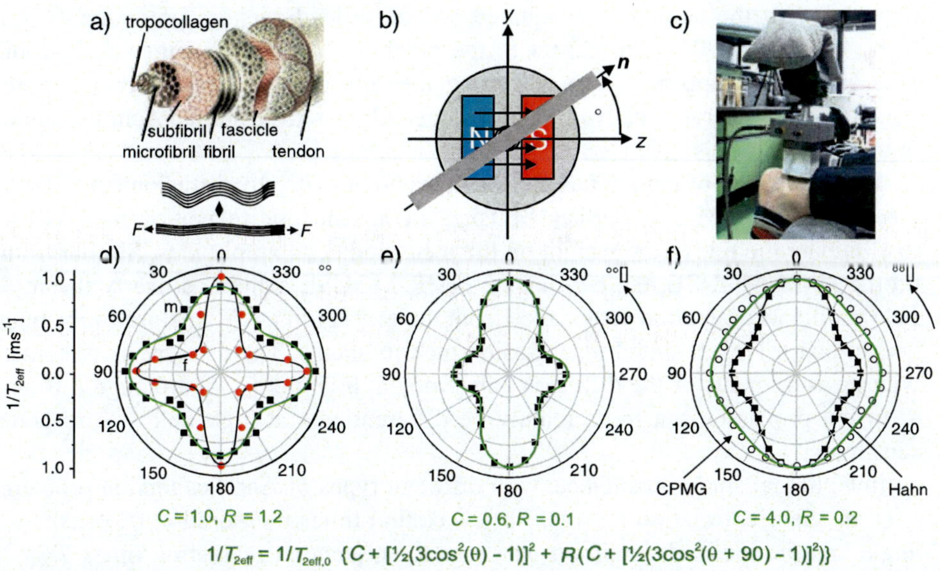

Fig. 8.6 Angle dependences of the transverse relaxation rates in tendon. (**a**) Schematic drawing of the hierarchical structure of tendon and the crimps which provide elasticity. (**b**) Definition of the orientation angle between the B_0 field direction of the NMR-MOUSE and the tendon orientation. (**c**) Setup for measuring human Achilles tendon in vivo with the NMR-MOUSE. (**d**) Transverse relaxation rates normalized to 1 ms at $\theta = 0°$ of the Achilles tendons of a male (m) and a female (f) volunteer. (**e**) Normalized CPMG relaxation rates for sheep Achilles tendon in vitro. (**f**) Normalized CPMG relaxation rates of a rat tail acquired with the CPMG sequence at an echo time of 0.1 ms and with the Hahn echo sequence. The points are measured data, and with the exception of the inner curve in (**f**), the drawn lines are fits with the equation and parameters given in the figure. All measurements were conducted in 4–5 mm depth. The acquisition of the data for one orientation lasted 5 min

The magic angle effect has first been observed with the NMR-MOUSE on porcine Achilles tendon in vitro and human Achilles tendon in vivo [20, 40, 42]. The orientation dependence of the relaxation rate was found to follow the square of the second Legendre polynomial in agreement with high-field studies [45–47]. The investigations were subsequently extended to compare the relaxation anisotropy of human Achilles tendon in vivo (Fig. 8.6d) with that of sheep Achilles tendon (Fig. 8.6e) and a rat tail in vitro (Fig. 8.6f). A smaller transverse relaxation rate is observed near the magic angle. As the relaxation in collagen is dipolar in nature [62] and the spins near resonance are locked in a CPMG sequence [23, 63–65], the dipolar interaction is partially averaged by the CPMG excitation but not by the Hahn echo excitation, so that the apparent anisotropy is reduced in CPMG measurements (Fig. 8.6f). Furthermore, the anisotropy of the effective relaxation rate may be reduced due to signal attenuation by diffusion. Yet the CPMG sequence is preferred for in vivo studies as the measurements are much faster than with Hahn echoes [2].

While the angle dependence of $1/T_{2\text{eff}}$ observed in vitro for rat tail and sheep Achilles tendon follows the square of the second-order Legendre polynomial, the angle dependence observed in two humans is more complex. A good fit could be obtained with the sum of two squared second-order Legendre polynomials, one being rotated by 90° with respect to the other (cf. formula in Fig. 8.6, bottom). Both orientations appear to arise with about the same amplitude, indicating a twisted structure of the hierarchically ordered collagen fibers for the human Achilles tendon in vivo (Fig. 8.6a) [16, 22].

The anisotropy of tendon has also been probed in vitro by diffusion studies with the bar-magnet NMR-MOUSE [18] using a constant-time relaxation method [66] to eliminate the relaxation weight on the measured echo amplitudes. Similar to the profile NMR-MOUSE, the bar-magnet NMR-MOUSE exhibits a nearly linear B_0 field profile as a function of distance from the magnet in a region close to the magnet, so that it is well suited for diffusion measurements. The data are in agreement with measurements of the diffusion anisotropy at high field [51] although different diffusion processes can more readily be differentiated at high field due to better sensitivity.

In a clinical study on subjects with different types of Achilles tendon problems [21], no clear correlation between the relaxation times measured at 0° orientation angle and the medical condition was found. The tendon relaxation times $T_{2\mathrm{eff}}$ of athletes and healthy young adults were all in the same range including that of an Achilles tendon healed after a fracture. This indicates that the measurement of relaxation times in Achilles tendon at 0° orientation angle is not a useful medical diagnostic tool and that alternative approaches need to be explored such as the relaxation measurements at a different orientation angle, measurement of anisotropy parameters, relaxation-weighted signals, and double-quantum filtered signals possibly in combination with volume selection to eliminate signal from surrounding tissue [22, 49, 52, 67].

8.3 Mummies and Bones

Mummies and bones consist of biological tissues which in some cases are of interest for non-destructive analysis by mobile NMR, as they may not be allowed to leave the museum or need to be chilled as the 5,300-year-old glacier mummy in the Archeological Museum in Bozen (Fig. 8.7). In Egyptian mummies, depth profiles can discriminate between textile wrapping from tissue and bone, and in the glacier mummy the thickness of the ice layer, tissue layers and the bone structure can be examined. Interestingly, the signal from the bone region of the glacier mummy is higher than that of the new skull and the model cadaver, which had been artificially dried down to the same weight as the iceman. Furthermore, the signal amplitude scales with the bone density and decreases with progressive states of decay.

Bone density can be determined quickly and non-destructively with the NMR-MOUSE in contrast to the conventional and destructive way of determining bone density by mercury injection. It is an important indicator for the state of conservation of the bone material. This is corroborated by the observation that an Egyptian mummy head and a 900-year-old skull give NMR depth profiles with amplitudes

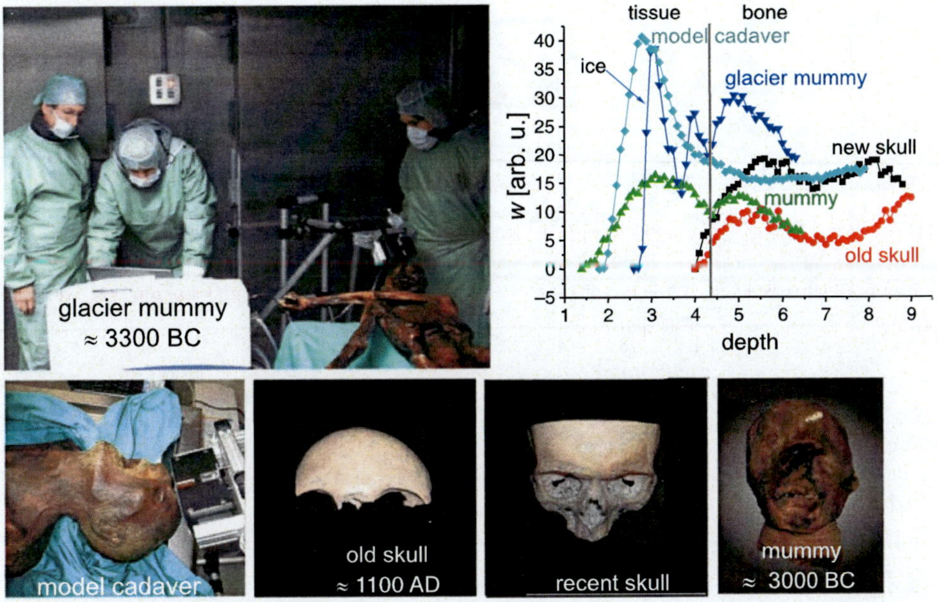

Fig. 8.7 Depth profiling of foreheads [16, 28] and photos of mummies and skulls including a measurement setup for analysis of the frozen glacier mummy from the Archaeological Museum in Bozen, Tyrol. The profile NMR-MOUSE is mounted near the foreheads of the mummies to measure depth profiles

significantly lower than those of profiles through a modern skull. The state of bone conservation is a quantity in demand in the search for historic DNA: the higher the bone density, the higher the chance to find DNA [16, 28].

To interpret the signal measured for the iceman at about 0°C, bone sections taken from the forehead of the recent skull and of an old skull were measured at room temperature, frozen in the dry state, in the wet state, and frozen in the wet state (Fig. 8.8). The bone of the new skull gives a higher signal in the dry state than that of an old skull, both at room temperature and frozen (Fig. 8.7). In the water-saturated, wet state the signal amplitude is higher due to the water filling the pores. The signal increase for the old skull is higher than that for the new skull in agreement with the interpretation that the bone density is lower and all of the available pore space is filled with water. Moreover, after prolonged exposure to temperatures of −30°C, the signal from the water-soaked, new skull is the same as at room temperature. This indicates that the water appears to be liquid inside the bone at this low temperature due to confinement effects. On the other hand the old skull shows some signal decrease upon cooling to −30°C, so that some water appears to be frozen. This water is expected to reside in the larger pores formed by the bone decay. As the profile through the forehead of the iceman measured at a temperature of just below 0°C is higher than that of the new skull, it can be concluded, that the water in the bone pores of the iceman is not frozen.

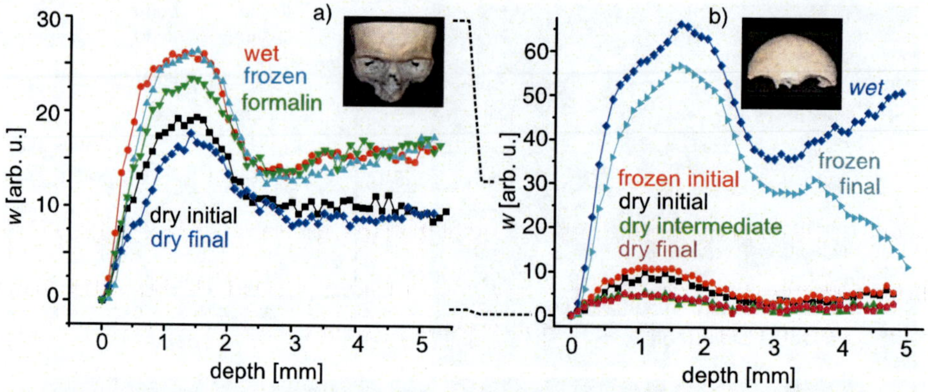

Fig. 8.8 Wetting, drying, and freezing study in terms of depth profiles through bone sections from foreheads. (**a**) New skull. It gives a higher signal in the dry state than the old skull. (**b**) Old skull from about AD1100 . Either water wet at room temperature or frozen to −30°C it gives much higher signal than the new skull with the same profiles in both states

8.4 Unilateral Imaging of Biological Matter

When dealing with patients and living species, the measurement times need to be short. This is why portable and more so, single-sided imaging has not yet found applications in the medical community [25]. The first 2D NMR images measured by unilateral NMR have been composed of single pixel measurements by displacing the NMR-MOUSE across a cut of a pork leg obtained from a butcher. The pixel information was acquired with a steady-state saturation recovery sequence which allows the simultaneous and rapid determination of T_1 and T_2 [23]. The spatial resolution was low, because it is determined by the extension of the sensitive volume, but the contrast is excellent.

To improve the spatial resolution and to cope with the field gradients inherent to unilateral NMR, single-point imaging in real space has to be abandoned in favor of single-point imaging in Fourier space [68–72]. Early Fourier images of medical tissue acquired with the unilateral NMR device are depicted in Fig. 8.9. Within an acquisition time of 9 min, one slice through a finger was acquired and three images with different contrast values were calculated from the data by adding different parts of the CPMG echo train for each pixel (b) [25]. The slice thickness is 1 mm and so is the lateral resolution. The image size is 32×32 pixels. In comparison with the first medical Fourier images [73], the quality is acceptable. For comparison with state-of-the-art medical MRI, a similar image acquired on a commercial 1.5 T tomograph is depicted in (a). Despite the low resolution of the images from unilateral MRI, the bone, the muscle, and the arteries can be identified.

Conventional MRI and single-sided MRI by single-point imaging with CPMG detection can produce images with similar contrast and quality as demonstrated in Fig. 8.10a for a chunk of raw bacon. The time to acquire a 32×32 pixel image with the unilateral scanner at a volume resolution of 1 mm^3 took about 10 min [25]. The images were acquired by phase-encoded single-point imaging with a CPMG train

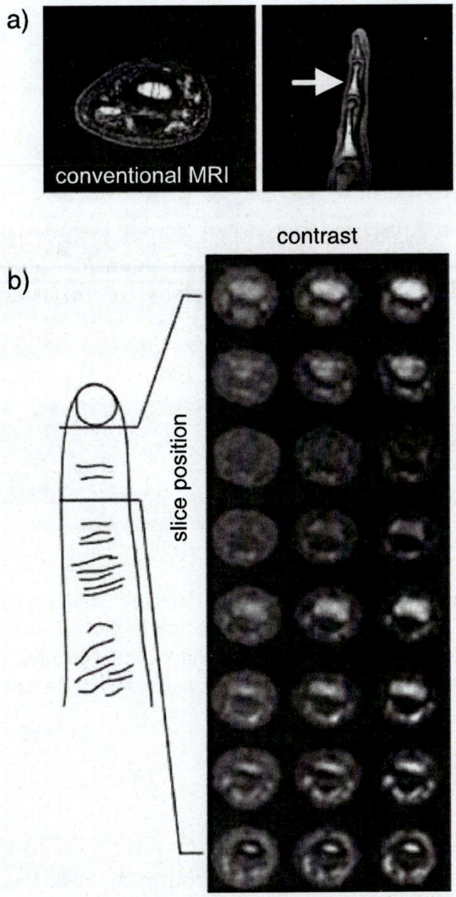

Fig. 8.9 NMR images through a human finger [25]. (**a**) NMR images of a finger acquired at 1.5 T with a clinical imager. (**b**) Two-dimensional images through different parts of a finger joint acquired in 9 min for each position with three different contrast values with a single-sided magnet by positioning the finger within the loop of the surface coil

attached for detection. The CPMG echoes can be used for either of two purposes: (1) extraction of parameters and parameter weights for contrast generation and (2) signal-to-noise enhancement by addition of echoes [19, 71]. T_1 contrast can be introduced by varying the repetition time t_R between scans. The effects of longitudinal and transverse relaxation weights on the image contrast have been studied by variation of the recycle delay t_R between scans (Fig. 8.10c) and the echo time t_{EE} of a spin echo encoding preceding the CPMG detection sequence (Fig. 8.10b)(see Chap. 5). Long echo times introduce a diffusion weight to the transverse magnetization decay from mobile molecules due to the presence of the strong background field gradient of the open sensor. The contrast between muscle and fat can be enlarged significantly and is best for large echo times and short recycle delays, which enhances the fat signal and suppresses the muscle signal. This is in agreement with the

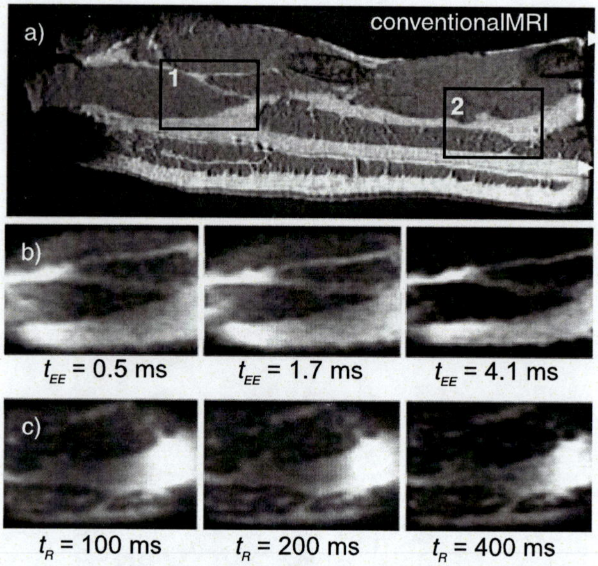

Fig. 8.10 (**a**) Image of a chunk of bacon measured in a clinical MRI. (**b**) Variation of T_2 contrast with the echo time t_{EE} of the spin-echo filter preceding the CPMG acquisition sequence. The images were measured in *region 1* shown in the conventional image. (**c**) Variation of T_1 contrast with the repetition time t_R. The images were measured in *region 2* shown in (**a**)

longitudinal relaxation times T_1 of muscle and fat being of the order of 400 and 225 ms, respectively.

Similar quality images were obtained from a rat head in vitro [26]. But the high gradient of the sensor along the depth direction affords the measurement of only thin slices parallel to the surface of the magnet. To shorten the measurement time, the slice thickness needs to be increased by shimming the stray field to lower gradients. Sufficiently homogeneous stray fields can be designed [74] that will enable frequency encoding of each of the three space directions using pulsed gradient fields generated by planar surface coils [75, 76]. Such a sensor could eventually be employed in remote areas or in an emergency vehicle, for example, to assist in the first aid of stroke victims.

8.5 Conclusions

Applications of unilateral NMR to in vivo studies of humans appear most promising for skin, in particular as depth profiles with a resolution of 10 μm and better can be acquired in a matter of a few minutes. Interesting applications arise in its use in clinical skin treatments, drug administration through the skin, allergy studies, wound healing, and in the development of skin-care products. The state of the skin could be assessed before treatment to help select an individual skin care program and to monitor the progress of treatment also underneath bandages. Tendon and muscle are more difficult to study, and an assessment of the orientation dependence of the signal requires long measurement times. But the wide possibilities of contrast generation have by far not yet been fully explored, and unilateral NMR without lateral spatial resolution may be not only of diagnostic value in medicine but also of practical interest in providing information at a level competitive to ultrasound measurements. In combination with pulsed gradient fields, the spatial selectivity of the NMR-MOUSE can be improved for better analysis of structures narrower than the sensitive volume, an issue which will help to improve, for example, the quality of tendon studies. The use of pulsed gradient fields for imaging in sufficiently homogeneous stray fields is most appealing so that mobile MRI machines may eventually be used in emergency vehicles and low-budget medical institutions. Given the progress of conventional medical imaging in the past, unilateral imaging may experience a similar evolution and become available in a few years in the offices of general practitioners for a first diagnostic analysis of skin, joints, and ligaments in competition with dedicated MRI scanners to be found in the office of the specialist.

References

1. Oppelt A (ed) (2006) Imaging systems for medical diagnostics. Publicis Corporate Publishing, Erlangen
2. Jackson JA, Burnett LJ, Harmon JF (1980) Remote (inside-out) NMR. 3. Detection of nuclear magnetic-resonance in a remotely produced region of homogeneous magnetic-field. J Magn Reson 41(3):411–421
3. Ferguson J (2001) The promise of portable MRI. Eng Sci 2:29–33
4. Esaote S.p.A. Via Siffredi 58, 16153 Genova, Italy. www.esaote.com
5. Odin Medical Technologies Inc. Coal Creek Circle, Louisville, CO. www.odinmed.com
6. MagneVu. Carlsbad, CA. www.magnevu.com.
7. Rokitta M, Rommel E, Zimmermann U, Haase A (2000, Nov) Portable nuclear magnetic resonance imaging system. Rev Sci Instrum 71(11):4257–4262
8. Trequattrini A, Coscia G, Pittaluga S (2000) Double-cavity open permanent magnet for dedicated MRI. IEEE Trans Appl Supercond 10:756–758
9. Kose K (2006) Compact MRI for chemical engineering. In: NMR in chemical engineering. Wiley-VCH, Weinheim
10. Kose K, Matsuda Y, Kurimoto T, Hashimoto S, Yamazaki Y, Haishi T, Utsuzawa S, Yoshioka H, Okada S, Aoki M, Tsuzaki T (2004, Aug) Development of a compact MRI system for trabecular bone volume fraction measurements. Magn Reson Med 52(2):440–444
11. Iita N, Handa S, Tomiha S, Kose K (2007, Feb) Development of a compact MRI system for measuring the trabecular bone microstructure of the finger. Magn Reson Med 57(2):272–277
12. Zur Y (2004, Nov) An algorithm to calculate the NMR signal of a multi spin-echo sequence with relaxation and spin-diffusion. J Magn Reson 171(1):97–106
13. Blank A, Alexandrowicz G, Muchnik L, Tidhar G, Schneiderman J, Virmani R, Golan E (2005, July) Miniature self-contained intravascular magnetic resonance (IVMI) probe for clinical applications. Magn Reson Med 54(1):105–112
14. Schneiderman J, Wilensky RL, Weiss A, Samouha E, Muchnik L, Chen-Zion M, Ilovitch M, Golan E, Blank A, Flugelman M, Rozenman Y, Virmani R (2005, June) Diagnosis of thin-cap fibroatheromas by a self-contained intravascular magnetic resonance imaging probe in ex vivo human aortas and in situ coronary arteries. J Am Coll Cardiol 45(12):1961–1969
15. Mitchell J, Blümler P, McDonald PJ (2006, July) Spatially resolved nuclear magnetic resonance studies of planar samples. Prog Nucl Magn Reson Spectrosc 48(4):161–181

16. Blümich B, Perlo J, Casanova F (2008, May) Mobile single-sided NMR. Prog Nucl Magn Reson Spectrosc 52(4):197–269
17. Eidmann G, Savelsberg R, Blümler P, Blümich B (1996, Sep) The NMR mouse, a mobile universal surface explorer. J Magn Reson Ser A 122(1):104–109
18. Blümich B, Anferov V, Anferova S, Klein M, Fechete R, Adams M, Casanova F (2002, Dec) Simple NMR-mouse with a bar magnet. Concepts Magn Reson 15(4):255–261
19. Perlo J, Casanova F, Blümich B (2005, Sep) Profiles with microscopic resolution by single-sided NMR. J Magn Reson 176(1):64–70
20. Haken R, Blümich, B (2000) Anisotropy in tendon investigated in vivo by a portable NMR scanner, the NMR-mouse. J Magn Reson 144:195–199
21. Miltner O, Schwaiger A, Schmidt C, Bucker A, Kölker C, Siebert CH, Zilkens KW, Niethard FU, Blümich B (2003, Mar) Portable NMR-mouse (r): a new method and its evaluation of the Achilles tendon. Z Orthop Ihre Grenzgeb 141(2):148–152.
22. Navon G, Eliav U, Demo DE, Blümich B (2007, Feb) Study of order and dynamic processes in tendon by NMR and MRI. J Magn Reson Imaging 25(2):362–380
23. Guthausen A, Zimmer G, Eymael R, Schmitz U, Blümler P, Blümich B (1998) Soft-matter relaxation by the NMR-MOUSE. In: Spatially resolved magnetic resonance. Wiley-VCH, Weinheim
24. Casanova F, Perlo J, Blümich B (2005) Depth profiling by single-sided NMR. In: NMR in chemical engineering. Wiley-VCH, Weinheim
25. Blümich B, Kölker C, Casanova F, Perlo J, Felder J (2005) Kernspintomographie für medizin und materialforschung: Ein mobiler und offener kernspintomograph. Phys Unserer Zeit 36:236–242
26. Goga NO, Pirnau A, Szabo L, Smeets R, Riediger D, Cozar O, Blümich B (2006, Aug) Mobile NMR: applications to materials and biomedicine. J Optoelectron Adv Mater 8(4):1430–1434
27. Perlo J, Casanova F, Blümich B (2006) Advances in single-sided NMR. In: Modern magnetic resonance. Springer, Berlin
28. Ruhli FJ, Boni T, Perlo J, Casanova F, Baias M, Egarter E, Blümich B (2007, July) Non-invasive spatial tissue discrimination in ancient mummies and bones in situ by portable nuclear magnetic resonance. J Cult Heritage 8(3):257–263
29. Kruger M, Schwarz A, Blümich B (2007, Feb) Investigations of silicone breast implants with the NMR-mouse. Magn Reson Imaging 25(2):215–218
30. Richard S, Querleux B, Bittoun J, Idyperetti I, Jolivet O, Cermakova E, Leveque JL (1991, July) In vivo proton relaxation-times analysis of the skin layers by magnetic-resonance-imaging. J Invest Dermatol 97(1):120–125
31. Richard S, Querleux B, Bittoun J, Jolivet O, Idyperetti I, Delacharriere O, Leveque JL (1993, May) Characterization of the skin in vivo by high-resolution magnetic-resonance-imaging – water behavior and age-related effects. J Invest Dermatol 100(5):705–709
32. Querleux B, Richard S, Bittoun J, Jolivet O, Idyperetti I, Bazin R, Leveque JL (1994, July) In-vivo hydration profile in skin layers by high-resolution magnetic-resonance-imaging. Skin Pharmacol 7(4):210–216
33. Mirrashed F, Sharp JC (2004, Aug) In vivo morphological characterisation of skin by MRI micro-imaging methods. Skin Res Technol 10(3):149–160
34. McDonald PJ (1997, Mar) Stray field magnetic resonance imaging. Prog Nucl Magn Reson Spectrosc 30:69–99
35. Bennett G, Gorce JP, Keddie JL, McDonald PJ, Berglind H (2003, Apr) Magnetic resonance profiling studies of the drying of film-forming aqueous dispersions and glue layers. Magn Reson Imaging 21(3–4):235–241
36. Dias M, Hadgraft J, Glover PM, McDonald PJ (2003, Feb) Stray field magnetic resonance imaging: a preliminary study of skin hydration. J Phys D Appl Phys 36(4):364–368
37. Backhouse L, Dias M, Gorce JP, Hadgraft K, McDonald PJ, Wiechers JW (2004, Sep) Garfield magnetic resonance profiling of the ingress of model skin-care product ingredients into human skin in vitro. J Pharm Sci 93(9):2274–2283

38. McDonald PJ, Akhmerov A, Backhouse LJ, Pitts S (2005, Aug) Magnetic resonance profiling of human skin in vivo using garfield magnets. J Pharm Sci 94(8):1850–1860
39. Doughty P, McDonald PJ (2006) Drying of coatings and other applications with GARField. In: NMR in chemical engineering. Wiley-VCH, Weinheim
40. Eymael R (2001) Methoden und Anwendungen der Oberflächen-NMR: Die NMR-MOUSE. PhD thesis, RWTH Aachen University
41. Heller F, Schwaiger A, Eymael R, Schulz-Wendtland R, Blümich B (2005, Oct) Measuring the contrast-enhancement in the skin and subcutaneous fatty tissue with the NMR-mouse (r): a feasibility study. Rofo-Fortschr Geb Der Röntgenstr Bildgeb Verfahr 177(10):1412–1416
42. Schwaiger A (2002) Biomedizinische Anwendungen der NMR-MOUSE. PhD thesis, RWTH, Aachen University
43. Heywang-Köbrunner SH (1990) Contrast-enhanced MRI of the breast. Karger-Verlag, Basel
44. Henkelman RM, Stanisz GJ, Kim JK, Bronskill MJ (1994, Nov) Anisotropy of NMR properties of tissues. Magne Reson Med 32(5):592–601
45. Xia Y (1998, June) Relaxation anisotropy in cartilage by NMR microscopy (μm MRI) at 14-μm resolution. Magn Reson Med 39(6):941–949
46. Xia Y (1998) Relaxation anisotropy as a possible marker for macromolecular orientations in carticular cartilage. In: Spatially resolved magnetic resonance, vols 351–362. Wiley-VCH, Weinheim
47. Xia Y, Moody JB, Alhadlaq H (2002, Sep) Orientational dependence of t-2 relaxation in articular cartilage: a microscopic MRI (mu MRI) study. Magn Reson Med 48(3):460–469
48. Fechete R, Demco DE, Blümich B (2002, July) Segmental anisotropy in strained elastomers by h-1 NMR of multipolar spin states. Macromolecules 35(16):6083–6085
49. Fechete R, Demco DE, Blümich B (2003, Nov) Parameter maps of h-1 residual dipolar couplings in tendon under mechanical load. J Magn Reson 165(1):9–17
50. Fechete R, Demco DE, Blümich B, Eliav U, Navon G (2003, May) Anisotropy of collagen fiber orientation in sheep tendon by h-1 double-quantum-filtered NMR signals. J Magn Reson 162(1):166–175
51. Fechete R, Demco DE, Eliav U, Blümich B, Navon G (2005, Dec) Self-diffusion anisotropy of water in sheep Achilles tendon. NMR Biomed 18(8):577–586
52. Wellen J, Helmer KG, Grigg P, Sotak CH (2004, Sep) Application of porous-media theory to the investigation of water ADC changes in rabbit Achilles tendon caused by tensile loading. J Magn Reson 170(1):49–55
53. Kastelic J, Galeski A, Baer E (1978) Multicomposite structure of tendon. Connect Tissue Res 6(1):11–23
54. Fullerton GD, Cameron IL, Ord VA (1985) Orientation of tendons in the magnetic-field and its effect on t2 relaxation-times. Radiology 155(2):433–435
55. Erickson SJ, Cox IH, Hyde JS, Carrera GF, Strandt JA, Estkowski LD (1991 Nov) Effect of tendon orientation on MR imaging signal intensity – a manifestation of the magic angle phenomenon. Radiology 181(2):389–392
56. Timins ME, Erickson SJ, Estkowski LD, Carrera GF, Komorowski RA (1995, July) Increased signal in the normal supraspinatus tendon on MR-imaging – diagnostic pitfall caused by the magic-angle effect. Am J Roentgenol 165(1):109–114
57. Chenevert TL, Brunberg JA, Pipe JG (1990, Nov) Anisotropic diffusion in human white matter – demonstration with MR techniques in vivo. Radiology 177(2):401–405
58. Basser PJ, Mattiello J, Lebihan D (1994, Jan) Diffusion tensor spectroscopy and imaging. Biophys J, 66(1):259–267
59. Knauss R, Schiller J, Fleischer G, Karger J, Arnold K (1999, Feb) Self-diffusion of water in cartilage and cartilage components as studied by pulsed field gradient NMR. Magn Reson Med 41(2):285–292
60. Wiesmath A, Filip C, Demco DE, Blümich B (2002, Jan) NMR of multipolar spin states excitated in strongly inhomogeneous magnetic fields. J Magn Reson 154(1):60–72

61. Wiesmath A, Filip C, Demco DE, Blümich B (2001, Apr) Double-quantum-filtered NMR signals in inhomogeneous magnetic fields. J Magn Reson 149(2):258–263
62. Mlynarik V, Szomolanyi P, Toffanin R, Vittur F, Trattnig S (2004, Aug) Transverse relaxation mechanisms in articular cartilage. J Magn Reson 169(2):300–307
63. McDonald PJ, Newling B (1998, Nov) Stray field magnetic resonance imaging. Rep Prog Phys 61(11):1441–1493
64. Guthausen A, Zimmer G, Blümler P, Blümich B (1998, Jan) Analysis of polymer materials by surface NMR via the mouse. J Magn Reson 130(1):1–7
65. Blümich B, Blümler P, Eidmann G, Guthausen A, Haken R, Schmitz U, Saito K, Zimmer G (1998, June) The NMR-mouse: construction, excitation, and applications. Magn Reson Imaging 16(5–6):479–484
66. Klein M, Fechete R, Demco DE, Blümich B (2003, Oct) Self-diffusion measurements by a constant-relaxation method in strongly inhomogeneous magnetic fields. J Magn Reson 164(2):310–320
67. Alhadlaq HA, Xia Y (2004, Nov) The structural adaptations in compressed articular cartilage by microscopic MRI (μm MRI) T2 anisotropy. Osteoarthr Cartil 12(11):887–894
68. Emid S, Creyghton JHN (1985) High-resolution NMR imaging in solids. Physica B Physica C, 128(1):81–83
69. Balcom BJ, MacGregor RP, Beyea SD, Green DP, Armstrong RL, Bremner TW (1996, Nov) Single-point ramped imaging with T-1 enhancement (sprite). J Magn Reson Ser A 123(1): 131–134
70. Casanova F, Blümich B (2003, July) Two-dimensional imaging with a single-sided NMR probe. J Magn Reson 163(1):38–45
71. Casanova F, Perlo J, Blümich B, Kremer K (2004, Jan) Multi-echo imaging in highly inhomogeneous magnetic fields. J Magn Reson 166(1):76–81
72. Perlo J, Casanova F, Blümich B (2004, Feb) 3D imaging with a single-sided sensor: an open tomograph. J Magn Res 166(2):228–235
73. Edelstein WA, Hutchison JMS, Johnson G, Redpath T (1980) Spin warp NMR imaging and applications to human whole-body imaging. Phys Med Biol 25(4):751–756
74. Perlo J, Casanova F, Blümich B (2007, Feb) *Ex situ* NMR in highly homogeneous fields: H-1 spectroscopy. Science 315(5815):1110–1112
75. Godward J, Ciampi E, Cifelli M, McDonald PJ (2002, Mar) Multidimensional imaging using combined stray field and pulsed gradients. J Magn Reson 155(1):92–99
76. Paulsen JL, Bouchard LS, Graziani D, Blümich B, Pines A (2008) Volume-selective magnetic resonance imaging using an adjustable, single-sided, portable sensor. Proc Natl Acad Sci USA 105(52):20601–20604

Chapter 9
Applications in Material Science and Cultural Heritage

Jürgen Kolz

The diversity of available techniques has made magnetic resonance a valuable tool in medicine, in chemistry, as well as in material science [1–3]. In spite of the complexity of obtaining high-resolution spectra and the lower sensitivity compared to NMR with superconducting magnets this is also true for single-sided NMR, particularly considering that the technique is truly non-destructive due to the open geometry of the sensor [4, 5]. The first U-shaped sensor was conceived to measure bulk properties close to the surface of the object under investigation. Later refinements of the magnet geometry opened up the possibility to achieve spatial resolution, either by exploiting the static gradient of the sensor in order to achieve depth resolution or by equipping the magnet with gradient coils in order to achieve spatial resolution in the lateral directions. Since spectral resolution is hard to achieve in the inhomogeneous field of a conventional single-sided sensor the main parameters used to characterize materials are spin density and relaxation times. The spin density reflects the number of hydrogen nuclei inside the sensitive volume and can be used to indicate inhomogeneities or voids, changes in the material density, or interfaces between different materials. The longitudinal and transverse relaxation times T_1 and T_2 can be used to obtain information about the molecular mobility inside the sample. These go often hand in hand with macroscopic mechanical properties and therefore relaxation times can be used as quality parameters for the production control. In most of the cases a CPMG sequence is applied, since it is the fastest accessible way to measure transverse relaxation. However, the obtained decay is not governed by pure transverse relaxation and the decay time is therefore called effective transverse relaxation time $T_{2\mathrm{eff}}$. The inhomogeneities of the static magnetic field B_0 and the rf-field B_1 cause off-resonance effects and a flip-angle distribution resulting in an incomplete conversion of z-magnetization into transverse magnetization. Therefore, the measured decay has also some inherent T_1 contribution (see Chap. 2).

In this chapter an overview is given of how relaxation times and spin densities extracted from measurements with single-sided sensors can be used to characterize a variety of different materials ranging from soft solids, like elastomers to solids, like

J. Kolz (✉)
Institut für Technische Chemie und Makromolekulare Chemie, RWTH Aachen University, D-52074 Aachen, Germany

F. Casanova et al. (eds.), *Single-Sided NMR*,
DOI 10.1007/978-3-642-16307-4_9,

semi-crystalline polymers or painting layers. Furthermore, the possibility to combine single-sided NMR with high spatial resolution is addressed in several applications, like the characterization of multilayer structures, the detection of failures, and the monitoring of the solvent ingress into polymer materials.

9.1 Elastomers

Due to their soft mechanical properties elastomers have a wide variety of applications in the industry, e.g., tyres, sealings, and tubings [6]. They consist of polymers with glass transition temperatures below ambient temperatures, in which crosslinks are introduced to form a network. The presence of the crosslinks can be monitored by NMR-relaxation times, since the chain mobility strongly depends on the molecular mobility [7–10]. This is based on the fact that the transverse relaxation time T_2 in solids is sensitive to the strength of the dipolar couplings. Depending on the timescale and geometry of motion, the dipolar couplings of mobile molecules are partially averaged leading to a longer T_2. However, by introducing crosslinks into an elastomer matrix the molecular mobility is hindered resulting in weaker averaging and therefore shorter relaxation times. Being sensitive to molecular motion makes T_2 also a useful parameter to probe aging processes and strain in rubber parts at a given temperature, since in both cases the mobility inside the elastomer network is affected. The following examples show how single-sided NMR can be applied to measure crosslink densities, to follow aging processes, to detect failures, and to measure load distributions in rubber parts.

9.1.1 Crosslink Density

The crosslink density is one of the key parameters which determines the mechanical properties of elastomers [6]. Crosslinks are introduced during a curing process in which commonly sulfur is used to connect the polymer chains to form a network [11]. This is necessary to avoid that the polymer chains disentangle under strain and polymer materials irreversibly deform. Conventionally the crosslink density is determined in mechanical deformation tests based on elongation or shear [12]. However, these methods are destructive and cannot provide information about the spatial distribution of the crosslink density across the object. Single-sided NMR, on the other hand, is a non-destructive technique, which is sensitive to variations of crosslink density and it can be combined with imaging techniques to resolve inhomogeneities across the material [13].

In order to correlate the relaxation times measured by NMR with material properties, a set of four natural samples with sulfur concentrations ranging from 2 to 5 phr (parts per hundred rubber) was prepared. The crosslink densities of the samples were determined by applying the Mooney–Rivlin approximation [14] to measurements of the stress–strain dependence and by shear monitoring the shear modules during vulcanization. Figure 9.1a shows the dependence of the resultant crosslink densities as a

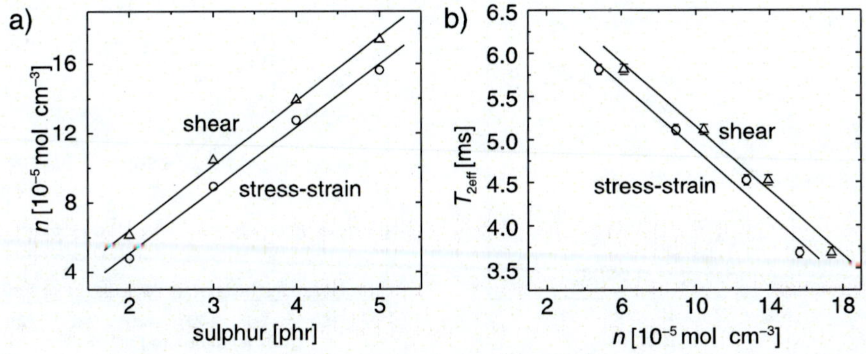

Fig. 9.1 (**a**) Dependence of the crosslink density determined by mechanical methods on the concentration of sulfur in the rubber mixture. (**b**) Dependence of the transverse relaxation time $T_{2\text{eff}}$ measured with a CPMG sequence vs. crosslink density

function of the amount of sulfur. For both methods a linear increase of the crosslink density is observed. The difference between the values produced by each method is due to the fact that rheology probes the shear modulus at the frequency of dynamic deformation and stress–strain measures the modulus of elasticity at zero frequency.

The same set of samples was analyzed by CPMG measurements with a bar-magnet NMR-MOUSE. The CPMG train consisted of 512 echoes with an echo time t_E of 0.1 ms. One hundred scans were averaged to increase the signal-to-noise ratio. Although the T_1 of the samples was about 50 ms a recycle delay of 3.5 s was chosen to avoid sample heating from power dissipation in the rf coil. Especially when dealing with elastomers a constant temperature of the sample has to be provided, since the molecular mobility and therefore the transverse relaxation time are strongly sensitive to temperature. A simple test to confirm constant temperature is to conduct the same CPMG measurement several times in succession at the same position. If an increase of the relaxation time as a function of the experiment number is observed, the sample is warming up and the recycle delay has to be increased. In order to reduce sample heating, it is also possible to guide an air stream between rf coil and sample, which helps to remove the generated heat by isolating the sample from the coil.

Figure 9.1b shows the $T_{2\text{eff}}$ values plotted as a function of the crosslink density determined by mechanical methods. $T_{2\text{eff}}$ was obtained by fitting the decays to stretched-exponential functions. In the observed range $T_{2\text{eff}}$ varies linearly with crosslink density. Using this calibration, the crosslink density can be measured at various lateral positions of arbitrarily sized objects. In many cases inhomogeneities in the crosslink density may appear in the direction perpendicular to the surface of the object under study, caused by non-uniform temperature distributions during the vulcanization process. Such inhomogeneities can be better identified using sensors like the profile NMR-MOUSE [15]. They select the signal from a thin flat-sensitive volume that is mechanically moved through the object, in order to achieve a profile along the depth direction of the sample (see Chap. 5).

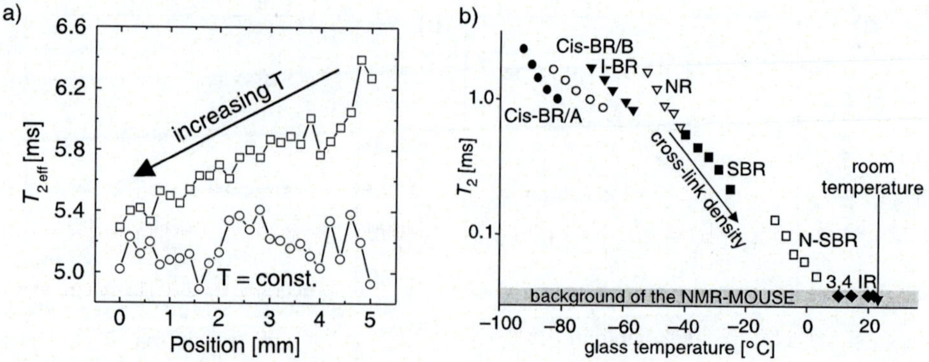

Fig. 9.2 (**a**) Profiles of $T_{2\mathrm{eff}}$ across rubber samples. One sample was cured at constant temperature, while for the other sample a temperature gradient during the curing process was introduced. The resultant gradient in crosslink density becomes visible in the $T_{2\mathrm{eff}}$ parameter. On the side with lower curing temperature a higher $T_{2\mathrm{eff}}$ value is obtained showing a lower crosslink density. (**b**) T_2 of different rubber samples as a function of the glass transition temperature measured with a Hahn-echo sequence

Figure 9.2a shows profiles of $T_{2\mathrm{eff}}$ measured through 5 mm thick elastomer samples. One sample was cured at constant temperature, while for the other one a temperature gradient across the sample was introduced during the vulcanization process. It can be seen that for the sample with uniform curing temperature $T_{2\mathrm{eff}}$ is uniform across the sample, while for the sample with temperature gradient $T_{2\mathrm{eff}}$ increases toward the side of lower curing temperature. This is due to the smaller rate of the curing reaction at lower temperatures resulting in a lower degree of crosslinking.

Since crosslinks change the mobility of the polymer chains also the glass transition temperature T_g is affected by the crosslink density. Herrman et al. [16] investigated the effect of varying T_g on T_2 by investigating various polymers, with side-groups characterized by different sterical hindrance and with different crosslink densities. The T_2 values were extracted from Hahn-echo decays measured with the NMR-MOUSE at room temperature, while the T_g was determined by dynamic-mechanical thermal analysis. Figure 9.2b shows T_2 values as a function of T_g. With increasing T_g the material becomes less mobile leading to a decrease of T_2. While the relaxations times for different crosslink densities within one polymer fall on a straight line, for different polymers do not, proving that the variation in crosslink density influences the glass transition in a way different from the sterical hindrance of the side groups.

9.1.2 Aging

Elastomers undergo aging processes during their life time. These processes are of chemical or physical nature and often alter the mobility of the chain network, which can be detected by measurements of NMR-relaxation times [17–20]. Chemical aging is caused by the attack of free radicals or oxygen on the polymer backbone leading, depending on the environmental conditions, to chain scission or additional crosslinks [12]. The formation as well as the activity of the attacking species can be triggered under certain environmental conditions like high temperature, high ozone concentration or by the exposure to light or UV-radiation. Such changes can be followed by measurements of transverse relaxation times using a single-sided sensor.

Figure 9.3a depicts $T_{2\text{eff}}$ as a function of aging time for two unfilled natural rubber samples, with and without a phenolic antioxidant. The initial $T_{2\text{eff}}$ is similar for both formulations, showing that the required quantity of antioxidant does not influence the relaxation of the unaged samples. With ongoing aging $T_{2\text{eff}}$ decreases for both formulations due to the formation of additional crosslinks. This results in a lower molecular mobility and therefore in stronger dipolar interactions. During the first days of aging the relaxation time of the sample with antioxidant decreases slower, proving the efficiency of the antioxidant. For longer aging times the relaxation times converge to the same value, caused by the consumption of the antioxidant. After 73 days the relaxation times are the same for both formulations, showing that the antioxidant is fully depleted. Furthermore, formulations consisting of different compositions can be compared in order to investigate which formulation shows the best aging resistance. Figure 9.3b shows $T_{2\text{eff}}$ values of fluorinated rubber with specific additives before and after aging at 120°C in air. It can be concluded that sample 4, which contains an ageing protectant but no carbon black, shows the highest aging resistance.

To obtain better insight into the mechanisms of the aging process profiles of $T_{2\text{eff}}$ can be measured [19]. Figure 9.4a shows $T_{2\text{eff}}$ profiles through a natural rubber sample, thermo-oxidatively aged for different times at 80°C in an oven. It can be seen that the sample shows a non-uniform $T_{2\text{eff}}$ profile already for the unaged sample, caused by a temperature gradient during the vulcanization process. With ongoing aging, $T_{2\text{eff}}$ decreases uniformly across the sample, while the initial $T_{2\text{eff}}$ gradient

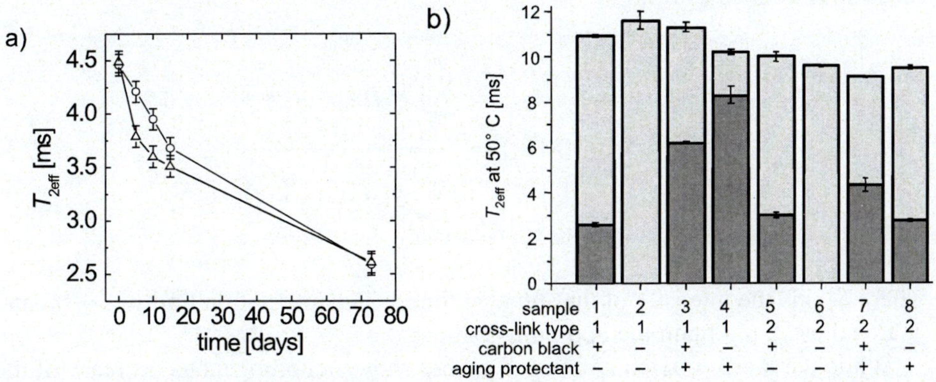

Fig. 9.3 (**a**) Time dependence of $T_{2\text{eff}}$ for natural rubber samples with (○) and without (△) antioxidant. (**b**) Relaxation times of different fluorine rubber sheets before (*light bars*) and after aging at 120°C (*dark bars*). Sample *No. 4* shows the best aging resistance

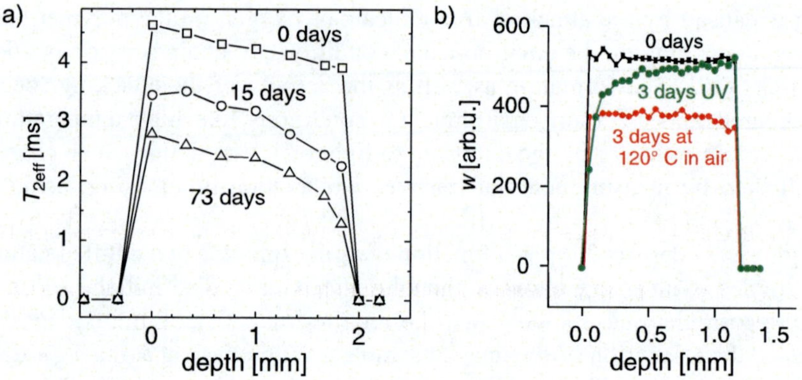

Fig. 9.4 (**a**) Depth profiles across natural rubber sheets new (□), after 15 days (○) and 73 days (△) of aging in air at 80°C. A $T_{2\mathrm{eff}}$ gradient originating from heterogeneous curing is observed during the whole aging period. (**b**) Depth profiles across natural rubber sheets new, aged in air at 120°C, and aged under UV irradiation

remains. This means that further formation of crosslinks by post curing, that could be expected for the regions of lower crosslink density, does not take place.

The effect of different aging mechanisms can also be visualized using the profiling technique. Two natural rubber samples were investigated, one aged for 3 days at 120°C in a cabinet dryer and one exposed to UV radiation for the same time. Figure 9.4b shows profiles of values of a weighting function for the two aged samples as well as a reference profile for a non-aged sample.

For the evaluation of the weighting function the echo amplitudes of two successive time intervals of the CPMG echo train are added and the ratio of the two sums is calculated to obtain a relaxation-weighted spin density number w. This option can be preferable over the fitting function when the decay has a complicated time dependence, as no fit function needs to be defined, or when $T_{2\mathrm{eff}}$ or at least one component of the decay is comparable with the echo time t_E. The weighting function is defined as follows:

$$w(i_\mathrm{i}, i_\mathrm{f}, j_\mathrm{i}, j_\mathrm{f}, t_\mathrm{E}) = \frac{\sum_{j=j_\mathrm{i}}^{j_\mathrm{f}} S(jt_\mathrm{E})/(j_\mathrm{f} - j_\mathrm{i})}{\sum_{i=i_\mathrm{i}}^{i_\mathrm{f}} S(it_\mathrm{E})/(i_\mathrm{f} - i_\mathrm{i})} \tag{9.1}$$

where $S(t)$ is the intensity of the signal at time t. The integration limits i_i, i_f, j_i, and j_f are adjusted to obtain the optimum contrast.

While the thermo-oxidatively aged rubber shows a homogenous decrease of the relaxation time across the whole sample, the response of the UV-aged sample is different [21]. At the exposed surface a strong decrease of the relaxation time is observed. This is attributed to the high amount of radicals formed at this location, which attack the polymer chains and give rise to additional crosslinks that result in an embrittlement of the material. With increasing depth the relaxation time increases exponentially to the value of the unaged sample. This can be explained by a spatial dependence of the radical concentration inside the polymer, which follows a Lambert–Beer behavior [21].

9.1.3 Imaging

As shown in Chap. 5, when equipped with additional gradient coils, single-sided sensors can also be used to measure 2D images [22]. Since parameters like spin density and relaxation times are responsible for contrast in NMR images, this technique can be useful to resolve structures, to detect failures or to map the load distribution in rubber parts [4, 19]. In order to achieve images in a reasonable time, a pure-phase encoding sequence is combined with a subsequent CPMG echo detection to increase the sensitivity (see Chap. 5). Figure 9.5 shows how the technique can be applied to monitor the position of crossed textile fibers embedded in layers incorporated for reinforcement of rubber products, like air spring bellows and high-performance rubber tubes.

An air spring bellow is an air-pressured rubber balloon used to suspend vehicles like trains and cars (Fig. 9.5a). The wall of an air spring bellow (Fig. 9.5b) of a passenger car is about 2 mm thick and contains two layers of crossed textile fibers (Fig. 9.5c). The fiber spacing and the layer spacing are about 1 mm. The sensor described in Chaps. 4 and 5 was used to produce slice-selective images across the

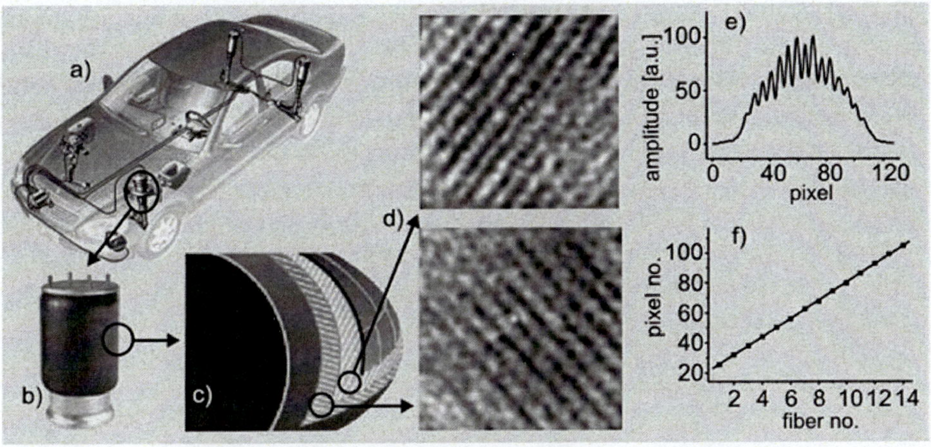

Fig. 9.5 Air springs bellows are used in intelligent suspension systems (**a**) that control (**b**) the hardness of the shock absorber by adjusting the air pressure in the rubber bellows. The rubber matrix of the device is reinforced by crossed layers of textile fibers (**c**). The fibers of each layer can be resolved by NMR imaging (**d**). The fiber spacing can be analyzed in a projection of the 2D images along the fiber direction (**e**). The positions of the fibers determined from the 1D profile depend linearly on space (**f**)

wall with a slice thickness of 0.2 mm. Two-dimensional images of each fiber layer with a field of view of 2 cm × 2 cm and an in-plane resolution of 0.2 mm were obtained within an acquisition time of 2 h per image (Fig. 9.5d). The signal of the textile fibers is negligible compared to that of the rubber matrix, so that fibers appear dark in the images. By projecting the 2D images along the fiber direction, a 1D projection is obtained (Fig. 9.5e). From the minima in the profile the position of the fibers can be determined with good precision (Fig. 9.5f), so that defects in the fiber positioning can be identified.

Another application concerns the defect analysis of fiber-reinforced rubber tubes. Figure 9.6a shows a rubber tube with a defect located at the inner rubber layer of the tube. A slice-selective image with a layer thickness of 0.66 mm was taken with a spatial resolution of 0.2 mm × 0.2 mm and a FOV of 4 cm × 4 cm. In order to decrease the measurement time to 120 min the complete echo train generated in the detection period of the sequence were added. The NMR image (Fig. 9.6b) reveals the textile fibers and shows a hole in the inner rubber layer. While the extension of the defect in the lateral direction can be mapped in a 2D image, the extension across the tube wall can be detected by measuring slice-selective images at different depths.

Besides its use in structure and failure analysis, NMR imaging also provides the possibility to map physical parameters like temperature and strain via their influence on relaxation times. This is of particular interest to identify critical high tensions during the design of rubber parts, which are intended to work under load during their usage. Compared to the failure detection in the previous example, where the spin density is used as contrast parameter, tensions inside the elastomer network are reflected in the change of relaxation times. Figure 9.7a shows that almost no $T_{2\mathrm{eff}}$ differences are observed in the CPMG decays for a rubber band at different elongation ratios, where $\lambda = l/l_0$ is the elongation ratio, l the length under stretching, and l_0 the initial length. However, a strong decrease of the relaxation time is observed in the Hahn-echo decay with increasing elongation ratio (Fig. 9.7b). When the sequence described in Chap. 5 is used to obtain a 2D image, the contrast by $T_{2\mathrm{Hahn}}$ can be controlled by varying the echo time t_{EE} in the encoding period, while

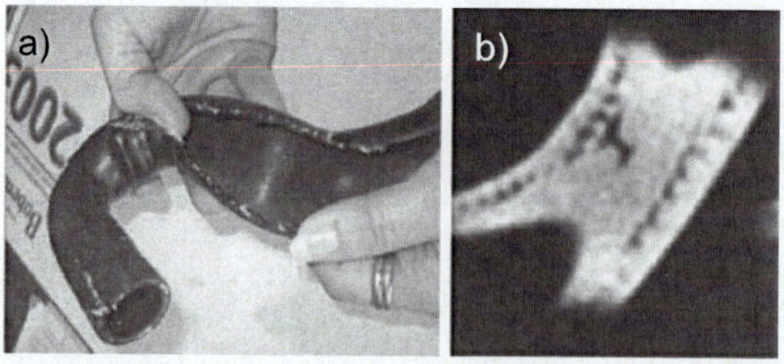

Fig. 9.6 Failure analysis. (**a**) Photograph of a fiber-reinforced rubber tube. The failure located at the inner rubber layer originates from the production. (**b**) Cross section taken with a high spatial resolution (0.2 mm)^2. The image reveals the defect as well as the textile fibers

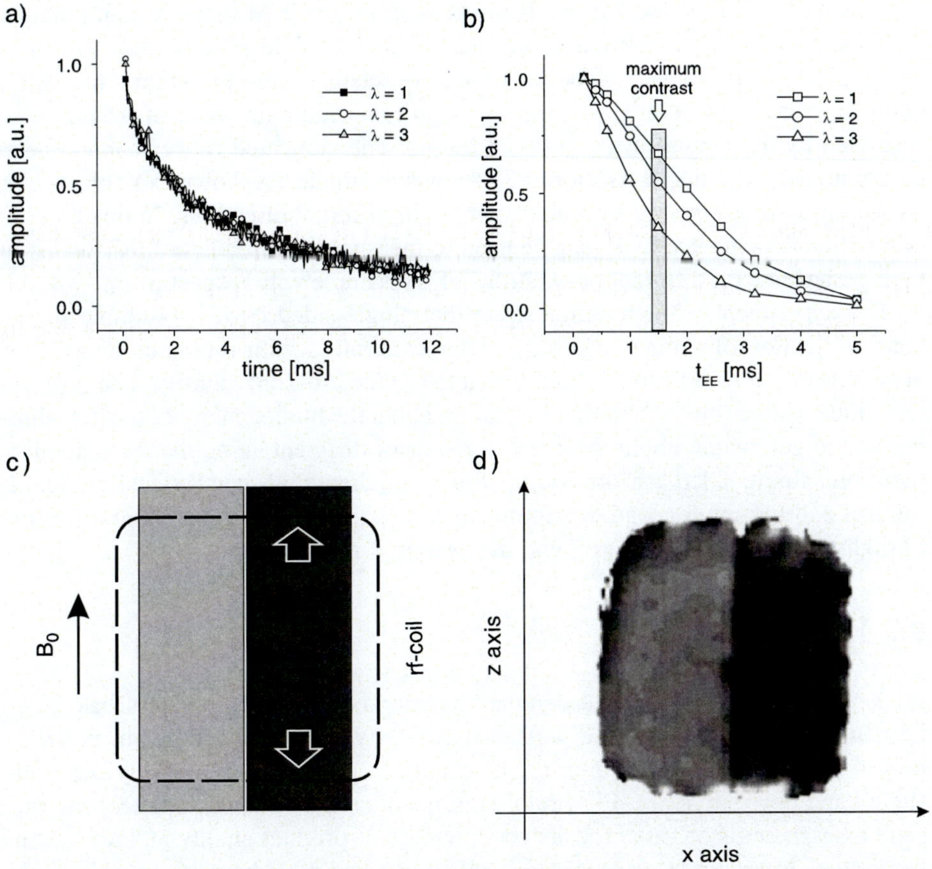

Fig. 9.7 Signal decay measured with a CPMG (**a**) and a Hahn-echo sequence (**b**) for different elongation ratios. Maximum contrast is achieved with the Hahn-echo sequence by setting the echo time of 1.5 ms. (**c**) Drawing of the sample setup. Two rubber bands, one stretched, and one relaxed, are placed in the sensitive volume of the sensor. (**d**) Map revealing the elongation of the rubber bands obtained as the ratio between an image taken with an encoding time 1.5 ms and an image with an encoding time of 0.5 ms

all echoes in the detection period are co-added to improve the SNR. In the present case t_{EE} was set to 1.5 ms to achieve maximum contrast. By calibration against λ the resultant pixel intensity can directly be translated into the elongation ratio, while for the single-sided setup the B_1 inhomogeneity as well as the change of the amount of sample in the sensitive volume caused by stretching has to be taken into account. This can be achieved by normalizing the image to an image acquired for short encoding time, where the difference in the Hahn-echo intensity for different elongation ratios is minimized.

Figure 9.7c shows a schematic drawing of a setup used to test the performance of the method to measure stretching. The sample consists of two rubber bands (10 mm wide) positioned next to each other parallel to the B_0 direction. While the right one is elongated to $\lambda = 3$, the left one is unstressed. Figure 9.7d shows a map obtained with the normalization method described before. The field of view was 30 mm × 30 mm, while the longer encoding time was set to 1.5 ms to achieve maximum contrast, and the short encoding time was set to a minimum value of 0.5 ms being restricted by the length of the gradient pulses. The elongated rubber band appears darker in the image due to its shorter T_2. However, unlike the shown example, rubber parts may also be stressed by forces that act in different directions. In this case the orientation dependence of T_2 can be used to measure the direction of the forces, by taking images with different orientations of the sample with respect to B_0.

The experiments shown demonstrate that single-sided NMR is suitable for the characterization of rubber materials. After calibration with reference data, transverse relaxation times can be directly related to the crosslink density. The profiling technique enables one to obtain information about the homogeneity of the crosslinking and to get insight about the characteristics of different aging mechanisms. Furthermore, single-sided sensors equipped with gradient coils can be used to identify failures and to monitor load distributions in rubber parts. Such information can be of high value for quality control and the design of rubber parts.

9.2 Hard Polymers

In the second half of the last century the market for hard polymers has grown tremendously. More and more parts that formerly were produced from metal are nowadays produced from polymers. Particularly pipes from poly(vinyl chloride) (PVC) and poly(ethylene) (PE) are of significant commercial importance, and their state assessment is of value for the optimization of product quality and for lifetime prediction. In this case single-sided NMR can be of great benefit, since it is non-destructive and sensitive to changes in the polymer morphology [23–28]. On the other hand, the measurements are more challenging than for elastomers, because the rigid chain morphology results in much shorter $T_{2\text{eff}}$ values. Since single-sided NMR sensors detect the NMR signal stroboscopically at the times of the echoes in multi-echo sequences, the number of points that can be detected is strongly limited for such materials. However, by constructing rf probes with shorter dead times more echoes can be acquired during the signal decay to improve the accuracy of data evaluation.

Poly(ethylene), which is mostly used for the production of pipes, is a semi-crystalline polymer. It is composed of crystalline domains where polymer chains are ordered and continues amorphous phase with disordered chains. Due to the different morphology of the two phases, two relaxation times are present in the decay of the transverse magnetization, a short one (A_{short}), corresponding to the rigid crystalline phase and a longer one (A_{long}) arising from the more mobile amorphous region. When fitting the CPMG decay with a bi-exponential function

$$S(t) = A_{\text{short}} \exp\left[\frac{-t}{T_{2\text{eff,short}}}\right] + A_{\text{long}} \exp\left[\frac{-t}{T_{2\text{eff,long}}}\right] \tag{9.2}$$

an NMR crystallinity parameter can be derived as $\alpha_{NMR} = A_{short}/(A_{short} + A_{long})$. This parameter differs from the one measured by X-ray scattering since NMR probes the molecular mobility and not the state of order. The NMR crystallinity also includes the fraction of molecules forming the interface between the crystalline and the amorphous phases. By proper calibration α_{NMR} can be used to monitor changes of morphology involved in the production process or caused by annealing and aging. Furthermore, this information can be acquired spatially resolved in different depths and at different lateral positions. Figure 9.8a shows the NMR crystallinity measured as a function of the depth of a poly(ethylene) pipe [26]. It can be seen that the crystallinity increases with increasing depth. This is due to the cooling process after the extrusion of the polymer melt into the pipe mould. The temperature gradient associated with the cooling process results in different crystallinities across the pipe wall.

Not only the variation in different depths can be monitored, but also the variations in the lateral directions can be measured by repositioning the sensor [28]. In this case the spatial resolution is determined by the lateral size of the sensitive volume, which is given by the dimensions of the rf coil. Following this concept the inner circumference of a PE pipe was mapped using the experimental setup shown

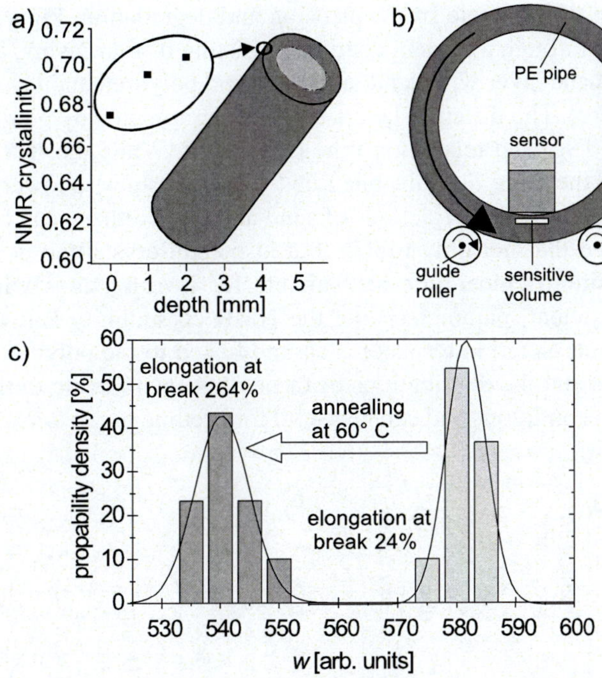

Fig. 9.8 Measurements of PE pipes with the NMR-MOUSE. (**a**) Variation of the NMR degree of crystallinity across the pipe wall. (**b**) Scheme of the positioning device used for mapping the inner pipe wall with the NMR-MOUSE. (**c**) Statistical distribution of transverse relaxation times measured at the inner pipe wall before and after annealing at 60°C for 24 h. The annealing has an effect on the PE morphology resulting in dramatic change of the mechanical properties

in Fig. 9.8b. A U-shaped MOUSE was placed inside a PE tube. In order to scan the pipe, the position of the sensor was fixed and the pipe was rotated by a stepper motor. The sensor was moved along the pipe to measure at different circumferences, in order to acquire 2D maps. In this way a matrix of 10 cm × 10 cm was mapped point by point with a step width of 1 cm along both dimensions. The pipe was scanned before and after annealing at 60°C for 24 h. The CPMG decays were analyzed using a weighting function similar to Eq. (9.1). The first echoes of the train, which correspond to the crystalline phase, were added and divided by the sum of the long components, which corresponds to the amorphous phase. The resultant parameter w scales with the crystallinity of the polymer. The statistical evaluation of these data, before and after annealing, is shown in Fig. 9.8c. It can be seen that even annealing 55°C below the glass transition temperature has a significant effect on the morphology. This proves that single-sided NMR can be used as a method to follow morphological changes. Furthermore, the detection of morphology changes at temperatures far below the glass transition suggest that for lifetime predictions of such semi-crystalline polymers the conditions used to accelerate the aging have to be chosen very carefully, when an extrapolation from elevated temperatures to the working temperature needs to be done.

The profiling technique is also of interest, when looking at multilayer structures of hard polymers. For example, concrete walls are often coated by different polymer layers, to protect the concrete from corrosion and degradation. Figure 9.9 shows a profile measured through a cement coating consisting of four layers, a surface with sand, a polyurethane layer with sand, a pure elastic polyurethane layer, and a layer of epoxy resin [29]. The different layers can be discriminated by plotting the signal amplitude (Fig. 9.9a) and relaxation time (Fig. 9.9b). While the two polyurethane layers show both the same $T_{2\text{eff}}$, the one filled with sand shows a lower signal amplitude, from which the volume fractions of sand and polyurethane can be derived.

It is concluded that not only rubber, but also semi-crystalline polymers below their glass transition temperature are suitable for investigations with the NMR-MOUSE. In particular, parameters like the NMR crystallinity and the segmental mobility can be measured in terms of $T_{2\text{eff}}$ and linked to the polymer morphology. The mean values and the distributions of these parameters relate to the processing and deterioration conditions and can be useful for lifetime predictions. Furthermore,

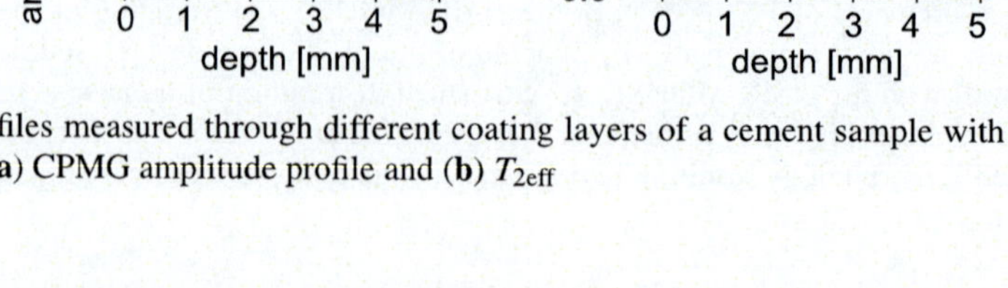

Fig. 9.9 Profiles measured through different coating layers of a cement sample with a resolution of 200 μm. (**a**) CPMG amplitude profile and (**b**) $T_{2\text{eff}}$

spatially resolved information can be achieved, when applying the profiling technique to multilayer structures.

9.2.1 Ingress of Solvents

Polymers used for casings, pipes, containers and packaging are often in contact with liquids that can ingress into the polymer matrix. In most cases this ingress should be kept to a minimum in order to hinder the release of toxic components to the exterior and to prevent the change of the morphology of the polymer material that can result in the weakening of its mechanical strength. The kinetics of such dynamic processes can be followed non-destructively by the profiling technique described in Chap. 5 [15]. One interesting application is the ingress of gasoline into a fuel tank wall [15, 19]. Fuel tank walls are normally composed of a multilayer polymer structure. It consists of a regrind and a polyethylene layer separated by a barrier layer, made for example from a ethyl-vinylalcohol copolymer, which is attached to both sides with an adhesive. The barrier is intended to prevent volatile compounds escaping to the exterior.

Figure 9.10a shows profiles measured through such a fuel tank wall before and after different times of exposure to gasoline. In cases where dynamic processes are to be followed, it is important to keep the measurements short enough to ensure that the changes of the system under investigation are negligibly small during the acquisition of one profile. In order to overcome the comparatively low sensitivity of the single-sided sensor, the addition of the echo amplitudes in the CPMG train has been proven to be an efficient way of increasing the SNR per unit of time.

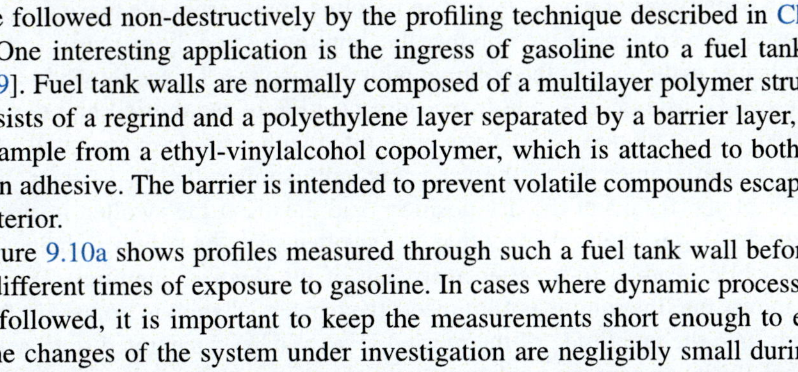

Fig. 9.10 (**a**) Series of profiles measured through a fuel tank wall before and at different times after exposure to regular-grade gasoline. Each point corresponds to the addition of the first 64 echo amplitudes of a CPMG train acquired with $t_e = 45\,\mu s$. The nominal resolution was set to $50\,\mu$ m by using an acquisition time of $20\,\mu s$ in the presence of a static gradient of 21 T/m. The profile taken before the exposure (0 min) shows that the different layers (polyethylene, resin, barrier, resin, regrind) can clearly be distinguished. The gasoline enters through the regrind layer and is then blocked at the barrier layer. (**b**) CPMG decays measured in the regrind layer, dry (○) and swollen (□). While the amplitude is similar for both decays, $T_{2\text{eff}}$ strongly increases by the ingressing solvent

However, the number of echoes that can be reasonably co-added depends on $T_{2\text{eff}}$ of the sample and the minimum applicable echo time. This in turn is a function of the chosen resolution, since it is defined by the acquisition time in the presence of the static field gradient. Because of this strong interdependence of the different acquisition parameters, it is crucial to determine the minimum resolution that is actually needed for a particular application. A change of the resolution by a factor of two may result in an increase of measurement time by an order of magnitude.

In this experiment a CPMG sequence with an echo time of 45 μs and an acquisition time of 20 μs resulting in a resolution of 50 μm was applied to measure the profiles. The amplitudes of the first 64 echoes of the CPMG train were added and plotted as a function of the position. Hence, the plotted amplitude corresponds to the spin density weighted by $T_{2\text{eff}}$. The gasoline enters from the regrind side. With increasing fuel content of the polymer the amplitude strongly increases reflecting a large change in the $T_{2\text{eff}}$ of the polymer with the gasoline. This is shown in Fig. 9.5b, where two CPMG decays, one from a dry position of the regrind and the other at a wetted one are plotted. The curves were fitted with a single-exponential function. While the initial intensity of the decays is quite similar, the $T_{2\text{eff}}$ values differ by a factor of six, 0.4 ms at the dry position, and 2.4 ms at the swollen position. The increase in $T_{2\text{eff}}$ is due to the fact that in a swollen state the polymer chains become more mobile, leading to a better averaging of the dipolar couplings. The solvent front moves into the sample until it meets the barrier, where it is stopped. After long enough time the amplitude of the regrind layer reaches a saturated state, while the external PE layer remains unaltered. It is noted that $T_{2\text{eff}}$ of the resin and of the barrier layer are increasing, reflecting that also these layers are partially swollen.

Besides organic solvents, water can also ingress into polymers. While in some cases polymers are specially designed for high water uptake, e.g., hydrogels, in most cases this is an unwanted property. The car industry, for example, is interested in substituting metal parts of the car body by polymer parts in order to reduce the weight and therefore the fuel consumption. Since these parts are exposed to all kinds of weather conditions, their moisture content continuously changes, resulting in changes of the mechanical properties as well as the clearance between body parts, caused by the swelling of the parts. Such processes of water penetration can be monitored using the profiling technique [29]. In this example the ingress of water into PVC (polyvinyl chloride) was followed by applying the profiling technique as a function of time to a sample immersed in water. Two cases were investigated, both conducted at room temperature: a first one where the sample was immersed in water and a second one where the sample was exposed to room humidity. In order to remove residual water the samples were dried for 48 h at 100°C prior to the start of the experiments. The profiles were measured using a solid-echo sequence with an echo time of 30 μs and an acquisition time of 6 μs per echo defining a nominal spatial resolution of 200 μm. Due to the extremely short T_2 (200 μs) only the amplitudes of the first eight echoes were added for signal-to-noise improvement. Figure 9.11a shows a series of profiles measured for different immersion times. The profiles clearly show how the water ingresses from both sides to the center of the sample reaching an equilibrium state after 65 days. In a different experiment

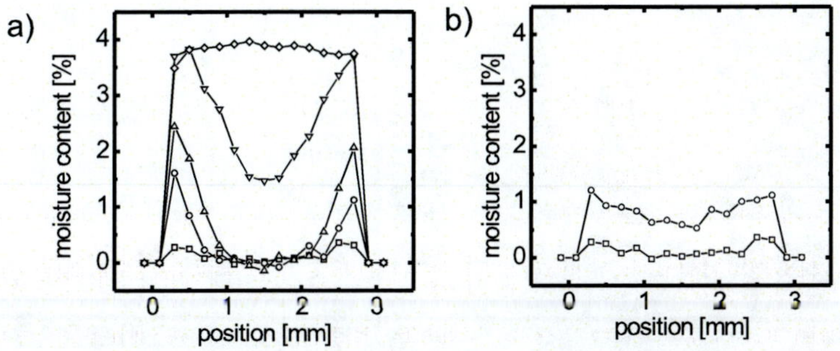

Fig. 9.11 (**a**) Profiles measured across a sheet of PVC: dry (□) and after 3 (○), 6 (△), 27 (▽), and 65(◇) days of immersion in water. The ingress reaches a saturation state after 65 days. (**b**) Moisture profiles measured after drying (□) and after exposition to room humidity for 4 weeks (○)

a dried sample was exposed to room humidity. Figure 9.11b shows that a lower saturation value is observed in this case. As for the previous example, also here the change in amplitude is due to a $T_{2\text{eff}}$ variation. However, the changes in $T_{2\text{eff}}$ are much smaller compared to the ones observed for the fuel tank, where the amount of absorbed liquid is larger. In order to obtain the quantitative information presented in the figures, the dry sample as well as the samples saturated in water and exposed to room humidity were weighed. The weights were plotted vs. the corresponding NMR amplitude and the obtained linear relation was used for the calibration. It is pointed out that even small quantities in a range of 1% can be well detected.

These examples prove that the profiling technique is also suitable to monitor the ingress of solvents into polymer materials in a non-destructive way. This can be of great value for the design of polymer parts, which are in contact with liquids or exposed to air humidity. The mobility and the non-invasiveness make the technique especially suited for applications outside the laboratory.

9.3 Cultural Heritage

Single-sided NMR has also proven to be a valuable tool in the field of cultural heritage, where non-destructive techniques are especially in demand considering the uniqueness of the objects. Furthermore, also the transport of the objects is often not possible or very expensive so that mobile equipment which can be brought to the site of the object is required. Single-sided NMR has been applied to study different kinds of objects of cultural heritage like frescoes [30], mummies [31], or ancient paper [32–35]. Here the studies of old master paintings by the profile technique are featured [36] as the structure and complexity of paintings is especially challenging. Depending on the style, paintings are made of several layers on a support, normally a wooden panel or canvas. Several paint layers are applied to produce color gradients, small details, shadow regions, and highlights. Usually a varnish layer is applied after drying to protect and saturate the colors.

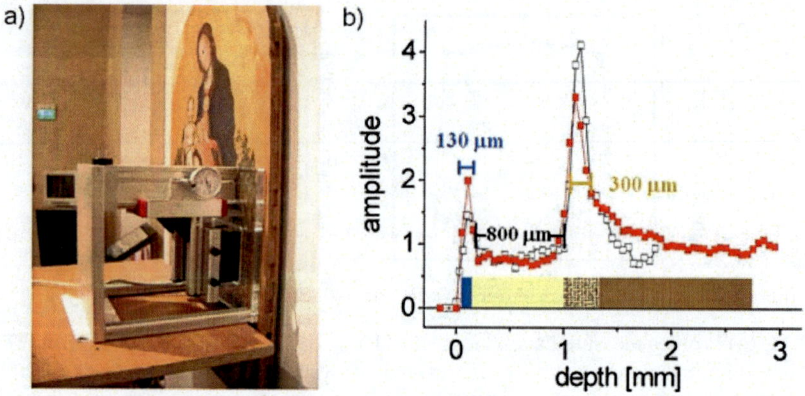

Fig. 9.12 (**a**) The NMR-MOUSE setup to measure depth profiles of the Virgin and child by Gentille da Fabriano (1411). The acrylic plate of the lift is aligned parallel to the painting surface but does not touch the paint. A gap of about 1 mm is left between them. (**b**) Measurements at two different spots show the multilayer structure of the painting consisting from *left* to *right*: paint, primer, canvas, and wood

Different old master paintings have been studied in the Galleria Nazionale dell'Umbria in Perugia (Italy). Figure 9.12a shows the setup used to measure profiles through the different layers of the paintings; in this case Virgin and Child by Gentille da Fabriano (1411). The sensor was placed on a movable table to control the depth position, while the lateral position was controlled simply by displacing the sensor. Since the lateral resolution is of about 1 cm in both directions (up–down and left–right), given by the dimension of the rf coil, the positioning precision was sufficient. More critical was the positioning along the depth direction, where a resolution of 20 μm was required. This could be done by carefully aligning the acrylic plate of the lift parallel to the painting. Due to the size ratio between plate and sensitive volume an error of 0.5 mm at the border of the plate represents a deviation of only 15 μm across the sensitive volume. Notice that there is no need to have any contact with the painting. Once the plate was aligned, the sensor was moved with respect to the plate by the high-precision lift. Figure 9.12b shows two depth profiles measured at two different positions. The first peak (depth 0 mm) corresponds to the paint layer which has a thickness of about 130 μm. This is followed by a flat region of about 800 μm that corresponds to the primer layer. The second peak at about 1 mm depth is assigned to canvas that is glued on the wood support.

The same structure, paint–preparation–canvas–wood, was observed in an Altar frontal, Virgin and child, Adoration of the Magi, and Triptych of Sant'Antonio, but the thickness of those layers was different from painting to painting. Interestingly, in the case of Adoration of the Magi, a region with a canvas layer much broader than normal was found. Figure 9.13 shows that in region 1, a thin primer layer is followed by a canvas that reaches a thickness of about 1 mm. A possible explanation could be that the wood support is composed of a number of pieces that are joined together and reinforced using several canvas layers. Close to the measured spot one of these joints was found.

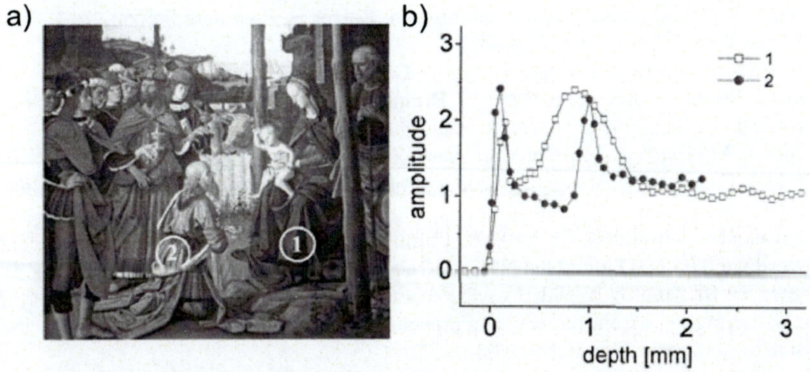

Fig. 9.13 (**a**) Adoration of the Magi by Perugino (1470s). (**b**) The same structure as the one described in Fig. 9.12 was observed. However, the canvas layer in spot 1 is much larger than in all other cases. A possible explanation could be that several canvas layers are used for reinforcement in places above wooden pieces are joined together

The analysis of old mater paintings shows that single-sided NMR can give new insight into the structure of objects of cultural heritage in a contact-free way. Different layers can be distinguished, their thickness can be determined, and even conclusions about the preparation technique of the painting can be drawn. Furthermore, the possibility to apply relaxation time contrast to the profiles is helpful in accessing information about the pigments, the aging of the binder, and possibly the authenticity of a painting [36].

References

1. Ernst RR, Bodenhausen G, Wokaun A (1987) Principles of nuclear magnetic resonance in one and two dimensions. Clarendon Press, Oxford
2. Blümich B (2000) NMR imaging of materials. Clarendon Press, Oxford
3. Callaghan PT (1991) Principles of nuclear magnetic resonance microscopy. Clarendon Press, Oxford
4. Blümich B, Perlo J, Casanova F (2008) Mobile NMR. Progr. Nuc Magn Reson Spec. 52: 197–269
5. Blümich B (2004) NMR-Bildgebung in den Materialwissenschaften, Nordrhein-Westfälische Akademie der Wissenschaften, Vorträge zur 483. Sitzung, Verlag Ferdinand Schöningh, München, pp 33–54
6. Cowie JMG (1991) Polymers, chemistry and physics of modern materials, 2nd edn. Blackie Academic & Professional, London
7. Litvinov VM, Dee P (eds) (2002) Spectroscopy of rubber and rubbery materials. Rapra Press, Sheffield
8. Blümler P, Blümich B (1997) NMR imaging of elastomers: a review. Rubber Chem Tech (Rubber Reviews) 70:468–518.
9. Blümich B, Demco DE (2002) NMR imaging of elastomers. In: Litvinov VM, De PP (eds) Spectroscopy of rubbers and rubbery materials. Rapra Technology Limited, Shawbury, pp 247–289
10. Blümich B, Blümler P, Gasper L, Guthausen A, Göbbels V, Laukemper-Ostendorf S, Unseld K, Zimmer G (1999) Spatially resolved NMR in polymer science. Macromol. Symp. 141: 83–93

11. Dinges K (1984) Kautschuk und Gummi. In: Batzer H, Polymere Werkstoffe 3. Band Georg Thieme Verlag, Stuttgart, pp 330–387
12. Gent AN (1992) Engineering with rubber. Carl Hanser Verlag, München
13. Kolz J, Martins J, Kremer K, Mang T, Blümich B (2007) Investigation of the elastomer-foam production with single-sided NMR, Kautschuk Gummi Kunststoffe 60:179–183
14. Mooney M (1940) A theory of large elastic deformation. J Appl Phys 11:582.
15. Perlo J, Casanova F, Blümich B (2005) Profiles with microscopic resolution by single-sided NMR. J Magn Reson 176:64–70
16. Herrmann V, Unseld K, Fuchs H-B, Blümich B (2002) Molecular dynamics of elastomers investigated by DMTA and the NMR-MOUSEÒ. Colloid Polym Sci 280:758–764
17. Anferova S, Anferov V, Adams M, Fechete R, Schroeder G, Blümich B (2004) Thermo-oxidative aging of elastomers: a temperature control unit for operation with the NMR-MOUSE®. Appl Magn Reson 27:361–370
18. Wolter B, Köller E (2001) Prüfung von Elastomeren mit NMR, DGzfP-Jahrestagung, 21–23 May, Berlin
19. Kolz J, Goga N, Casanova F, Mang T, Blümich B (2007) Spatial localization with single-sided NMR sensors. Appl Magn Reson 32:171–184
20. Somers AE, Barstow TJ, Burgar MI, Forsyth M, Hill AJ (2000) Quantifying rubber degradation using NMR. Polym Degrad Stab 70:31–37
21. Goga NO, Demco DE, Kolz J, Ferencz R, Haber A, Casanova F, Blümich B (2008) Surface UV aging of elastomers investigated with microscopic resolution by single-sided NMR. J Magn Reson 192:1–7
22. Perlo J, Casanova F, Blümich B (2004) 3D imaging with a single-sided sensor: an open tomograph. J Magn Reson 166:228–235
23. Blümich B, Casanova F, Perlo J, Anferova S, Anferov V, Kremer K, Goga N, Kupferschläger K, Adams M (2005) Advances of unilateral, mobile NMR in nondestructive materials testing. Magn Reson Imaging 23:197–201
24. Blümich B, Casanova F (2006) Mobile NMR. In: Webb GA (ed) Modern magnetic resonance. Springer, Berlin, pp 369–378
25. Blümich B, Casanova F, Buda A, Kremer K, Wegener T (2005) Mobile NMR for analysis of polyethylene pipes. Acta Physica Polonica A 108:13–23
26. Blümich B, Buda A, Kremer K (2006) Non-destructive testing with mobile NMR. RFP 1:34–37
27. Blümich B, Casanova F, Buda A, Kremer K, Wegener T (2005) Anwendungen der mobilen NMR zur Zustandsbewertung von Bauteilen aus Polyethylen. 3R Int 44:349–354
28. Blümich B, Adams-Buda A, Baias M (2007) Alterung von Polyethylen: Zerstörungsfreies Prüfen mit mobiler magnetischer Resonanz. Gas Erdgas 148:95–98
29. Casanova F, Perlo J, Blümich B (2006) Depth profiling by single-sided NMR. In: Stapf S, Han S (eds.) NMR in chemical engineering. Wiley-VCH, Weinheim, pp 107–123
30. Proietti N, Capitani D, Lamanna R, Presciutti F, Rossi E, Segre AL (2005) Fresco paintings studied by unilateral NMR. J Magn Reson 177:111–117
31. Rühli F, Böni T, Perlo J, Casanova F, Baias M, Egarter E, Blümich B (2007) Noninvasive spatial tissue discrimination in ancient mummies and bones by in situ portable nuclear magnetic resonance. J Cult Heritage 8:257–263
32. Capitani D, Emanuele MC, Bella J, Segre AL, Attanasio D, Focher B, Capretti G (1999) 1H NMR relaxation study of cellulose and water interaction in paper. TAPPI J 82:117–124
33. Blümich B, Anferova S, Sharma S, Segre A, Federici C (2003) Degradation of historical paper: nondestructive analysis by the NMR-MOUSE®. J Magn Reson 161:204–209
34. Casieri C, Bubici S, Viola I, De Luca F (2004) A low-resolution non-invasive NMR characterization of ancient paper. Solid State Nucl Magn Reson 26:65–73
35. Viola I, Bubici S, Casieri C, De Luca F (2004) The codex major of the collectio altaempsiana: a non-invasive NMR study of paper. J Cult Heritage 5:257–261
36. Presciutti F et al (2008) Noninvasive nuclear magnetic resonance profiling of painting layers. Appl Phys Lett 93:033505

Chapter 10
Spectrometer Hardware

Jörg Felder

The purpose of this chapter is to familiarize the reader with the particular hardware requirements of single-sided nuclear magnetic resonance (NMR) tomographs. A good starting point is the comparison with a clinical tomograph as shown in Fig. 10.1. The following sections assume a basic knowledge of NMR hardware since the detailed description of the individual components is without the scope of this book. The reader is referred to standard textbooks, e.g., [1, 2], for an in-depth treatment.

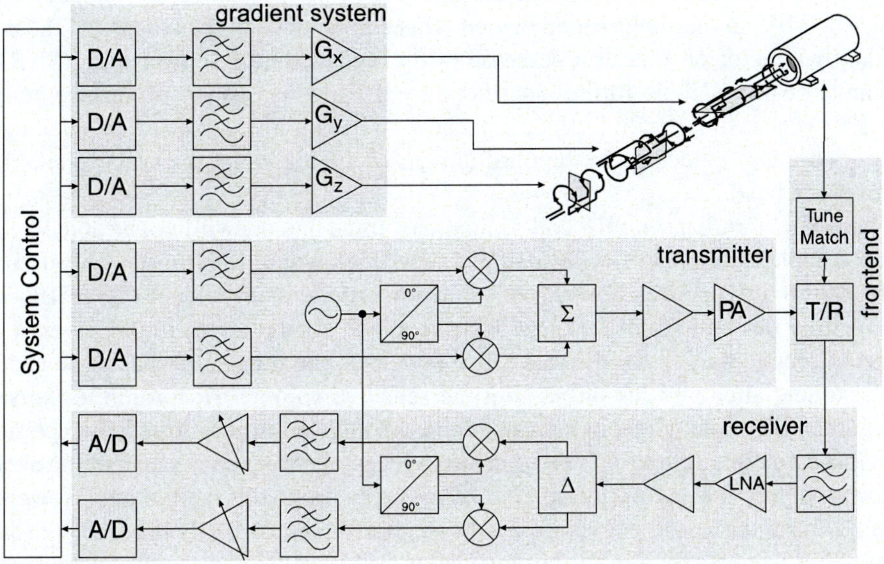

Fig. 10.1 Block diagram of a clinical MRI system employing a whole body transmit/receive birdcage resonator with linear polarization

J. Felder (✉)
Institute of Neuroscience and Medicine 4, Forschungszentrum Jülich, 52425 Jülich, Germany

F. Casanova et al. (eds.), *Single-Sided NMR*,
DOI 10.1007/978-3-642-16307-4_10, © Springer-Verlag Berlin Heidelberg 2011

10.1 Single-Sided vs. Conventional: Systematic Differences

The fundamental difference between a conventional tomograph and a single-sided device lies in the realizable magnetic field strength and homogeneity. Superconducting solenoidal coils in whole body systems achieve magnetic flux densities between 1.5 and 11.7 T with a typical homogeneity of 10–50 ppm over a 30–50 cm spherical volume [3]. Characteristic switchable field gradients range from 60 mT/m in wide bore magnets up to almost 1,000 mT/m in animal magnetic resonance imaging (MRI) systems [4]. In contrast, the permanent magnet of a single-sided tomograph generates a maximum magnetic induction of around 0.25 T and gives rise to a *static* field gradient as high as 2.5 T/m in direction of penetration (Chap. 4) [5].

Thus, an analysis of the implications caused by the inhomogeneous $\mathbf{B}_0$ field for the radio frequency (RF) hardware becomes necessary. Assume a system with a narrowband transmitter capable of amplifying signals which fall within a bandwidth (BW) of 1 MHz centered around the proton resonant frequency.[1] At a field gradient of 66 mT/m in slice direction, this transmitter allows selective excitation anywhere within a range of 35.60 cm. With the single-sided tomograph the strong field gradient reduces the field of view (FOV) to just 1 cm in the slice direction. The other way round, the NMR signal originating from a slice 1 cm thick is contained within a BW of 28.05 kHz in a clinical tomograph, while it is spread over a frequency range of 1.06 MHz in the single-sided device. This corresponds to an increase of the noise BW by a factor of 37.8 or a decrease of the signal-to-noise power ratio (SNR) by almost 16 dB. SNR is further sacrificed by the low sensitivity of surface coils at higher penetration depth. Surface coils have to be employed in a single-sided tomograph because extension in the third dimension is limited by the space available in the probe head.

Readers familiar with RF coil design may argue that one will not be able to cover the complete 1 cm FOV with a single shot. This would require not only tremendously high transmitter power but also a coil with a quality factor (Q) of 10 – and hopefully our coil is not that bad. But the above arguments still hold. Even if one has to retune the coil for different slice positions, the transmitter has to be capable of exciting slice profiles throughout the sensitive volume. According to the same argument, the analog receiver frontend must display an equally broad characteristic. In this case digital filters can be adjusted to accept the frequency range of the excited slices only, so that the optimum SNR of the receive signal is maintained. However – as has been discussed above – for a given slice thickness the BW is much broader and SNR is worse than in the homogeneous case.

In essence, optimization of RF frontend hardware is of uttermost importance during construction of a single-sided tomograph and is crucial for successful system application [6]. The examples given above intend to emphasize this fact. An in-depth

---

[1] A BW of 1 MHz does not seem to be that great but already represents a relative BW of 10% or requires that the Q of the amplifier, filters, and matching circuits be equal or less than 10 for a 0.25 T system.

treatment of the major components shown in Fig. 10.1 can be found in the next sections.

10.2 Frontend Design

Simply stated, the frontend of an NMR tomograph includes all circuitry between transmitter and receiver; compare Fig. 10.1. It includes the NMR coil itself, the impedance matching network, and the transmit/receive (T/R) switch. Depending upon the system configuration, the T/R switch may alternatively be a conventional antenna switch in a single coil setup or employ PIN diodes to actively decouple multiple coils.[2] In addition, many commercial systems integrate signal combiners for quadrature coil operation and support modules. These may be impedance measurement equipment, hardware detection to automatically modify the system setup according to the connected components, and specific absorption rate (SAR) surveillance devices.

10.2.1 Matching and Balancing

Industry standard tomographs usually employ "two-capacitor" matching because of its simplicity and ease of tuning. The susceptibility of this circuit to electromagnetic interference can easily be disregarded, because the complete magnet is installed inside an RF-shielded room. Obviously, a portable tomograph has to live without this luxury. Another way of minimizing external (or man-made) noise and of removing interfering signals that may easily swamp the receive signal[3] (cf. [1] for a good example) has to be sought. A good solution is matching circuits which balance the receive coil with respect to ground. Advantages of balanced coils are as follows:

- Reduced capacitive coupling to ground or to the sample under investigation which is usually assumed to be a lossy path to ground. This increases the current flow through the coil by minimizing the shunt effect of ground loops. Furthermore, capacitors at the coil terminals shorten the electrical length of the coil and produce a more homogeneous current distribution along the coil. This is important even at low frequencies because, as a rule of thump, only components with linear dimensions (including the length of the wire from which the coil is wound) less than $\lambda/16$ can be assumed to have lumped element characteristics.
- Reduced proneness to external interference. Interfering sources produce common mode voltages at the coil terminals which cancel. This concept is well known to radio amateurs who frequently connect unbalanced-to-balanced transformers (baluns) to their antennas in order to improve system performance.

---

[2] A technique also known as "*Q*-factor spoiling."

[3] The noise temperature of an antenna operating at 10 MHz is approximately 2×10^5 K [7]!

- Reduced far-field radiation. Current coupled from a balanced coil tends to produce a quadrupole field distribution in the dielectric volume under investigation instead of an electric dipole which arises from an unbalanced current distribution[8]. This greatly reduces radiation and minimizes the associated losses.

Before analyzing alternative matching circuits, a quantitative measure for the balancing properties is required. The theory of even and odd mode wave propagation provides the required tools [7]. Designating the complex voltages at the top and bottom of the NMR coil as $\underline{v}_{\text{top}}$ and $\underline{v}_{\text{bottom}}$, respectively, balanced (odd) and unbalanced (even) voltages can be expressed as

$$\underline{v}_{\text{even}} = \frac{1}{2}\left(\Re\left\{\underline{v}_{\text{top}}\right\} + \Re\left\{\underline{v}_{\text{bottom}}\right\}\right) + \jmath\frac{1}{2}\left(\Im\left\{\underline{v}_{\text{top}}\right\} + \Im\left\{\underline{v}_{\text{bottom}}\right\}\right), \quad (10.1)$$

$$\underline{v}_{\text{odd}} = \frac{1}{2}\left(\Re\left\{\underline{v}_{\text{top}}\right\} - \Re\left\{\underline{v}_{\text{bottom}}\right\}\right) + \jmath\frac{1}{2}\left(\Im\left\{\underline{v}_{\text{top}}\right\} - \Im\left\{\underline{v}_{\text{bottom}}\right\}\right). \quad (10.2)$$

A graph of the even and odd voltages at the coil terminals over frequency allows to analyze the balancing properties of a given matching circuit.

Practical matching circuits – apart from the conventional "two-capacitor" matching mentioned above – include inductive coupling [9, 10], lumped element baluns [11], transmission line matching [12], and several solutions borrowed from radio amateurs who have embraced the concept of symmetric antenna feeding for a long time [13].

To the extent of balancing, inductive coupling seems to present the optimum choice. In the confined space available in the probe head of a single-sided tomograph, however, it is difficult to arrange a second coil that has negligible coupling to the NMR signal itself but still be coupled with the primary coil at the same time. An additional disadvantage of this arrangement is its proneness to ringing which, in an extreme scenario, may by a patient safety hazard. Space limitations also render the idea of counterloops difficult to realize. Systems with two loops connected in opposite polarity have been reported to cancel external interference, which can be considered to be uniform in space over the region of interest (ROI). Again, one coil has to be arranged so as to have no coupling with the NMR signal.

Let us take a closer look at the two selected matching circuits shown in Fig. 10.2a, b. Both realizations have improved balancing properties compared to the standard matching scheme. Figure 10.2c, d displays the absolute value of the even and odd mode voltages calculated according to Eq. (10.2). Another discrimination between the two circuits is the way their input impedance changes in vicinity of the NMR resonance frequency; compare Fig. 10.2e, f. The impedance variation of the circuit in Fig. 10.2a is greater because it resonates at a frequency different from the NMR resonance. Thus circuit (Fig. 10.2b) achieves a greater BW for a given NMR coil.

For comparison, the performance of standard "two-capacitor" matching employing a shunt capacitor C_1 across the coil and a connection to the transmission line via

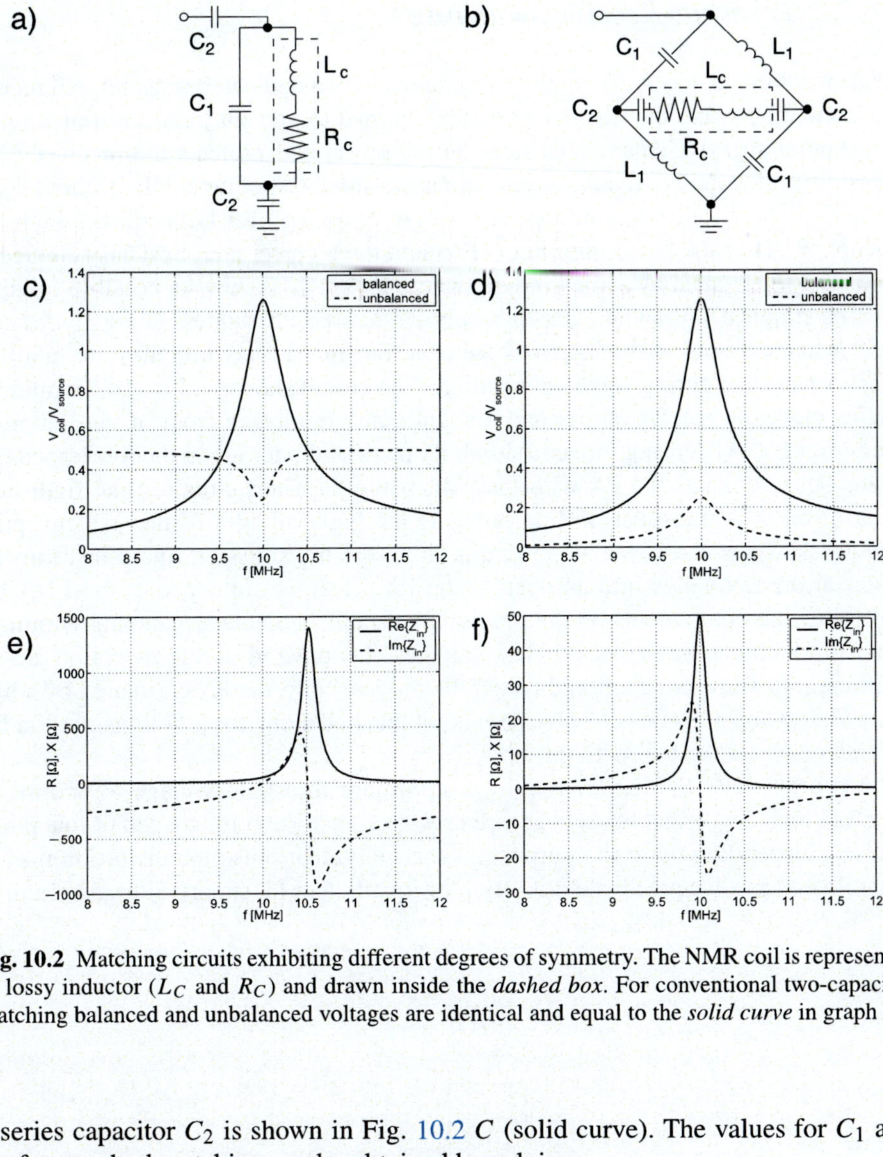

Fig. 10.2 Matching circuits exhibiting different degrees of symmetry. The NMR coil is represented as lossy inductor (L_C and R_C) and drawn inside the *dashed box*. For conventional two-capacitor matching balanced and unbalanced voltages are identical and equal to the *solid curve* in graph (**c**)

a series capacitor C_2 is shown in Fig. 10.2 C (solid curve). The values for C_1 and C_2 for standard matching can be obtained by solving

$$S_1 = \frac{Q \pm \sqrt{(Q^2+1)k - 1}}{Q^2+1}, \tag{10.3}$$

$$S_2 = \frac{k}{\pm\sqrt{(Q^2+1)k - 1}}, \tag{10.4}$$

where $S_1 = \omega_0 C_1 R_C$, $S_2 = \omega_0 C_2 R_C$, $k = \frac{R_C}{R_0} = (1 - QS_1)^2 + S_1^2$, and R_0 is the reference resistance.

10.2.2 Transmit-Receive Switching

T/R switching is necessary in all tomographs. When dedicated transmit and receive coils are employed, the inactive coils are detuned to prevent them from picking up substantial signals. This avoids field deterioration and protects the receiver hardware from destructive overvoltages. In single-sided tomographs, the limited space available usually dictates the use of a single NMR coil for both transmission and reception. Thus the power amplifier (PA) and the receiver are physically connected and have to be gated by some kind of switch. Figure 10.3 shows a possible solution.

The simplest T/R switch employs pairs of crossed PN diodes. These diodes conduct when exposed to the high voltage of a transmit pulse while they are in a high impedance state during signal acquisition. The circuit shown in Fig. 10.3 would use series diodes at the PA output and a shunt configuration in front of the low-noise amplifier (LNA). During transmission both pairs of diodes are in the low impedance state. This connects the PA with the coil, while the short circuit at the right-hand side of the $\lambda/4$ transmission line prevents the high voltages of the transmit pulse from destroying the LNA. At the same time, the transmission line transforms the short at the LNA gate into an open at the PA. Thus the influence of the LNA and all following components on the previously established power matching within the transmit path can be neglected. The voltages encountered during reception are not sufficient to allow conduction through the diodes. Now, the PA is isolated by a high impedance and a low loss receive path is established along the $\lambda/4$ transmission line which connects the coil with the LNA.

Unfortunately, PN diodes require a minimum threshold voltage to commence conduction. This effect leads to severe crossover distortion for shaped or low power pulses. Depending upon the purpose of the individual pulse the distortion may be manifested as slice profile deterioration or insufficient fat suppression. Alternatives

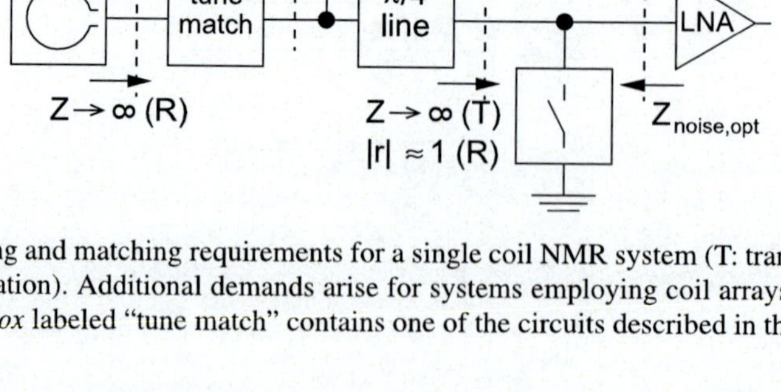

Fig. 10.3 Tuning and matching requirements for a single coil NMR system (T: transmit operation, R: receive operation). Additional demands arise for systems employing coil arrays and have been included. The *box* labeled "tune match" contains one of the circuits described in the last section

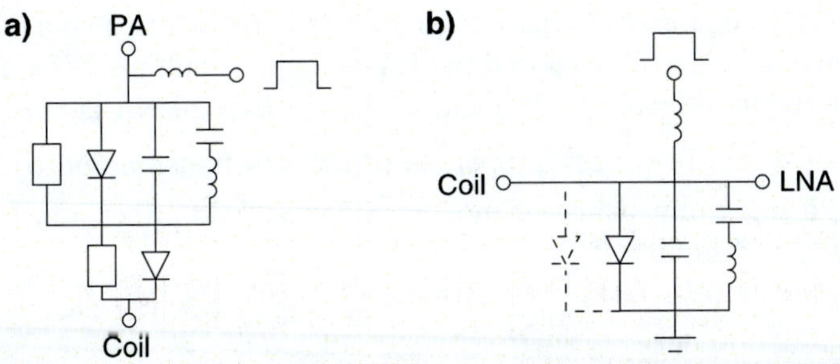

Fig. 10.4 Detailed view of the PIN diode switches. (**a**) A series switch as employed in the transmitter path, and (**b**) a shunt switch used to protect the preamplifier input. The switches are integrated in a parallel resonant circuit to increase isolation. The high impedance resistor in (**a**) is required to provide reverse bias conditions for both diodes. The shunt switch in (**b**) may employ multiple diodes – although parallel diodes potentially cause problems when the individual characteristics differ significantly – in order to decrease the forward-biased resistance

to PN diodes are PIN diodes.[4] Under the influence of a forward direct current (DC) or reverse voltage they act as controlled RF resistance which exhibits a low or high impedance, respectively. The series switch and the shunt switch configuration, both of which could be used in the frontend shown in Fig. 10.3, are reproduced in detail in Fig. 10.4.

During transmission both PIN diode switches are forward biased, while a reverse voltage is applied when acquiring the NMR signal. Apart from simplifying the drive circuitry this configuration has the advantage of avoiding shot noise generation since there is practically no DC current flow during signal acquisition. Thus the SNR of the receive signal is maintained as best as possible. Another advantage is that the reverse-biased PIN diodes keep noise emitting from the quiescent PA – there is always some noise even if the PA manufacturer has already included means for noise blanking – from reaching the receiver.

One should think that the low frequency of operation of a single-sided tomograph – which is of course due to the low main field strength – simplifies RF system design. This is generally true; however, there are some exceptions. First the encountered frequencies are notoriously neglected by semiconductor device manufacturers who are focusing on mass markets like ISM, GSM, GPS, which cover different frequency bands. Second, there is a lower frequency limit for proper operation of PIN diodes. It depends on the minority carrier lifetime τ of the specific diode. Diodes with longer carrier lifetimes are suited for low-frequency operation but exhibit longer switching times. Finally, many circuits easily and reproducibly implemented using transmission lines have to be approximated by a lumped replica.

Included in Fig. 10.3 are the requirements for multiple coil operation in so-called phased arrays [14]. Multiple receive coils are widely used to speed up image acquisition or improve SNR. A fact which makes phased arrays an attractive choice

[4] PIN diodes have an additional intrinsic region between the P- and N-doped parts.

for single-sided tomographs where high averaging ratios often cause long imaging times. A matching circuit proposed in [15] decouples multiple coils if the absolute value of the input reflection coefficient of the connected LNA is close to 1. Due to the impedance transformation properties of the $\lambda/4$ transmission line $Z_{\text{in}} = \frac{Z_L^2}{Z_{\text{out}}}$ and the high input impedance of modern Si/GaAs field effect transistors (FETs)[5] this is readily accomplished.

10.3 Transmitter Design

Before going into details, the four most common figures of merit for PAs are presented. Once these are understood, they can be brought into relationship with the NMR experiment and a suitable amplifier topology may be selected. Going straight ahead, main performance criteria are as follows:

Output power It usually describes the peak output power which cannot be sustained in continuous wave (CW) operation. Hard pulses at peak output power are limited to the maximum pulse duration while their rate of repetition is limited by the maximum duty cycle.

Bandwidth A minimum requirement is that the transmitter covers the range of Larmor frequencies encountered in the tomograph. For proton imaging, narrow band amplifiers are sufficient while broadband PAs are employed when different nuclei are investigated.

Efficiency Power efficiency is of concern in a dual sense. It allows miniaturization of the transmitter by reducing the size of cooling equipment and heat sinks. Additionally, the requirements on the power supply can be relaxed.

Pulse rise time Especially for short, hard pulses a fast rise time is necessary in order to obtain the full output power. It is usually measured between 10 and 90% of the peak envelope power (PEP) for an input rectangular voltage waveform. Fast rise times require transistor output matching circuits with a low Q or, synonymously, a large BW.

Let us investigate these issues for PAs employed in single-sided NMR tomographs. Output power required to excite small sensitive volumes close to the transmit

[5] Note that commercially available GaAs HEMTs are difficult to employ in high-field applications because they display a strong Hall effect.

surface coil is relatively low, albeit sufficient power has to be available to reduce 180° pulse length to the low microsecond range. Depending upon the desired FOV in direction of penetration, about 50–500 W should be sufficient.

As stated above, the BW has to cover the range of expected Larmor frequencies. Keep in mind that this may be rather broad even for proton imaging because of the strong decay of $\mathbf{B}_0$ with penetration depth. Finally, efficiency is of major concern in mobile devices. Generally, MRI PAs tend to be bulky and quite heavy because they contain large toroidal transformers, capacitor banks for pulse power supply, and huge heat sinks[6]. Thus, efficient amplifiers with small heat sinks are advantageous in mobile, single-sided devices. For modern applications, especially transmit SENSE (TSENSE) which has the capability or reducing the effects of $\mathbf{B}_1$ inhomogeneities [16], multiple small and light-weight amplifiers have further benefits.

10.3.1 Conventional Power Amplifiers

Conventional amplifiers are operated in class A or class AB (push–pull). Inherent advantage of these classes is their linearity. The output follows the input voltage waveform at a higher power level without significant distortion. Typically, nonlinearities in these classes give rise to harmonics which are 30–35 dB below the carrier signal. The amplifier is said to have an intermodulation distortion (IMD) of -30 to -35 dBc.

The major disadvantage of a linear PA is its wasteful handling of the supplied power. Ideal class A amplifiers achieve a maximum efficiency of 50% when driven by a CW signal at PEP. Push–pull operation increases theoretical efficiency to 78.5%. However, most (e. g., shaped) NMR pulses are far from displaying a constant envelope and efficiency for the linear amplifiers decreases rapidly. Table 10.1 shows the efficiencies obtained for some common-shaped pulses – assuming that the active devices are ideal and do not dissipate power themselves. Columns 2–4 imply that the transmit PA is gated off (or blanked) by removing its supply voltage during inactivity. This is a common method to keep noise emitting from a quiescent PA from disturbing the NMR signal. In rare occasions, e. g., when applying fast RF sequences, removing the amplifier bias voltage is not possible because it would take too long to get the PA back to operation. This further reduces overall efficiency.

Table 10.1 Average efficiencies of linear PAs for some common NMR pulses. Columns 2–4 imply transistor shutdown during inactivity while column 5 has been obtained for continuous PA operation and hard pulsing. In-depth treatment of PA efficiencies as a function of the statistical properties of the input envelope can be found in [17, 18]

	Sinc3 (%)	Sinc5 (%)	Gauss (%)	Continuous operation (%)
Class A	11.9	8.1	16.6	0.5
Class AB	28.6	21.2	36.6	2.5

[6] Many high-power amplifiers in the kilowatt range even require water cooling.

The last column shows the result obtained for the theoretical case of a spin-echo experiment (TR = 2 s and TE = 20 ms) without amplifier gating.

An important aspect for PAs employed in single-sided NMR systems is the high required BW and fast pulse rise time. It is desirable that the amplifier is capable of exciting slices at either end of the sensitive volume. Additionally, the strong $\mathbf{B}_0$ inhomogeneity causes fast signal dephasing and therefore requires short excitation pulses and fast pulse rise times. Thus, in most cases, the amplifier to be chosen cannot employ narrowband circuitry and the output filter must display a suitably low Q. The last demand is often met by adding resistive elements into the output circuit which clearly reduces the efficiency of the amplifier even further.

10.3.2 Alternative Amplifier Designs

In contrast to linear amplifiers, switching amplifiers yield high efficiencies. Trade-off is their gross nonlinearity. Information contained in the signal envelope is lost and the unfiltered output spectrum contains significant intermodulation (IM) products. A comparison of typical PA linearities and efficiencies for some common modulation formats can be found in [18].

Figure 10.5 shows a classification of amplifier modes of operation. The linear amplifiers are replenished by switching amplifiers in which the active device/devices operate alternatively in cutoff or saturation. On the right-hand side, the systematic is expanded by two alternative applications of amplification systems. They might be distinguished as "efficiency boosting" and "amplification systems employing signal processing". Both achieve high efficiencies and a linear transfer function at the same time.

Efficiency boosting systems are designed to operate a PA closer to its PEP. This clearly improves efficiency but does not change the linearity of the underlying

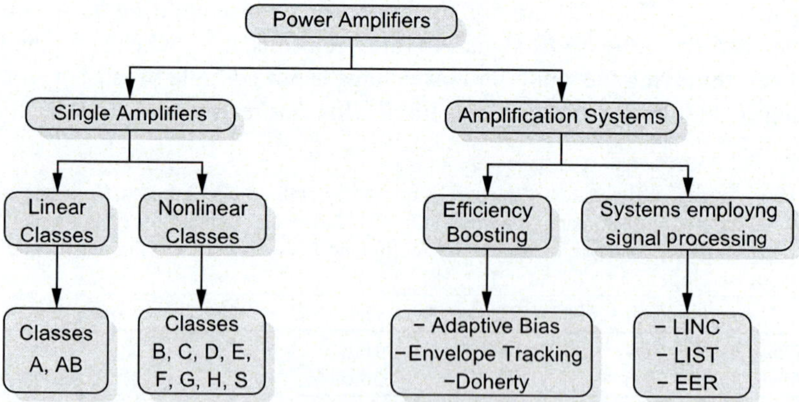

Fig. 10.5 Classification of PAs and amplification systems. For a detailed description of the different amplification methods confer to [18–21]

amplifier. For this reason, linear amplifier classes have to be employed – which sets the theoretical limit of efficiency at 78.5%, e. g., that of ideal class B operation. However, efficiencies obtained in reality are generally much lower. For 20 dB power backoff, which is slightly less than the average power backoff of a sinc pulse with two or four sidelobes, ideal class A achieves less than 5% efficiency and ideal class B 24.8%. This value is increased to around 43% for Doherty systems with ideal class B active devices [22]. Both adaptive bias and envelope tracking reside in approximately the same range of efficiencies.

Systems employing signal processing are constructed of nonlinear but highly efficient amplifiers. They are combined to form linear amplification systems which are potentially more efficient than the approaches described above. Efficiencies greater than 50% are reported in literature even for high power backoff operation [23, 24].

Inherent advantages and potential technological difficulties of the individual amplification methods are summarized in Table 10.2. As far as efficiency is concerned, systems employing signal processing promise superior results as compared to efficiency boosting systems. Within the first group, envelope elimination and restoration (EER) presents the system with the highest potential, even more so since its efficiency decreases only gradually with increasing power backoff levels. The reason why linear amplification employing nonlinear components (LINC) and linear amplification employing sampling techniques (LIST) perform worse is that they require a signal combiner or transformer located at the output of the two PAs. Efficiency is sacrificed in these components either by dissipating the out-of-phase power in the combiner resistor or by load pulling.

As far as BW is concerned, frequency-selective components like $\lambda/4$ transmission lines are usually the bottleneck. They limit the operating frequency range of the amplifier and thus penetration depth of the single-sided tomograph. Amplification systems requiring frequency-selective components include Doherty, Chireix outphasing, and parallel amplification. Although EER also employs a resonant circuit, it achieves suitable results. A BW greater than 1.2 MHz in a PA complying with the IS-95 standard has been realized [25].

Whether EER achieves the required linearity for NMR experiments is difficult to answer at once. Primarily because literature seems to pay little attention to PA linearity[7] and, second, the inhomogeneity of $\mathbf{B}_0$ of a single-sided magnet is not encountered in conventional tomographs. A state-of-the-art EER system has been constructed and the obtained efficiency has been measured [6]. For the pulses given in Table 10.1, greater than 50% efficiency can be maintained. At the same time gating the amplifier output during transmit pauses is obsolete since supply voltage is effectively removed from the final stage at zero output power. Figure 10.6 shows a graph of the efficiency obtained with the constructed EER system.

[7] Albeit a treatise on required transmitter amplifier linearity can be found [26].

Table 10.2 Advantages and disadvantages of individual efficiency improvement approaches from [6]

Technique	Potential advantages	Potential disadvantages
Adaptive bias control	• Low system complexity	• Limited efficiency improvements • Transistor gain variations with changing bias cause AM/AM distortion
Envelope tracking	• Low system complexity	• Limited efficiency improvement • No linearization of underlying amplifier
Doherty	• No bandwidth broadening due to nonlinear signal processing • No high-power modulators required • High efficiency	• Limited bandwidth due to $\lambda/4$-transmission line • Variable load impedances (load pulling) • IMD performance relatively poor
LINC	• Power dissipation occurs external from active device • Low system complexity	• Efficiency degradation by signal combiner • Broad bandwidth cancelation necessary because of greater bandwidth of PM signals • Good load match required
LIST	• Power dissipation occurs external from active device • Reduced gain/phase balance requirements compared to LINC	• Efficiency degradation by signal combiner • High delta-coding frequency • Complex reconstruction filter architecture • Slope overload of delta coder causes ACI
EER	• High efficiency over wide range of power levels • Good linearity performance	• High-side modulation required • High bandwidth of signal components • Distortions at low signal levels

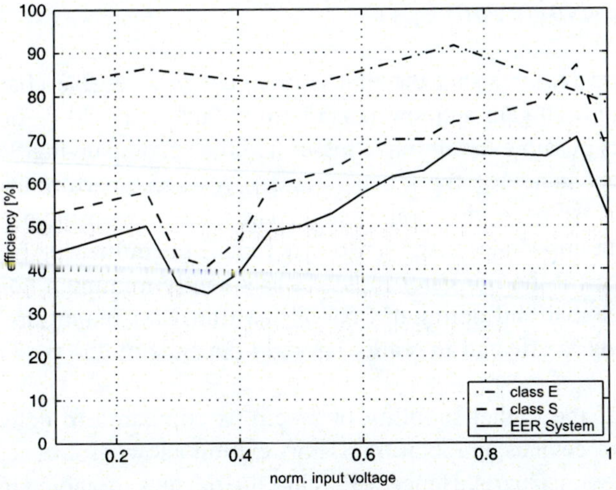

Fig. 10.6 Measured efficiencies of the class E and class S amplifier. The calculated efficiency of the complete EER system obtained by multiplying the individual amplifiers' linearities has been added for comparison

10.4 Receiver Design

The strong inhomogeneity of $\mathbf{B}_0$ in single-sided NMR devices requires frontend components with a high BW. Consequently, a receiver has to cover at least the range of Larmor frequencies encountered within the designated FOV. Since noise is acquired from within the complete analog BW, even if only thin slices are excited, noise performance is of uttermost importance. In a well-designed system, the component governing the overall noise figure (F)[8] is the LNA. It should be located as close to the receive coil as possible for optimum performance.

Because of inhomogeneous flip angles and low sensitivity for high penetration depths, the received amplitude is significantly lower in a single-sided NMR system than in a conventional tomograph. At receive amplitudes of a few microvolts only, gains greater than 100 dB are required to drive a 1 V analog-to-digital converter (ADC) at full range. Thus, the second consideration must be how to achieve high receiver gain. Certainly no stable amplification of this magnitude is realizable without frequency separation. The difficulties of implementing a truly balanced phase-sensitive detector (PSD) suggests down-conversion to an intermediate frequency (IF) where the receive signal can be acquired with fast ADCs. At the same time, this solution avoids DC coupling of the final ("audio") amplifiers and the associated problem of bias drift.

[8] F is defined as ratio of SNR at the amplifier output to SNR at its input.

10.4.1 Low-Noise Amplifier

Friis' equation requires that the first component in a receive chain ideally be an amplifier with high gain and low-noise figure. And, it should be placed as close to the antenna as possible to minimize losses and the accompanying SNR degradation. Because power matching the transistor sacrifices noise performance for amplifier gain, the LNA is operated in noise match. The underlying principle is to present an optimum input impedance in the sense as to reduce the influence of noise current and voltage sources at the transistor input. Consequently minimum noise deterioration of the input signal and optimum SNR at the transistor output are achieved. How noise matching is effected in reality strongly depends on the type of active device employed.

Because of the higher mobility of electrons compared to defects ("holes") in semiconductor devices, unipolar transistors exhibit a lower intrinsic noise level than their bipolar counterparts. Hence, FETs are the devices of choice in LNAs. A multitude of FETs is available differing in the semiconductor materials employed, their operating principles and last but not least their geometric layout. For frequencies below approximately 30 MHz, conventional silicon (Si) metal oxide semiconductor FETs (MOSFETs) achieve lower noise figures since the $1/f$ noise edge frequency of gallium arsenide (GaAs) FETs and high electron mobility transistors (HEMTs) is higher [7].

Assuming a simplified MOSFET equivalent circuit [7] the optimum noise impedance $Z_{\mathrm{noise,opt}}$ at the transistor input is given by

$$Z_{\mathrm{noise,opt}} = 0.67\omega C_{\mathrm{GS}} - \jmath\omega C_{\mathrm{GS}}, \tag{10.5}$$

where C_{GS} is the transistor gate-to-source capacitance. $Z_{\mathrm{noise,opt}}$ of FETs is generally in the range of several hundred ohms to kilo-ohms parallel to an imaginary reactance.

The trade-off between highest gain (power matching) and optimum SNR performance (noise matching) is best illustrated with the aid of a transmission line chart. In Fig. 10.7 circles of equal gain (*dotted*) and equal F (*dashed*) are plotted. All source impedances lying inside these circles will at least yield the given gain, respectively, noise performance. The impedances of some coil matching networks transformed via a $\lambda/4$ transmission line are added in the Smith chart to illustrate the influence of matching networks. Obviously, the less impedance variation the matching circuit displays over the frequency range of interest the better noise performance inside the complete FOV can be achieved.

Note that optimum noise matching will sometimes not be feasible for a given device. High-performance transistors tend to be susceptible for positive feedback loops which may cause unwanted oscillation. This is why stability should be checked for every design, e.g., by calculating the Linvill factor and eventually adding the regions of potential instability in Fig. 10.7.

A prototype design employing the dual-gate BF998 MOSFET achieves a noise figure of 0.7 dB with the bridge matching circuit shown in Fig. 10.2b. Dual-gate

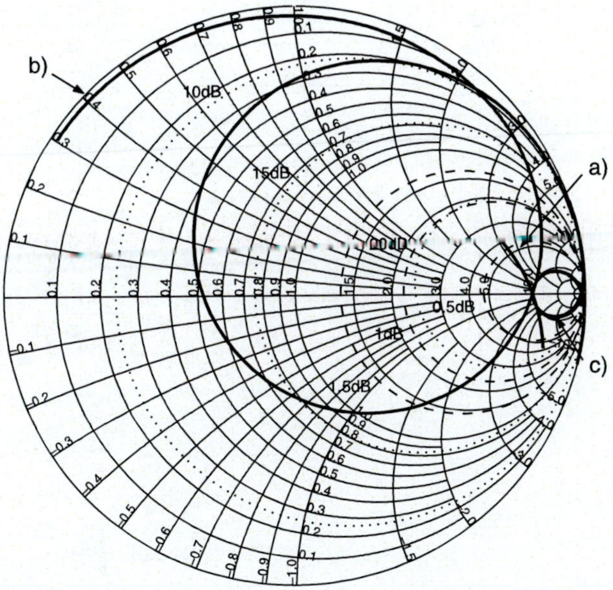

Fig. 10.7 Smith chart displaying circles of equal gain (*dotted*) and equal noise figure (*dashed*) for a BF998 MOSFET. The transistor is unconditionally stable so that no region of potential instability is marked. The *solid lines* show characteristic impedance variations over a range of 4 MHz transformed through a $\lambda/4$ transmission line ($Z_L \approx 158\,\Omega$) for a fixed 50 Ω load (**a**), conventional two-capacitor matching (**b**), and lumped balun matching according to Fig. 10.2b (**c**)

MOSFETs have the advantage of simplifying bias circuit implementation. At the same time they display high gains and have similar noise performance compared to their single gate counterpart. Noise matching is achieved with the aid of a Collins filter, which transforms the 50 Ω source impedance of the matched NMR coil into a high impedance at the transistor input.

10.4.2 Frequency Generation and Mixing

NMR is a resonant phenomena and therefore a high-resolution and highly stable frequency reference is necessary. This is driven to the extent that many spectrometers employ frequency locking – a control mechanism which lets the frequency generator output track small changes in the static $\mathbf{B}_0$ field as they arise from noise in shim or gradient power supplies or from main field drift.

Another important issue is control of the reference oscillator phase. It becomes relevant for signal averaging, a common method to improve the receive signal SNR. Phase control is also required for imaging sequences employing phase cycling. The application of phase cycling allows the removal of artifacts generated from imperfections in the PSD and suppresses ringing of the filters caused by pulse breakthrough. Generally, the higher the order of distortion to be corrected, the more cycles with smaller phase steps are required.

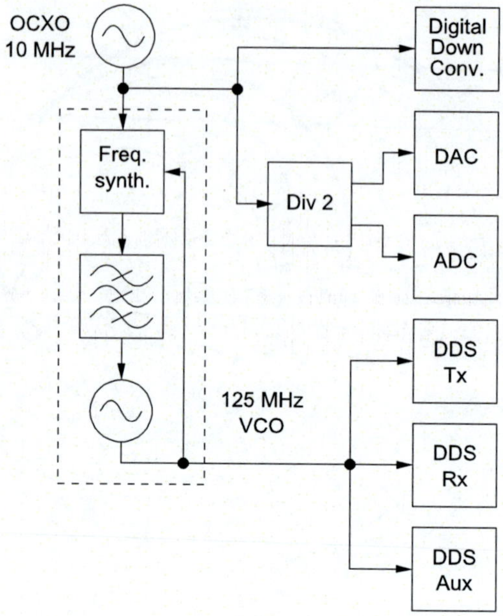

Fig. 10.8 Derivation of the LO from one stable reference. The design allows stable frequency generation and establishes phase coherence between all LO signals

The third requirement a spectrometer frequency generator has to satisfy is low phase noise. The free induction decay (FID) or echo signals exhibit $1/f$ noise which mixes with the phase noise from the local oscillator and leads to a reduction of SNR. Furthermore, phase noise correlates signal and noise and reduces the effectiveness of signal averaging. Good frequency sources exhibit a phase noise of at least 80 dB below the carrier signal in a frequency band 10 Hz to 1 kHz away from the center frequency [1].

Figure 10.8 shows a possible realization of the frequency generator. All clock frequencies are derived from a stable oven-controlled crystal oscillator (OCXO) hence displaying the same frequency stability. Frequency deviation of OCXOs is usually below 5×10^{-9} per day. The reference frequency is divided to drive the ADC and digital-to-analog converter (DAC) clocks, which are synchronized by this measure. For derivation of the remaining local oscillator (LO) signals, the OCXO output is first up-converted with the aid of a phase-locked loop (PLL). The PLL signal in turn drives direct digital synthesis (DDS) circuits which generate the required LO frequencies. DDS employs a digital counter to address a read-only memory (ROM) in which the sine function is stored in digital form. The ROM output is converted to analog via DACs. By changing the increment added to the counter at each clock cycle or loading an offset, DDS allows fast, high-resolution frequency variations of the output signal and gives direct control over the output phase. Another advantage is that the phase noise of the input clock is reduced by the noise transfer function $|\Theta_{\mathrm{DDS}}|^2 = 20 \log \frac{f_{\mathrm{out}}}{f_{\mathrm{clock}}}$.

The prototype design [6] built according to the block diagram in Fig. 10.8 displays a phase noise floor of -105 dBc/Hz and drops below -95 dBc/Hz at a frequency offset of less than 1 kHz from the carrier. Frequency resolution is better than 1.0 Hz. The phase register allows to set initial phase in increments of 11.25°.

Sampling of the receive signal at an IF simplifies the receiver architecture. It avoids the problems associated with analog baseband detection [27] such as quadrature gain balance and DC drift. However, because of the high gain required in the receiver at least one analog frequency conversion is required. The mixer should employ a single-sideband (SSB) architecture.[9] SSB conversion employs double-balanced mixers preceded by a mirror frequency filter and concluded by a low-pass filter. The first filter removes noise from the mirror frequency range which would otherwise be folded into the NMR signal and the second filter eliminates the sum term generated during the mixing process. Both analog filters should deviate from linear phase as little as possible to avoid distortion caused by dispersion. Tabulated filter coefficients [28] as well as modern electronic design and automation (EDA) software allow fast filter design and simple performance evaluation.

10.5 Digital Hardware

The multitude of tasks to be carried out during an NMR experiment is usually scheduled to several processors which are chosen according to performance requirements. Time critical and computationally extensive routines require digital signal processors (DSPs) or field programmable gate arrays (FPGAs). Less crucial work can be scheduled to microprocessors or standard personal computer (PC) systems. Signal synthesis for RF and gradient channels as well as receive signal processing and image reconstruction require high-performance processors, while only limited effort is required for tasks like hardware supervision and status control. Last but not least, user interaction usually employs of the shelf PC hardware for ease of implementation.

Returning to small mobile NMR units, PC-based architecture offers the advantage that fast DSP and FPGA boards are available which can be plugged into the extension slots of the host system. A basic MRI tomograph can hence be build with a powerful FPGA plugged into a standard PC a small rack for external hardware and the magnet setup. With efficient amplifiers and small heatsinks, the external hardware easily integrates into a rack of about the same size as your PC.

10.5.1 Frontend Signal Processor Selection

The selection of the primary data processor strongly depends upon the expected data rate. While this is usually relatively low for single channel systems – common receiver bandwidths are a few hundred kilohertz allowing high downsampling rates – multiple receive channels require immense data handling capabilities. Since

---

[9] The NMR signal is contained in a frequency band centered around the Larmor frequency. This makes it an SSB signal from an engineering point of view.

the SNR of a single-sided NMR system is certainly lower than that of conventional tomographs, multiple receivers are a highly potent method to reduce scan time and should be considered from the beginning.

Because of the strong $\mathbf{B}_0$ inhomogeneity in a single-sided device, a minimum receiver BW of 1 MHz[10] seems appropriate. A suitable (low) IF for digital quadrature detection could also be 1 MHz. Thus the receive signal is contained inside the interval 500–1.5 MHz. Clocking the ADCs at 5 MHz achieves an oversampling rate of 3.3 for the highest possible frequency.

Another important aspect which influences the data rate is resolution. State-of-the-art receivers employ ADCs with a hardware resolution of 12–16 bits. Oversampling and parallel acquisition of one receive channel with multiple converters can increase the effective resolution to around 20 bit giving a dynamic range of 120 dB. In a single-sided system dynamic range is of true importance for an additional reason. Because of their proximity to the receive coil, slices close to the objects surface induce a large signal. In contrast, signal power is low at high distances. To increase dynamic range, gain switching during acquisition as well as compression circuits may be employed. An in-depth treatment is without the scope of this book. Further information regarding compression circuits can be found in [29].

To derive an estimate of the required data rate, let us investigate a fictive single-sided NMR system with four receive channels each of which having a physical resolution of 16 bits and being sampled at 5 MHz. The data rate for this configuration is

$$r = 4 \times 16\,\text{bit} \times 5\,\text{MHz} \approx 38.15\,\text{MB/s}. \tag{10.6}$$

Note that the convention $1\,\text{MB} = 1,024^2$ bytes has been used. For the output channels – they include at least the RF pulse data and three independent gradient data streams – a similar data handling capability is required. Clocking gradient outputs at 5 MHz may seem to be overdone but modern imaging sequences employ more and more complex waveforms so the assumption may well be justified. For transmission and reception in combination a processor has to sustain burst data rates in excess of 80 MB/s.

Many DSPs are overwhelmed with this task, even more so if real-time filtering of the receive signals and digital PSD are taken into account. Currently, a better choice for the described tasks is FPGAs. They allow parallel processing of the individual data streams and can therefore operate at lower clock frequencies. Trade-off for this solution are a lower flexibility and higher expenditures in programming an FPGA. This can be alleviated to some extent by employing a standard PC for pulse sequence programming and storing the preprocessed output data in random access memories (RAMs) with a fast bus interconnection to the FPGA. A possible top-level schematic of the described architecture is shown in Fig. 10.9.

10 Compare the specifications for the single-sided magnet given earlier in this chapter.

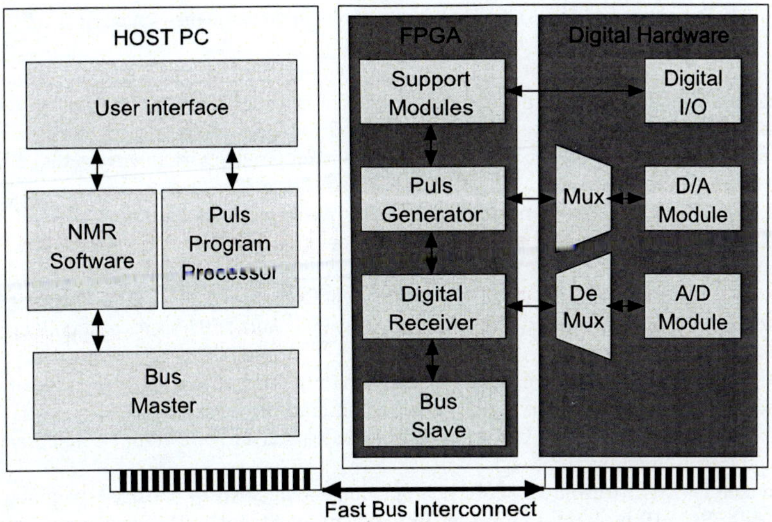

Fig. 10.9 Possible task sharing between a PC and a dedicated processor, e.g., an FPGA. The architecture arises from a trade-off between flexibility and fast processing speeds. Connections between the FPGA and digital hardware are usually realized as first-in-first-out (FIFO) buffers, while the connection between PC and processor module might well use a standard PCI architecture

10.5.2 Digital Phase Sensitive Detector

Digital baseband detection offers the advantage of high gain and phase balance and can be implemented in two different ways. The IF is either multiplied by a single complex LO with subsequent low-pass filtering or it is shifted to baseband by means of the Hilbert transformation. The later solution is generally given preference because it results in lower distortion of the receive signal. The Hilbert filter approximates a filter of linear phase in a broad frequency range and consequently minimizes dispersion because of constant group delay [30]. The common compromise of fourth-order Butterworth filtering is avoided.

The Hilbert filter is best implemented as type III finite impulse response (FIR) filter by invoking the Parks–McClellan algorithm. Asymmetry and vanishing even filter coefficients allow a resource-efficient mapping of the filter structure on digital hardware. The numeric complex oscillator for conversion to baseband can be made phase coherent with the analog hardware by feeding the reference signal – usually the OCXO output signal – into the digital processor and using this signal to drive the numeric oscillator. Thus phase coherence is maintained throughout both the analog and the digital receiver sections.

References

1. Chen CN, Hoult DI (1989) Biomedical magnetic resonance engineering. Adam Hilger, Bristol
2. Vlaardingerbroek MT, den Boer JA (1996) Magnetic resonance imaging. Springer, Berlin

3. Liang Z-P, Lauterbur PC (2000) Principles of magnetic resonance imaging. IEEE Press, New York, NY
4. BioSpec – MR Imaging and in vivo Spectroscopy at High Magnetic Fields. Bruker BioSpin MRI GmbH. http://med.cornell.edu/research/cbic/facilities/biospec_apps.pdf
5. Popella H (2003) Auslegung und Optimierung des magnetischen Kreises eines mobilen Kernspintomographen sowie Entwicklung eines planaren Gradientenspulensystems. Ph.D. thesis, RWTH Aachen University
6. Felder J (2004) Design and optimisation of RF frontend components for unilateral and mobile MR tomographs employing efficient. Linear Power Amplifiers. Ph.D. thesis, RWTH Aachen University
7. Zinke O, Brunswig H (eds) (1999) Hochfrequenztechnik, vol 2, 5th edn. Springer, Berlin
8. Peterson DM, Duensing GR, Fitzsimmons JR (1997, Jan) MRI basics and coil design principles. RF design, 56–64
9. Suddarth S (1998, Aug) A method for matching high-temperature superconductor resonators used for NMR signal pickup at 400 MHz. IEEE Trans Biomed Eng 45(8):1061–1066
10. Raad A, Darasse L (1992) Optimization of NMR receiver bandwidth by inductive coupling. Magn Reson Imaging 10:55–65
11. Peterson DM, Wolverton BL (2002) Simulation and analysis of balanced matching circuits at 3 Tesla. Proc 10th scientific meeting Intl Society for Magnetic Resonance in Medicine, May 2002, Hawaii, USA, page 885
12. Hirata H, Ono M (1997, Sep) Impedance-matching system for a flexible surface-coil-type resonator. Rev Sci Instrum 68(9):3528–3532
13. Texas Instruments (2001, May) Technical application report. Radio frequency identification systems
14. Roemer PB et al (1990) The NMR phased array. Magn Reson Med 16:192–225
15. Reykowski A, Wright SM, Porter JR (1995) Design of matching networks for low noise preamplifiers. Magn Reson Med 33:848–852
16. Ullmann P et al (2005) Experimental verification of transmit SENSE with simultaneous RF-transmission on multiple channels. Proc 13th scientific meeting Intl Society for Magnetic Resonance in Medicine, May 2005, Miami Beach, USA, page 15
17. Krauss HL, Bostian CW, Raab FH (1980) Solid state radio engineering. Wiley, New York, NY
18. Kenington PB (2000) High-linearity RF amplifier design. Artech House, Bristol
19. Cripps SC (1999) RF Power amplifiers for wireless communications. Artech House, Bristol
20. Cripps SC (2002) Advanced techniques in RF power amplifier design. Artech House, Bristol
21. Breed GA (1993, Aug) Classes of power amplification. RF Design
22. Raab FH (1987, Sep) Efficiency of doherty RF power-amplifiers. IEEE Trans Broadcast 33(3):77–83
23. Raab FH (1985, Oct) Efficiency of outphasing RF power amplifier systems. IEEE Trans Commun 33(10):1094–1099
24. Raab FH, Rupp DJ (1994) High-efficient single-sideband HF/VHF transmitter based upon envelope elimination and restoration. International Conference on HF radio systems and techniques, pp 21–25, July
25. Hannington G, Chen PF, Asbek PM, Lawrence LE (1999, Aug) High-efficiency power amplifier using dynamic power supply voltage for CDMA applications. IEEE Trans Microw Theory Tech 47(8):1471–1476
26. Chan F, Pauly J, Macovski A (1992) Effects of RF amplifier distortion on selective excitation and their correction by prewarping. Magn Reson Med 23:224–238
27. Crols J, Steyart SJ (1998, Mar) Low-IF topologies for high-performance analog front ends of fully integrated receivers. IEEE Trans Circ Syst II 45(3):269–282
28. Zverev AI (1967) Handbook of filter synthesis. Wiley, New York, NY
29. Gittinger NC (1991, June) Analog signal compression circuit. United States Patent 5023490
30. Oppenheim AV, Schafer RW (1989) Discrete-time signal processing. Prentice Hall, Upper Saddle River, NJ

Index

A
ADC, 238
Aging profiles, 208
Application of D-T_2 distribution function, 74
Archeology, 194
Average velocity measurements, 136

B
Bar magnet, 5, 89
  B_0 dependence on size, 91
  gradient coils, 108
Barrel magnet, 100
Bias of solution, 67
Bi-exponential fitting, 212
Bone density measurements, 194
Bones, 194
Building materials, 214
Butterfly coil, 103
B_1 variations, 103

C
Capacitive coupling, 223
Carbonate rock, 77
Chemical composition, 78
Chemical shift imaging, 156
Chirp shim pulses, 160
Coherence pathways, 11, 14
  CPMG, 29
  expanding tree, 17
  Hahn-echo, 27
  13-interval sequence, 132
  phase, 16
  selection, 17
Composite z-rotations, 147
Contrast in skin, 191
CPMG/CP sequence, 121
CPMG - diffusion attenuation, 33
CPMG - diffusion effect on relaxation, 61
CPMG - echo shape, 30
CPMG - effective relaxation time, 31, 60
CPMG - quadrature loss, 118
CPMG - sequence, 28, 59
CPMG - setting rf amplitude, 34
CPMG - shortest echo time, 34
CPMG - signal decay, 31
CPMG - T_1 contamination, 31
CPMG - T_2 error, 32
CPMG - timing, 29
Cream ingress into skin, 192
Crosslink density, 204
  gradient, 206
Crude oil, 78
C-shaped magnet, 88
Cultural heritage, 217

D
Dairy product, 75
Data inversion, 65
Depth profiles, 112
  of skin, 189
Depth resolution, 111
Detection efficiency, 20
Detection-period, 122
Diagnostic imaging, 187
Diffusion, 58
  measurement, 42, 62
    Hahn and stimulated echoes, 43, 63
  relaxation distribution function, 69
Diffusion attenuation
  13-interval sequence, 132
  CPMG, 33,61
  Hahn-echo, 28, 43
  stimulated echo, 43
Digital signal processor, 237
2D imaging
  gradient coil, 108
  sequence, 115
3D imaging, 128

F. Casanova et al. (eds.), *Single-Sided NMR*,
DOI 10.1007/978-3-642-16307-4, © Springer-Verlag Berlin Heidelberg 2011

Dipolar magnetic field, 3
Displacement encoding, 129
Distribution function, 64

E

Earth field, 2
Echo condition, 17
Echo-time limit, 34, 216
Effective magnetic field, 13
Elastomers, 204
Emulsion, 78
Encoding-period, 122
Excitation bandwidth, 22
  CPMG, 30
  Hahn-echo, 25
  single-pulse, 22
Ex situ matching, 147
Ex situ NMR, 144
Ex situ shim pulse, 158

F

Failure detection, 210
Far-field radiation, 224
Field lines, 90
Figure-8 coil, 102
Filtering, 60
Finger images, 197
Flip angle distribution, 11, 21
Flow encoding, 129
Flow imaging, 137
Fluid saturated rock, 75
Fluorinated rubber, 207
Food products, 75
Fourier imaging, 113
Frequency generation, 235
Frequency mixing, 235

G

Gels, 78
Glass transition temperature, 206
Gradient coils, 107
Gradient strength, 216

H

Hahn-echo, 23
  attenuation, 28
  bandwidth, 25
  decay of stretched rubber, 211
Hard polymers, 212
High order shimming, 177
High-resolution NMR, 144
Homogeneity B_1, 103
HOMOGENEIZED, 145
Homogeneous Stray field, 165
Horseshoe geometry, 5
^{1}H spectroscopy - NMR MOUSE, 177
Humidity in polymers, 217

I

Iceman, 195
Imaging
  air spring bellow, 209
  biological matter, 196
  diffusion contrast, 198
  materials, 209
  maximizing sensitivity, 125
  T_1 contrast, 198
  T_2 contrast, 121, 126
Inductance rf coil, 105
Inside-out NMR, 1
Intermolecular zero-quantum coherence, 145
13-Interval sequence, 131
  off-resonance distortion, 133
Inverse Laplace transform, 65
Inversion recovery - sampling times, 38
Inversion recovery - sequence, 35

J

Jackson geometry, 2, 87

L

Laplace transformation, 65
Lateral resolution, 113
Layer thickness determination, 217
Low-noise amplifier, 234

M

Magnetization evolution
  calculus, 12
  CPMG, 29
  Hahn-echo, 25
  single pulse, 22
Miscible fluids, 77
Mobile NMR, 3
Movable magnets, 170
Multi-dimensional methods, 6, 35, 43, 63
Multi-echo acquisition, 117
Multi-echo detection - relaxation and diffusion contrast, 126
Multi-echo detection - sensitivity improvement, 125
Multilayer building materials, 214
Multilayer painting structures, 218
Multiple quantum coherence, 144
Multiple receiver channels, 237
Mummies, 194

N
Natural rubber, 205
NMR-MOLE, 102
NMR-MOUSE, 4
NMR-MOUSE - profiling, 98
NMR-MOUSE - spectral resolution, 177
NMR-MOUSE - spectroscopy, 165, 176
NMR-MOUSE - Temperature compensation, 181
NUMAR tool, 87
Nutation echoes, 147

O
Off-resonance excitation, 21
One-sided access, 4
Optimal acquisition bandwidth, 27
Optimal regularization parameter, 66

P
Permeability, 81
Phase cycle
    add/subtract, 18
    CPMG, 29
    CPMG/CP sequence, 121
    Hahn-echo, 28
    Hahn-echo diffusion, 45
    13-interval sequence, 135
    inversion recovery, 38
    saturation recovery, 38
    stimulated echo, 19
    stimulated echo diffusion, 47
Phase sensitive detectors, 239
Phase-encoding imaging, 114
Pin diodes switch, 227
Poly(ethylene), 212
Polymer annealing, 213
Polymer crystallinity, 212
Polymer morphology, 212
Pore connectivity, 80
Pore size, 79
Profile NMR-MOUSE, 98
Profile NMR-MOUSE- gradient coils, 108
Profiling, 95
    concrete walls, 214
    field profile, 99
    gasoline ingress, 215
    paintings, 217
    plastic fuel tank walls, 215
    reduced gradient magnets, 97
    rf coil, 103
    water ingress in PE, 216
Pulse sequence analysis, 21
Pulsed field gradients, 130

Q
Quadrature multi-echo detection, 121

R
RARE-sequence, 117
Receiver design, 233
Reciprocity principle, 20, 48
Regularization, 67
Relaxation, 57
Relaxation enhanced images, 121
Resolution
    lateral imaging, 108
    profiling, 95
    spectrum, 177
Rf amplifiers, 228
Rf circuit bandwidth, optimum, 51
Rf frontend hardware, 222
Rf matching and balancing, 223
Rf-coils, 102
Rubber tubes, 210

S
Sample heating, 205
Saturation recovery
    recycling delay, 40
    sequence, 39
Scaled adiabatic pulses, 150
Scaling law, 92
Sensitive volume
    flat, 98
    spectroscopy, 176
Sensitivity
    rf coil, 106
    Sweet-Spot magnets, 101
    bar magnet, 93
Shim
    lineal terms, 171
    second order terms, 173
    magnets, 150
    mechanical, 165
    pulses, 157
        bandwidth, 158
    unit, 168
        temperature compensation, 181
Signal calculation, 20
Signal processing, 237
Signal-to-noise ration, 47
    analytical solution, 55
    numerical calculation, 49
Single point imaging, 196
Single pulse excitation, 22
Single-loop rf coil, 102
Skin studies, 188
Skulls, 195

Solenoid and Bar magnet equivalence, 90
Solvent ingress, 215
Space-dependent phase corrections, 150
Spatial encoding, 113
Spatially dependent z-rotations, 145
Spatially matched rf field, 146
Spatially resolved velocity distributions, 137
Spatial resolution for short signals, 216
Spectral resolution, 177
Spectroscopy, 165
  temperature compensation, 181
Spin density contrast, 210, 216
Spin dynamics in inhomogeneous fields, 12, 59
Spin-lattice relaxation, 35
Square - rf coil, 102
Stability of solution, 67
Stimulated echo, 17
STRAFI, 62
Strain, 210
Stray field
  profiling, 95
  spectroscopy, 165
Stretching ratio of rubber, 210
Structured fluids, 78
Surface relaxation, 58
Sweet-Spot magnets, 99
  rf coil, 105

T

T_1 measurements, 35, 64
T_1-T_2 distribution function, 69
T_1/T_2 ratio, 76
T_2 contrast, 210
  in hard polymers, 214
  in swollen polymers, 215
T_2-mapping in PE pipes, 213
Temperature coefficient magnetic material, 181
Temperature compensation, 181
Tendon studies, 192
Thermo-oxidative aging, 207
Transmit-receive switch, 226
Two component systems, 75
Two-dimensional distribution functions, 58
Two-dimensional Laplace transform, 71
Two-dimensional pulse sequences, 63

U

U-shaped magnet, 88
  B_0 dependence on size, 95
  field profile, 96, 99
  gradient coils, 108
  mechanical shim, 166
  profiling, 96
  spectroscopy, 166
  tailored geometry, 96
  yoke, 94
Ultrafast spectroscopy, 160
Uncertainty in distribution function, 68
UV aging, 208

V

Velocity distributions, 135
Velocity maps, 137

W

Water ingress into skin, 192
Waxing, 78
Weighting function, 208
Well-logging, 2
Wettability, 76